D1626111

WITHDRAWN
FROM STOCK
QMUL LIBRARY

FOR RETURN

Between Nilpotent
& Solvable

Between Nilpotent and Solvable

Henry G. Bray
W. E. Deskins
David Johnson
John F. Humphreys
B. M. Puttaswamaiah
Paul Venzke
Gary L. Walls
edited by Michael Weinstein

Polygonal Publishing House
80 Passaic Avenue
Passaic, NJ 07055 USA

222973

Copyright © 1982 by Polygonal Publishing House
All rights of publication reserved

Library of Congress Cataloging in Publication Data
  Main entry under title:

Between nilpotent and solvable.

  Bibliography: p.
  Includes indexes.
  1. Finite groups. 2. Solvable groups. 3. Groups,
Nilpotent. I. Bray, Henry G. II. Weinstein, Michael.
QA171.B48   512'.22   82-539
ISBN 0-936428-06-6   AACR2

Manufactured in the United States of America
by Braun-Brumfield

Contents

Preface: *What the Reader should Know before Reading and while Reading this Book*

The importance and utility of the distinction between *solvable* and *nonsolvable* in finite group theory can hardly be overestimated. Solvable groups, built up as they are from Abelian groups by repeated extensions, still bear some family resemblances to their Abelian ancestors, even if these resemblances are often vague. Nonsolvable groups, on the other hand, are so wildly non-Abelian that the theories necessary to study them have to be radically different from those needed in the solvable case.

This book is about solvable groups. It focusses on particular classes of finite solvable groups (supersolvable, M-group, etcetera) giving, in each case, at least the basic results about that class. We have chosen to consider only classes which contain $\mathfrak{N}$, the class of all finite nilpotent groups. This choice is not as capricious as it first appears. Several results proven in Chapter 5 show that $\mathfrak{N}$ is contained in all classes which are "well-behaved" in various senses (Most notable of these results is that $\mathfrak{N}$ is contained in all formations which can be locally defined by a system of nonempty formations).

The class $\mathfrak{N}$ itself is not given an exposition in the book. This seemed unnecessary as the basic results about nilpotent groups are so well-known, being covered in many texts on general group theory. It is assumed then that the reader is familiar with the concepts of ascending (upper) central series, descending (lower) central series, and with the following facts about nilpotent groups:

THEOREM *For a finite group G, the following are equivalent*:
(1) *G is nilpotent.*
(2) *All Sylow subgroups of G are normal.*
(3) *All maximal subgroups of G are normal.*
(4) *All subgroups of G are subnormal.*
(5) *If $H < G$, then $H < N_G(H)$.*
(6) *G is the direct product of its Sylow subgroups.*

THEOREM *If $H \leq Z(G)$ and G/H is nilpotent, then G is nilpotent.*

THEOREM *If G is nilpotent and $\{1\} \neq N \triangleleft G$, then $Z(G) \cap N \neq \{1\}$.*

THEOREM *Finite p-groups are nilpotent.*

The class $\mathfrak{S}$ of finite solvable groups is not assumed to be so well-known to the reader. Fundamental concepts of the theory of solvable groups are presented in the appendices. The reader is advised to have a look at these appendices before starting Chapter 1 and to either work through the appendix material at that time or at least to familiarize himself with what the appendices contain, with the intention of returning to them if and when necessary.

Briefly, let us mention some of the basic concepts of the theory of finite solvable groups. First there is the idea of a Hall subgroup (a subgroup whose index and order are coprime) and Phillip Hall's theorem about the existence and conjugacy of Hall subgroups in solvable groups. This theorem and related results are referred to as extended Sylow theory since they extend (for solvable groups) Sylow's theorems. An account of this theory is given in Appendix A.

Appendix B discusses chief factors, covering and avoidance, and the induced automorphism groups $\text{Aut}_G(H/K)$. The importance of these concepts to the study of solvable groups cannot be stressed too highly. Indeed, an inspection of the contents of a good many of the theorems throughout the book will reveal a heavy dependence on these concepts. Let us, therefore, take a minute to define chief factor and point out its most important property. If G is a group and H and K are normal subgroups of G with $K \leq H$, then H/K is called a **chief factor** (or principal factor) of G if there is no normal subgroup T of G such that $K < T < H$. Stated differently, a chief factor of G is a minimal normal subgroup of a quotient group of G. The single most important property of chief factors—and something that the reader should keep in mind throughout the book—is that chief factors of solvable groups are elementary Abelian p-groups for some prime p.

In addition to the above, there is one further isolated fact about solvable groups the reader should know and keep in mind at all times: namely, in a finite solvable group, the index of a maximal subgroup must be a prime-power. This is proven as part of Lemma 1.11 in Chapter 5.

A final point that the reader is asked to keep in mind throughout the book is that all groups under consideration here are finite and solvable. This requirement will be dropped occasionally—but only occasionally—and it will be painfully obvious where this is being done (e.g., in theorems, such as Theorem 2.3 of Chapter 1, whose point it is to establish that a group meeting certain conditions is solvable). Other than these few occasions then, all groups considered are finite and solvable. Despite this blanket assumption, it has seemed desirable to explicitly use the phrase "finite solvable" whenever it was wished to *emphasize* that the group under consideration was such. As this was done for emphasis, it does not appear in every case.

Michael Weinstein

Supersolvable Groups

W. E. Deskins
and Paul Venzke

1.   Basic Results

DEFINITION   A finite group is said to be **supersolvable** whenever its chief factors are all cyclic.

It is easily seen that the chief factors of a supersolvable group are not only cyclic but of prime order as well. It is also an immediate consequence of the definition that supersolvable groups are solvable.

THEOREM 1.1   (a) *Every subgroup of a supersolvable group is supersolvable.*
   (b) *Every homomorphic image of a supersolvable group is supersolvable.*
   (c) *The direct product of two supersolvable groups is supersolvable.*
   (d) *If the group G has two normal subgroups H and K with both G/H and G/K supersolvable, then G/H ∩ K is supersolvable.*
Proof: Let G be a supersolvable group and

$$\{1\} = G_0 \lhd G_1 \lhd G_2 \lhd \ldots \lhd G_r = G$$

be a chief series of G. For the subgroup S of G,

$$\{1\} = S \cap G_0 \lhd S \cap G_1 \lhd S \cap G_2 \lhd \ldots \lhd S \cap G_r = S$$

is a normal series of S in which the factor $(S \cap G_i)/(S \cap G_{i-1})$ is isomorphic to the subgroup $(S \cap G_i)G_{i-1}/G_{i-1}$ of G_i/G_{i-1}. For each i, G_i/G_{i-1} is of prime order, so that $(S \cap G_i)/(S \cap G_{i-1})$ is either trivial or of prime order. Hence the above normal series for S will yield, after deleting redundant terms, a chief series of S with cyclic factors. Therefore S is supersolvable. If α is a homomorphism of G onto the group T, then

$$\{1\} = G_0^\alpha \triangleleft G_1^\alpha \triangleleft G_2^\alpha \triangleleft \ldots \triangleleft G_r^\alpha = T$$

is a normal series of T. The factor $G_i^\alpha/G_{i-1}^\alpha$ is a homomorphic image of the factor G_i/G_{i-1} and hence is either trivial or of prime order. It is now seen that the above normal series for T will provide, after the deletion of redundant terms, a chief series of T with cyclic factors. Therefore T is supersolvable.

Let H also be a supersolvable group with chief series

$$\{1\} = H_0 \triangleleft H_1 \triangleleft H_2 \triangleleft \ldots \triangleleft H_t = H.$$

The series

$$\{1\} = H_0 \triangleleft H_1 \triangleleft \ldots \triangleleft H_t = H \triangleleft G_1 \oplus H \triangleleft \ldots \triangleleft G_r \oplus H = G \oplus H,$$

is a chief series of $G \oplus H$ with cyclic factors. Therefore $G \oplus H$ is supersolvable.

Suppose now that G has normal subgroups H and K with G/H and G/K supersolvable. The mapping $g \to (Hg, Kg)$ is a homomorphism of G into $G/H \oplus G/K$ with kernel $H \cap K$. Thus $G/H \cap K$ is isomorphic to a subgroup of the supersolvable group $G/H \oplus G/K$ and is therefore supersolvable.

If a normal subgroup H and its associated quotient group G/H are both supersolvable, G itself need not be supersolvable. We do have, however

THEOREM 1.2 *Let G have a normal cyclic subgroup H with G/H supersolvable. Then G is supersolvable.*
Proof: Let

$$\{1\} = H_0 \triangleleft H_1 \triangleleft H_2 \triangleleft \ldots \triangleleft H_r = H = G_0 \triangleleft G_1 \triangleleft G_2 \triangleleft \ldots \triangleleft G_s = G$$

be a chief series for G which passes through the normal subgroup H. Let $\overline{A}$ denote the image of the set A in the quotient G/H. The series

$$\{1\} = \overline{G_0} \triangleleft \overline{G_1} \triangleleft \overline{G_2} \triangleleft \ldots \triangleleft \overline{G_s} = \overline{G}$$

is a chief series of the supersolvable group $\overline{G} = G/H$. The factor G_i/G_{i-1} of G is isomorphic with the factor $\overline{G_i}/\overline{G_{i-1}}$ of $\overline{G}$ and is therefore cyclic. Since H is cyclic, each factor H_i/H_{i-1} is cyclic and therefore G is supersolvable.

We are approaching an important result (Theorem 1.4) which gives a criterion for a chief factor to be cyclic of prime order. This theorem and its immediate corollary, Corollary 1.5—a necessary and sufficient condition for supersolvability, will be used heavily throughout the book. The proof of Theorem 1.4 will be made to depend on a lemma about vector spaces. This lemma, then, is supporting a considerable amount of the theory of super-solvable groups. Because of its importance, two proofs will be given.

LEMMA 1.3 *Let V be a vector space of dimension $n \geqslant 1$ over $GF(p)$. Let G be an Abelian group of linear transformations of V of exponent dividing $p - 1$. If G acts irreducibly on V, then $n = 1$ and G is cyclic.*

Proof: Let g be an element of G. Since G has exponent dividing $p - 1$, g satisfies the polynomial $x^{p-1} - 1$. Over $GF(p)$ this polynomial is the product of linear factors and hence g has a characteristic root $\lambda \neq 0$ in $GF(p)$. Since λ is a characteristic root of g, the subspace $W = \{v : vg = \lambda v\}$ of V is nonzero. Let $x \in G$ and $v \in W$, then $vxg = vgx = \lambda vx$ and we conclude that $vx \in W$. W is thus a nonzero G-invariant subspace of V and hence by the irreducibility of G, $V = W$. It is now clear that each element of G induces a scalar multiplication on V, so that in order for G to act irreducibly on V we must have $n = 1$. Furthermore, since $n = 1$, G must be isomorphic with a subgroup of the multiplicative group of $GF(p)$ and is therefore cyclic.

Alternate proof: Consider the endomorphism ring $E(V)$. For all $b \in GF(p) \backslash \{0\}$, let b^* be the endomorphism described by scalar multiplication by b. There are $p - 1$ of these; say $b_1^*, b_2^*, \ldots, b_{p-1}^*$. Let $C = \{t : t \in E(V)$ and t is a sum of products of elements in $G \cup \{b_1^*, b_2^*, \ldots, b_{p-1}^*\}\}$. Note that each b_i^* commutes with all endomorphisms. It is straightforward then to show that C is a commutative subring of $E(V)$. Moreover, C acts irreducibly on V because G does. Schur's lemma may now be applied to yield that $D = \{d : d \in E(V)$ and $dt = td$ for all $t \in C\}$ is a division ring. Since C is a subring of D and $E(V)$ is finite, C is a finite integral domain, hence a field. Thus the multiplicative group $C \backslash \{0\}$ is cyclic, hence so is its subgroup G.

Let $G = \langle \gamma \rangle$. Then γ and each b_i^*, $1 \leq i \leq p - 1$, are roots of the polynomial $x^{p-1} - 1 \in C[x]$ (γ by hypothesis; each b_i^* because $b_i^{p-1} = 1$ in $GF(p)$). Since a polynomial of degree $p - 1$ cannot have p roots and since $b_i^* \neq b_j^*$ for $i \neq j$, we conclude $\gamma = b_i^*$ for some i. Now let W be the subspace spanned by any nonzero vector $w \in V$. Then $\gamma = b_i^*$ shows that W is γ-invariant, hence G-invariant. Conclude $W = V$ so that V has dimension one.

We are now ready for Theorem 1.4 which involves the automorphisms that a group induces on its chief factors. See Appendix B for an explanation of chief factors and induced automorphisms.

THEOREM 1.4 *Let H/K be a p-chief factor of the group G. Then $|H/K| = p$ if and only if $\mathrm{Aut}_G(H/K)$ is Abelian of exponent dividing $p - 1$.*
Proof: If $H/K \cong \mathbb{Z}_p$, then $\mathrm{Aut}_G(H/K)$ can be embedded in $\mathrm{Aut}(\mathbb{Z}_p)$ $\cong \mathbb{Z}_{p-1}$.

To prove the converse, suppose $\mathrm{Aut}_G(H/K)$ is Abelian of exponent dividing $p - 1$. View H/K as a vector space of dimension n over $\mathrm{GF}(p)$ on which $\mathrm{Aut}_G(H/K)$ acts irreducibily as a group of linear transformations. Lemma 1.3 implies $n = 1$ so that $|H/K| = p$.

Theorem 1.4 immediately implies

COROLLARY 1.5 *The group G is supersolvable if and only if for every prime p and every p-chief factor H/K of G, $\mathrm{Aut}_G(H/K)$ is Abelian of exponent dividing $p - 1$.*

THEOREM 1.6 *The commutator subgroup of a supersolvable group is nilpotent.*
Proof: Let G be a supersolvable group with chief series

$$\{1\} = G_0 \lhd G_1 \lhd G_2 \lhd \ \ldots \ \lhd G_r = G.$$

By Theorem 1.4, $\mathrm{Aut}_G(G_i/G_{i-1}) \cong G/C_G(G_i/G_{i-1})$ is Abelian for each i, so that $G' \leq C_G(G_i/G_{i-1})$. We now consider the normal series

$$\{1\} = G_0 \cap G' \lhd G_1 \cap G' \lhd G_2 \cap G' \lhd \ \ldots \ \lhd G_r \cap G' = G'$$

of G'. Since $G' \leq C_G(G_i/G_{i-1})$, we see that $[G_i \cap G', G'] \leq [G_i, G'] \cap G' \leq G_{i-1} \cap G'$. Hence this is a central series for G' and therefore G' is nilpotent.

In the next theorem we will see that the maximal subgroups of a supersolvable group are always of prime index. In Section 3 we will show that the converse of this theorem is also true.

THEOREM 1.7 *Every maximal subgroup of a supersolvable group has prime index.*
Proof: Let G be a supersolvable group of smallest possible order for which the theorem fails. Let M be a maximal subgroup of G and $K = \mathrm{Core}_G(M)$, the largest normal subgroup of G contained in M. If $K \neq \{1\}$, then M/K is a maximal subgroup of the supersolvable group G/K and by the minimality of G, $[G:M] = [G/K : M/K]$ is prime. We may therefore assume that $K = \{1\}$. Let N be a minimal normal subgroup of G. As G is supersolvable, $|N| = p$ for some prime p. Since $K = \{1\}$, $M \cap N = \{1\}$ and from maximality of M we know that $G = MN$. It follows that $|G| = |M|p$ or equivalently that $[G:M] = p$. Thus every maximal subgroup of G has prime index so that G was no counterexample to the

theorem. Therefore, by induction, the theorem must be valid for all supersolvable groups.

DEFINITION Let G be a group and let $p_1 > p_2 > \ldots > p_r$ be the distinct primes dividing $|G|$. Then G is said to satisfy the **Sylow tower property** if there exist G_{p_1}, G_{p_2}, ..., G_{p_r} such that each G_{p_i} is a Sylow p_i-subgroup of G and $G_{p_1} G_{p_2} \ldots G_{p_k} \lhd G$ for $k = 1, 2, \ldots, r$.

THEOREM 1.8 *Supersolvable groups satisfy the Sylow tower property.*
Proof: Let G be supersolvable. It is enough to show that if p is the largest prime dividing $|G|$, then G_p, a Sylow p-subgroup of G, is normal in G, for the theorem will then follow by induction on the number of distinct prime divisors of $|G|$. We now induct on $|G|$. Let N be a minimal normal subgroup of G. By induction NG_p/N is a normal subgroup of G/N and hence $NG_p \lhd G$. Let $|N| = q$ where q is some prime. If $p = q$, then $NG_p = G_p$ and we are done; hence we may assume that $p > q$. By Theorem 1.4, $G/C_G(N)$ is Abelian of exponent dividing $q - 1$. Since $p > q$, we conclude that $C_G(N) \geq G_p$. It now follows that G_p is a normal subgroup of NG_p and, as G_p is also a Sylow subgroup of NG_p, that G_p is characteristic in NG_p. This and $NG_p \lhd G$ imply $G_p \lhd G$ which establishes the theorem.

The converse of this last theorem is not true, indeed, one can easily find groups which enjoy the Sylow tower property and yet are not supersolvable (we will give an example of order 200 at the end of this section; the reader is invited to find examples of smaller order). Our next goal will be to develop a criterion which can be applied to determine when a group enjoying the Sylow tower property is supersolvable. For this we need to introduce a very special type of supersolvable group.

DEFINITION Let p be a prime. A group G is said to be **strictly p-closed** whenever G_p, a Sylow p-subgroup of G, is normal in G with G/G_p Abelian of exponent dividing $p - 1$.

Elementary arguments can be used to show that subgroups and homomorphic images of strictly p-closed groups are again strictly p-closed. We also have

THEOREM 1.9 *Every strictly p-closed group is supersolvable.*
Proof: Let G be a strictly p-closed group of the smallest possible order for which the theorem could fail. If we show that G is supersolvable, the theorem will follow by induction. Let G_p denote the normal Sylow p-subgroup of G. Should $G_p = \{1\}$, then G will be Abelian and hence supersolvable. Should $G_p > \{1\}$, then G_p contains a minimal normal subgroup N of G which is elementary Abelian of order $p^n > 1$. Let Z denote

the center of G_p; Z is a normal subgroup of G and $N \cap Z \neq \{1\}$. By our choice of N it follows that $N \leq Z$, or equivalently, that $C_G(N) \geq G_p$. From the strict p-closure of G, we now know that $\text{Aut}_G(N) \cong G/C_G(N)$ is Abelian of exponent dividing $p - 1$. With the aid of Theorem 1.4, we conclude that N is cyclic. By our choice of G, the strictly p-closed group G/N is supersolvable and we may therefore appeal to Theorem 1.2 and conclude that G is supersolvable.

COROLLARY 1.10   *If G is a group of order qp^r, where p and q are primes and $q \mid (p - 1)$, then G is supersolvable.*
Proof: By the Sylow theory a Sylow p-subgroup G_p of G is normal in G. Hence G is strictly p-closed.

EXERCISE   Show that if G is a group of order $q^2 p^r$, where p and q are primes and $q^2 \mid (p - 1)$, then G is supersolvable.

The following well-known result on automorphisms of p-groups will be useful in the sequel. If α is an automorphism of the group G with subgroup H, then we let $[H, \alpha]$ denote the subgroup of G generated by all elements of the form $h^{-1}h^\alpha$ as h ranges over H.

LEMMA 1.11   *Let p be a prime, P a p-group, and α an automorphism of P whose order is prime to p. If P has a chain of subgroups*

$$\{1\} = P_0 \leq P_1 \leq P_2 \leq \ldots \leq P_r = P,$$

such that $[P_i, \alpha] \leq P_{i-1}$ for $i = 1, 2, \ldots, r$, then $\alpha = 1$.
Proof: Induce on r. The lemma is certainly valid for $r = 1$ and we may therefore assume $r \geq 2$. Since $[P_{r-1}, \alpha]$ is contained in P_{r-1}, the restriction of α to P_{r-1} is an automorphism of P_{r-1} which, by induction, fixes each element of P_{r-1}. Let g be an arbitrary element of P. By hypothesis $g^{-1}g^\alpha \in P_{r-1}$ and hence $g^\alpha = gh$ where h is some element of P_{r-1}. Simple induction can be used to show that for each natural number k, $g^{\alpha^k} = gh^k$. For if $g^{\alpha^{k-1}} = gh^{k-1}$, then $g^{\alpha^k} = (g^{\alpha^{k-1}})^\alpha = (gh^{k-1})^\alpha = g^\alpha(h^\alpha)^{k-1} = gh(h^\alpha)^{k-1}$. Since α acts trivially on P_{r-1}, $h^\alpha = h$ and $g^{\alpha^k} = gh^k$. We now see that $g = g^{\alpha^{o(\alpha)}} = gh^{o(\alpha)}$ and that $h^{o(\alpha)} = 1$. Since h is a p-element and $(o(\alpha), p) = 1$, it follows that $h = 1$ so that $g^\alpha = g$. Therefore $\alpha = 1$.

The following theorem is due to Reinhold Baer (see [Baer 1957]).

THEOREM 1.12   *The group G is supersolvable if and only if*
(a) G satisfies the Sylow tower property, and
(b) $N_G(G_p)/C_G(G_p)$ is strictly p-closed for every Sylow p-subgroup, G_p, of G.
Proof: Suppose first that G is supersolvable. We have already seen (Theorem 1.8) that condition (a) is valid. Thus we need only show the

validity of (b). Let p be a prime, G_p a Sylow p-subgroup of G, N the normalizer of G_p in G, and C the centralizer of G_p in G. As a subgroup of the supersolvable group G, N itself is supersolvable. If now, $\{1\} = N_0 \lhd N_1 \lhd \ldots \lhd N_r = G_p$ is a portion of a chief series of N passing through G_p; then, by Corollary 1.5, $N/C_N(N_i/N_{i-1})$ is Abelian of exponent dividing $p - 1$. Set $L = \bigcap\{C_N(N_i/N_{i-1}) : 1 \leq i \leq r\}$ so that N/L is also Abelian of exponent dividing $p - 1$ and, moreover, $L \geq C$. We claim that L/C is a p-group. Suppose to the contrary that some $Cx \in L/C$ has order n relatively prime to p. Let $\alpha \in \text{Aut}(G_p)$ be the automorphism induced by x, i.e., $\alpha(g) = x^{-1}gx$. Then the order of α in $\text{Aut}(G_p)$ divides n hence it too is relatively prime to p. Also note that $x \in L$ implies $[N_i, \alpha] \leq N_{i-1}$ for $1 \leq i \leq r$, so that Lemma 1.11 applies to show α is trivial. Hence so too is Cx, proving the claim. It follows that N/C is strictly p-closed with Sylow p-subgroup L/C.

Suppose now that G satisfies conditions (a) and (b). Let H/K be a chief factor of G; we wish to show that H/K is cyclic. Solvability of G implies H/K is an elementary Abelian p-group for some prime p. By condition (a) we know that G has a normal subgroup R of order prime to p such that G_pR is normal in G, where G_p denotes a Sylow p-subgroup of G. $H \cap R$ has order prime to p and since H/K is a p-group, we must have $K \geq H \cap R$. Now $HR/KR = H(KR)/KR \cong H/H \cap KR = H/K(H \cap R) = H/K$ and thus there is no loss of generality to assume that $K \geq R$. Now if R is nontrivial, then G/R is supersolvable by induction (this, of course, depends on the fact that conditions (a) and (b) are quotient-closed; verifying this is straightforward). Thus the chief factor $(H/R)/(K/R)$ is cyclic, hence so is H/K.

We may therefore assume $R = \{1\}$. This forces $G_p \lhd G$ and so $G/C_G(G_p)$ is strictly p-closed. Now $G_pK \lhd G$ and hence G_pK/K is a normal Sylow p-subgroup of G/K, showing that $H/K \leq G_pK/K$. Moreover, H/K is a minimal normal subgroup of G/K and so must lie in the center of G_pK/K and hence $G_p \leq C_G(H/K)$. Since $H \leq G_pK$, it also follows that $C_G(G_p) \leq C_G(H/K)$ and so $G/C_G(H/K)$ is strictly p-closed. But $G_p \leq C_G(H/K)$ means $G/C_G(H/K)$ has trivial Sylow p-subgroup so $G/C_G(H/K)$ is Abelian of exponent dividing $p - 1$. Theorem 1.4 may now be applied and we conclude that H/K is cyclic. Supersolvability of G now follows.

If one attempts to generalize the concept of the Fitting subgroup in the obvious way by considering the product of all normal supersolvable subgroups of a group G, then the result is not necessarily supersolvable.

EXAMPLE Let H be the direct product $\langle x \rangle \times \langle y \rangle$, where $o(x) = o(y) = 5$. The maps $\alpha : x \to x^2$, $y \to y^{-2}$ and $\beta : x \to y^{-1}$, $y \to x$ are automorphisms of H and generate a subgroup $A \leq \text{Aut}(H)$ of order 8 (A is isomorphic with the quaternion group). Take $G = H[A]$. Then $S = \langle H, \alpha \rangle$

and $T = \langle H, \beta \rangle$ are normal subgroups of G and both are supersolvable since S/H and T/H are Abelian of exponent $4 = 5 - 1$. Thus $G = ST$ is the product of normal supersolvable subgroups but is not itself supersolvable.

However, a result due to Baer tells us exactly when the product of normal supersolvable subgroups is supersolvable. We state this result for two subgroups; the extension to an arbitrary finite number is an easy corollary.

THEOREM 1.13 *Let $G = ST$, where S and T are normal supersolvable subgroups of G. Then G is supersolvable if and only if G' is nilpotent.*

Proof: If G is supersolvable, then we know (Theorem 1.6) that G' is nilpotent, so we must prove G supersolvable when G' is nilpotent. Let θ be a homomorphism of G. Then $G^\theta = S^\theta T^\theta$, S^θ and T^θ are normal supersolvable subgroups of G^θ, and $(G^\theta)' \leq (G')^\theta$ is nilpotent. So if $|G^\theta| < |G|$, then by induction G^θ is supersolvable. This means that if G contains two different minimal normal subgroups M and N, then G/M, G/N and hence $G/M \cap N \cong G$ are supersolvable. So we are left with the case G has a unique minimal normal subgroup H. This implies that $\mathrm{Fit}(G)$ is a p-group for some prime p. Moreover, since the Fitting subgroup is characteristic, $\mathrm{Fit}(S)$ and $\mathrm{Fit}(T)$ are contained in $\mathrm{Fit}(G)$. From Theorem 1.12 we conclude that S and T are strictly p-closed. Thus $S/\mathrm{Fit}(S)$ and $T/\mathrm{Fit}(T)$ both have exponent dividing $p - 1$. This together with $G/\mathrm{Fit}(G)$ being Abelian (since G' is nilpotent) imply $G/\mathrm{Fit}(G)$ has exponent dividing $p - 1$. Hence we see that G is strictly p-closed and conclude that G is supersolvable.

2. Equichained Groups

Let H denote a subgroup of a group G. A chain of subgroups, $H = M_0 \leq M_1 \leq M_2 \leq \ldots \leq M_r = G$, is called a **maximal chain** joining H to G of length r whenever M_{i-1} is a maximal subgroup of M_i for $i = 1, 2, 3, \ldots, r$. Should all of the maximal chains joining H to G have the same length, we say that H is **equichained** in G, moreover, should every subgroup of G be equichained in G, we simply say that G is an **equichained** group. In Section 1 we observed that every maximal subgroup of a supersolvable group is of prime index in the group and therefore every supersolvable group must be an equichained group, since the length of the maximal chains joining a subgroup H to the supersolvable group G is simply the number of prime divisors, counting repetitions, of $[G:H]$. K. Iwasawa ([Iwasawa 1941a]) was the first to show that the equichained groups are precisely the supersolvable groups. This section will be devoted to Iwasawa's theorem. The most difficult step in the proof is showing that equichained groups are solvable. For this we will need the theorem of Grün. The proof of Grün's theorem is not given here, the reader is directed to [Scott 1964, 13.5.4] for its proof.

Let p be a fixed prime. We say that the group G is **p-normal** whenever G_p and $G_p{}^*$ are Sylow p-subgroups of G such that $Z(G_p) \leq G_p{}^*$, then $Z(G_p) = Z(G_p{}^*)$. We now state Grün's theorem.

THEOREM 2.1 *Let G be a p-normal group and G_p a Sylow subgroup of G for some prime p. then the largest Abelian p-group which occurs as a factor group of G is isomorphic with the largest Abelian p-group which occurs as a factor group of $N_G(Z(G_p))$.*

If a group G is not p-normal, then for some Sylow p-subgroup G_p of G, $Z(G_p)$ is contained in some Sylow p-subgroup $G_p{}^*$, but $Z(G_p) \neq Z(G_p{}^*)$. We observe that $Z(G_p)$ is not normal in $G_p{}^*$. For assume $Z(G_p) \triangleleft G_p{}^*$, then $N_G(Z(G_p)) \geq \langle G_p{}^*, G_p \rangle$. Hence by Sylow's theorem there is an element $w \in N_G(Z(G_p))$ such that $G_p{}^w = G_p{}^*$ and therefore $Z(G_p) = Z(G_p)^w$ $= Z(G_p{}^w) = Z(G_p{}^*)$, contrary to $Z(G_p) \neq Z(G_p{}^*)$. Hence, if G is not a p-normal group, the center of a Sylow p-subgroup is always a nonnormal subgroup of some other Sylow p-subgroup. The following theorem of Burnside will be useful in working with groups which are not p-normal.

THEOREM 2.2 *Let p be a prime and H a p-subgroup of the group G such that H is a normal subgroup of some Sylow p-subgroup of G but is a nonnormal subgroup of some other Sylow p-subgroup of G. Then there exists a p-group L contained in G such that $N_G(L)/C_G(L)$ is not a p-group.*
Proof: Among all Sylow p-subgroups of G which contain H as a non-normal subgroup select P so that $[P:N_P(H)]$ is minimal. Set $L = N_P(H)$ and $M = N_P(L)$, thus $P \geq M > L > H$ and H is not a normal subgroup of M. Let P_1 be a Sylow p-subgroup of $N_G(H)$ with $P_1 \geq L$. Since, by hypothesis, $N_G(H)$ contains at least one Sylow p-subgroup of G, P_1 is in fact a Sylow p-subgroup of G and $P_1 > L$. Let R be a Sylow p-subgroup of $N_G(L)$ with $R \geq N_{P_1}(L)$ and let P_2 be a Sylow p-subgroup of G with $P_2 \geq R$. Then $H \triangleleft R$; for if not, then H is nonnormal in P_2 and, since $N_{P_2}(H) \geq N_{P_1}(L) > L$, we would have $[P_2 : N_{P_2}(H)] < [P_2 : L] = [P : N_P(H)]$, contrary to the choice of P. As $M \leq N_G(L)$, H is not a normal subgroup of $N_G(L)$ and, since $R \leq N_G(H)$, it follows that there is a Sylow q-subgroup, Q, of $N_G(L)$ such that Q is not contained in $N_G(H)$ (here q is a prime with $q \neq p$). Therefore Q is not contained in $C_G(L)$ and q divides $|N_G(L)/C_G(L)|$.

The next theorem is due to B. Huppert ([Huppert 1954]). It will be used in the proof of Iwasawa's theorem but it is also of independent interest.

THEOREM 2.3 *If every maximal subgroup of a group is supersolvable, then the group is solvable.*

Proof: Let G be a group of smallest possible order for which the theorem might fail. Thus every proper subgroup of G is supersolvable, hence solvable, and every proper quotient group of G will be solvable by induction. Hence the theorem will be established as soon as we find a nontrivial proper normal subgroup of G. Let p be the smallest prime dividing the order of G, we may certainly assume that G is not a p-group. Let $L \neq \{1\}$ be any p-subgroup of G. If $N_G(L) = G$, then we are done; so we may assume that $N_G(L)$ is proper and therefore supersolvable. This implies (Theorem 1.8) that $N_G(L)$ satisfies the Sylow tower property. Thus the elements in $N_G(L)$ of order prime to p form a normal subgroup R of $N_G(L)$ with $(|R|, p) = 1$. $[R, L] \leq R \cap L = \{1\}$ and hence $R \leq C_G(L)$, so that $N_G(L)/C_G(L)$ is a p-group. With the aid of Burnside's theorem (Theorem 2.2) and the discussion which preceded it, we see that G must be p-normal and hence Grün's theorem will apply. Let P be a Sylow p-subgroup of G. As before $N_G(Z(P))$ is a proper subgroup of G and hence supersolvable. Moreover, $N_G(Z(P)) = RP$ where $(|R|, p) = 1$ and R is a normal subgroup of $N_G(Z(P))$. Then $RP' \lhd N_G(Z(P))$ and $N_G(Z(P))/RP' \cong P/P'$ is a nontrivial Abelian p-group. Therefore, by Grün's theorem, G has a nontrivial Abelian quotient group. Thus G has a nontrivial proper normal subgroup and the theorem is established.

We can now prove Iwasawa's theorem.

THEOREM 2.4 *The group G is supersolvable if and only if G is an equichained group.*

Proof: We have already observed that a supersolvable group is an equichained group. Thus it suffices to show that an equichained group G is supersolvable. Certainly every subgroup and quotient group of an equichained group is also an equichained group, hence we may assume by induction that every proper subgroup and quotient group of G is supersolvable. By Theorem 2.3 we have that G is solvable. Let H be a minimal normal subgroup of G, so that H is an elementary Abelian p-group for some prime p say $|H| = p^a$. By induction, G/H is supersolvable. We may also assume that H is the unique minimal normal subgroup of G; for if $K \neq H$ is a minimal normal subgroup of G, then G/K is supersolvable and $G \cong G/H \cap K$ is supersolvable (Theorem 1.1). Since $H \cap Z(O_p(G)) \neq \{1\}$, we must have $H \leq Z(O_p(G))$. Let $M/O_p(G)$ be a minimal normal subgroup of $G/O_p(G)$. Since $G/O_p(G)$ is supersolvable by induction, $|M| = qp^b$, where $p^b = |O_p(G)|$ and q is a prime. Since $O_p(G)$ is the largest normal p-subgroup of G, $q \neq p$. Let Q be a Sylow q-subgroup of M. Thus $M = O_p(G)Q$ and, by virtue of the Frattini argument, $G = MN_G(Q) = O_p(G)N_G(Q)$. We now consider two cases.

Case 1. $HN_G(Q) \neq G$. In this case $NH_G(Q)$ is supersolvable by induction. Hence if L is a minimal normal subgroup of $HN_G(Q)$ contained in H, then $|L| = p$. Moreover, since $L \leq H \leq Z(O_p(G))$, $N_G(L) = G$ and $L = H$. The supersolvability of G now follows from Theorem 1.2.

Case 2. $G = HN_G(Q)$. As H is the unique minimal normal subgroup of G, $G \neq N_G(Q)$ and $H \cap N_G(Q) = \{1\}$. We claim that $N_G(Q)$ is a maximal subgroup of G. First note that if $G \geq B \geq N_G(Q)$, then $B = B \cap G$ $= B \cap HN_G(Q) = (B \cap H)N_G(Q)$. Since $G = HN_G(Q)$, it follows that $B \cap H$ is a normal subgroup of G. Hence either $B \cap H = \{1\}$, which implies $B = N_G(Q)$, or $B \cap H = H$, which implies $B = G$. Thus $N_G(Q)$ is a maximal subgroup as claimed. Let r be the length of a maximal chain joining $\{1\}$ to $N_G(Q)$. Since $N_G(Q)$ is a maximal subgroup of G, there is a chain of length $r + 1$ joining $\{1\}$ to G. On the other hand, since $G/H \cong N_G(Q)$, a maximal chain joining $\{1\}$ to G which contains H will have length $r + a$. The equichain condition implies $a = 1$, so that H is cyclic of order p. Therefore G is supersolvable by Theorem 1.2.

3.   Maximal Subgroups

In the first section it was shown that the maximal subgroups of a supersolvable group have prime index. Bertram Huppert ([Huppert 1954]) showed that this property characterizes supersolvable groups. This section will be devoted to Huppert's theorem and related topics. Certain properties of the Fitting and Frattini subgroups will be used; these are given in the relevant sections of Appendix C.

THEOREM 3.1   *If every maximal subgroup of the group G has prime index, then G is supersolvable.*
Proof: We first show that G is solvable. Let p be the largest prime dividing $|G|$ and let G_p be a Sylow p-subgroup of G. If G_p is not a normal subgroup of G, then there is a maximal subgroup M of G with $M \geq N_G(G_p)$. By Sylow's theorems $[G : M]$ is congruent to 1 modulo p, but $[G : M] = q$, where q is a prime with $q < p$. This certainly is impossible and hence G_p is a normal subgroup of G. By induction we can assume G/G_p is solvable and therefore G is solvable.

We now show that G is supersolvable. Let H be a minimal normal subgroup of G. Since G is solvable, H is an elementary Abelian p-group for some prime p and by induction we can assume G/H is supersolvable. If K is some other minimal normal subgroup, then we may assume by induction that G/K is supersolvable and hence by Theorem 1.1 $G \cong G/H \cap K$ would be supersolvable. We may therefore assume that H is a unique minimal normal subgroup and that $\text{Fit}(G)$ is a p-group. Should H not lie in $\Phi(G)$, then there is a maximal subgroup M of G with $G = MH$. Moreover, since $M \cap H$ is a normal subgroup of G, we have that $M \cap H = \{1\}$ and that $[G : M] = |H|$. Therefore, since $[G : M]$ is prime, it follows that $|H| = p$, so that H is cyclic. In this case the supersolvability of G follows from Theorem 1.2. Hence we can assume that H lies in $\Phi(G)$. Let $\overline{A}$ denote the image of the set A under the natural homomorphism of G onto $G/\Phi(G)$. By induction $\overline{G}$ is supersolvable. By Lemma 2.2 of Appendix C, $\text{Fit}(\overline{G}) = \overline{\text{Fit}(G)}$ and hence $\text{Fit}(\overline{G})$ is a p-group. From Theorem 1.6 we know that $\overline{G}'$ lies in $\text{Fit}(\overline{G})$ and hence for every chief factor $\overline{A}/\overline{B}$ of $\overline{G}$ of order coprime to p, we have $\overline{G} = C_{\overline{G}}(\overline{A}/\overline{B})$. This and Theorem 2.5 of Appendix C yield that $\text{Fit}(\overline{G})$ is the intersection of the centralizers of those chief factors of $\overline{G}$ whose order is p. Hence $\overline{G}/\text{Fit}(\overline{G})$ is Abelian of exponent dividing $p - 1$ and therefore $G/\text{Fit}(G)$ is also Abelian of exponent dividing $p - 1$. So G is strictly p-closed and hence supersolvable.

As an immediate corollary to this theorem we have

COROLLARY 3.2 *The group G is supersolvable if and only if $G/\Phi(G)$ is supersolvable.*

The next theorem, due to Otto-Uwe Kramer ([Kramer 1976]) is similar to Huppert's theorem. Notice that because of Kramer's theorem, the supersolvability of a solvable group can be concluded if only those maximal subgroups not containing the Fitting subgroup have prime index.

THEOREM 3.3 *Let G be a solvable group. Then G is supersolvable if and only if, for every maximal subgroup M of G, either $M \geq \text{Fit}(G)$ or $M \cap \text{Fit}(G)$ is a maximal subgroup of $\text{Fit}(G)$.*
Proof: Let G be supersolvable and M a maximal subgroup of G. By Theorem 1.7, $[G : M]$ is prime. Should M not contain $\text{Fit}(G)$, then $G = M\,\text{Fit}(G)$ and $[G : M] = [\text{Fit}(G) : M \cap \text{Fit}(G)]$. Therefore $M \cap \text{Fit}(G)$ has prime index in $\text{Fit}(G)$ and hence is a maximal subgroup of $\text{Fit}(G)$.

In order to show the sufficiency, let G satisfy the condition on maximal subgroups. Then so does $G/\Phi(G)$, by virtue of Lemma 2.2 of Appendix C. So if $\Phi(G) \neq \{1\}$, then $G/\Phi(G)$ is supersolvable by induction, which implies G is supersolvable by Corollary 3.2. We may, therefore, assume that $\Phi(G) = \{1\}$. We know by Lemma 2.3 of Appendix C that $\text{Fit}(G)$ is Abelian and is the direct sum of minimal normal subgroups of G, say $\text{Fit}(G) = H_1 \oplus H_2 \oplus \ldots \oplus H_r$. Since $\Phi(G) = \{1\}$, for each i, $i = 1, 2, \ldots, r$, there is a maximal subgroup M_i of G with $G = H_i M_i$ and $H_i \cap M_i = \{1\}$. Moreover, $\text{Fit}(G) = H_i(M_i \cap \text{Fit}(G))$ as is easy to see. By hypothesis, $M_i \cap \text{Fit}(G)$ is a maximal subgroup of $\text{Fit}(G)$ and, since $\text{Fit}(G)$ is nilpotent, $M_i \cap \text{Fit}(G)$ has prime index in $\text{Fit}(G)$. Since $M_i \cap H_i = \{1\}$, it follows that

H_i has prime order for $i = 1, 2, \ldots, r$. As $|H_i|$ is prime, we know that $\text{Aut}_G(H_i) \cong G/C_G(H_i)$ is an Abelian group and hence $C_G(H_i)$ contains the commutator subgroup G'. Thus $C = \bigcap\{C_G(H_i) : i = 1, 2, \ldots, r\}$ also contains G'. But C is the centralizer of $\text{Fit}(G)$ and hence by Theorem 2.6 of Appendix C we have $G' \leq C \leq \text{Fit}(G)$.

We are now in a position to show that maximal subgroups of G have prime index so that Huppert's theorem can be applied. Let M be a maximal subgroup of G. Then either $M \geq \text{Fit}(G)$ or $M\text{Fit}(G) = G$. If $M \geq \text{Fit}(G)$, then $M \geq G'$ and hence M is normal in G and $[G : M]$ is prime. If $M\text{Fit}(G) = G$, then $[G : M] = [\text{Fit}(G) : M \cap \text{Fit}(G)]$ and, since $\text{Fit}(G)$ is nilpotent, the hypothesis implies that $[G : M]$ is prime. We can now apply Huppert's theorem (Theorem 3.1) to conclude supersolvability of G.

COROLLARY 3.4 *Let G be a solvable group. Then G is supersolvable if and only if for each maximal subgroup M of G and each normal subgroup N of G either M contains N or $M \cap N$ is a maximal subgroup of N.*

4. *Existence of Subgroups*

From the fundamental theorem of Abelian groups it is easily seen that an Abelian group contains subgroups of every possible order, i.e., if G is Abelian and n is a factor of $|G|$, then G contains a subgroup N of order n (and naturally $N \triangleleft G$). In fact it is not difficult to show that a group is nilpotent if and only if it contains a normal subgroup of each possible order [Holmes 1966]. Ore and Zappa (and others) obtained a similar characterization for supersolvable groups.

THEOREM 4.1 *A group G is supersolvable if and only if each subgroup $H \leq G$ contains a subgroup of order d for each divisor d of $|H|$.*

The following alternate formulation, clearly equivalent, is more easily treated.

THEOREM 4.2 *A group G is supersolvable if and only if each subgroup $H \leq G$ contains a subgroup of index p for each prime divisor p of $|H|$.*
Proof: From Section 1 we recall that subgroups and factor groups of a supersolvable group are supersolvable. Also (Theorem 1.8) if G is supersolvable and p is the largest prime divisor of $|G|$, then a Sylow p-subgroup G_p of G is normal. Moreover, if $|G_p| = p^r$, then G_p contains a subgroup P of order p^{r-1} with $P \triangleleft G$.

To demonstrate the necessity of the condition in the theorem we proceed by induction on the order of G. If H is a proper subgroup of the supersolvable group G, then H is supersolvable and so by induction contains a subgroup of prime index for each prime divisor of $|H|$. Thus we need only show that G itself contains a subgroup of index q whenever q is a prime divisor of $|G|$. If $q < p$, the largest prime dividing $|G|$, then consider the factor group $\bar{G} = G/G_p$. Since $\bar{G}$ is supersolvable and $|\bar{G}| < |G|$, we conclude by induction that $\bar{G}$ contains a subgroup $\bar{K}$ with $[\bar{G} : \bar{K}] = q$. Clearly K, the pre-image of $\bar{K}$ in G, has index q in G.

Suppose on the other hand that $q = p$. From the Sylow tower property (or, more quickly from Theorem 1 of Appendix A) it follows that G_p has a complement G^p. Then $M = PG^p$ is a subgroup of index p in G, where P is the normal subgroup of G of order p^{r-1} mentioned above.

The sufficiency is proved by induction on $|G|$ also. Let q be the least prime divisor of $|G|$. By hypothesis G contains a subgroup K of index q in G, and by a standard exercise $K \vartriangleleft G$. Now we conclude by induction that K is supersolvable. Then a Sylow p-subgroup K_p, where p is the largest prime divisor of $|K|$, is normal in K and hence in G. If $p = q$, then G is a p-group and hence supersolvable, so we need only treat the case $q < p$. Then K_p is the Sylow p-subgroup of G.

Let N be a minimal normal subgroup of G lying in $Z(K_p)$. We shall show that $|N| = p$. By hypothesis G contains a subgroup M of index p in G. Then $M \cap N \vartriangleleft N$, since N is Abelian, so $M \cap N \vartriangleleft G$. Thus either $M \cap N = \{1\}$ or $M \cap N = N$ since N is minimal normal in G. If $M \cap N = \{1\}$, then $|N| = [G : M] = p$ so consider the case $M \cap N = N$. From induction M is a supersolvable group so we know that N contains a subgroup N_1 of order p which is normal in M. Now $N_G(N_1)$ contains M and K_p since $N_1 \leq N \leq Z(K_p)$. Since $[G : M] = p$, it follows that $G = K_p M$; hence $N_1 \vartriangleleft G$ and $N = N_1$. So G contains a normal subgroup N of order p in all cases.

Now we consider $G^* = G/N$ and show that each subgroup H^* of G^* contains a subgroup of prime index for each prime factor of $|H^*|$. If $H^* < G^*$, then its pre-image is a proper subgroup of G and hence super-solvable, so H^* is supersolvable and contains the required subgroups. Consequently we need only show that G^* contains the required maximal subgroups.

Let r be a prime divisor of $|G^*|$. Then r divides $|G|$ and G contains a subgroup R with $[G : R] = r$. If $R \geq N$, then $[G^* : R^*] = [G : R]$ where $R^* = R/N$ and R^* is the subgroup of G^* that we need. If R does not contain N, then $R \cap N = \{1\}$, $G = RN$, and $G^* = RN/N \cong R$. Since R is a proper subgroup of G, it is supersolvable, hence G^* is supersolvable (and contains the required maximal subgroups). So we conclude in all cases that G^* is supersolvable and it follows that G is supersolvable.

It should be noted that a group G is not necessarily supersolvable if the condition "each subgroup $H \leq G$ contains a subgroup of every possible order," of Theorem 4.1 is replaced by "G contains a subgroup of every possible order." The symmetric group S_4 satisfies the weaker condition but is not supersolvable. Groups satisfying this weaker condition will be discussed in Chapter 3.

The condition on a group G given in Theorem 4.2, namely

$$\text{for all } H \leqslant G \text{ and all primes } q \text{ dividing } |H|, \text{ there exists } K \text{ such that } K \leqslant G \text{ and } [H:K] = q \tag{1}$$

has the following dual formulation:

$$\text{for all } H \leqslant G \text{ and all primes } q \text{ dividing } [G:H], \text{ there exists } K \text{ such that } H \leqslant K \leqslant G \text{ and } [K:H] = q. \tag{2}$$

Groups satisfying condition (2) have been studied by a number of authors and have been christened with a number of different names; among these are A_1-groups, $\mathcal{Y}$-groups, GCLT-groups, and groups satisfying the overgroup condition. For consistency with Chapter 4 we will call them $\mathcal{Y}$-groups.

First we note that a $\mathcal{Y}$-group G also satisfies

$$\text{for all } H \leqslant G \text{ and all } d \text{ dividing } [G:H], \text{ there exists } K \text{ such that } H \leqslant K \leqslant G \text{ and } [K:H] = d. \tag{3}$$

Thus conditions (2) and (3) are equivalent and either may be taken as the definition of a $\mathcal{Y}$-group.

Conditions (2) and (3) turn out to be somewhat stronger than condition (1) which is to say that the class of $\mathcal{Y}$-groups is a proper subclass of the class of supersolvable groups. We will prove the following characterization:

THEOREM 4.3 $G \in \mathcal{Y}$ *if and only if G contains a normal Hall subgroup N such that*
(i) *N and G/N are both nilpotent*
(ii) *for all $H \leq N$, $G = NN_G(H)$.*

To prove that $\mathcal{Y}$-groups are supersolvable, let $G \in \mathcal{Y}$ and let M be a maximal subgroup of G. Let p be any prime factor of $[G:M]$. Then G has a subgroup K such that $[K:M] = p$, and K must be G by maximality. Consequently each maximal subgroup of G has prime index and so G is supersolvable by Huppert's theorem (Theorem 3.1). Now consider the following:

EXAMPLE Let G be the direct product of the symmetric group S_3 with the cyclic group $\mathbb{Z}_3$ of order 3. Let H be the subgroup of order 3 generated by the pair (x, y) where x is an element of order 3 in S_3 and y is an element of order 3 in $\mathbb{Z}_3$. Then $[G:H] = 6$ but there is no subgroup K of G with $K \geq H$ and $[K:H] = 2$. So we see that G is a supersolvable group which is not in class $\mathcal{Y}$.

We next note that if $G \in \mathcal{Y}$ and $N \lhd G$, then $G/N \in \mathcal{Y}$. For if H/N is a subgroup of G/N and p is a prime factor of $[G/N : H/N] = [G:H]$, then there is a subgroup K of G, $K \geq H$ with $[K : H] = p$. So $K/N \geq H/N$ and $[K/N : H/N] = [K : H] = p$.

Theorem 4.3 will be a consequence of the next two theorems.

THEOREM 4.4 *Let $G \in \mathcal{Y}$ and let $N = \gamma_\infty(G)$ be the nilpotent residual of G. Then*

(i) *N is a nilpotent Hall subgroup,*

(ii) *for all $H \leq N$, $G = NN_G(H)$.*

Proof: Since G is supersolvable, G' is nilpotent (Theorem 1.6), and since the nilpotent residual of G lies in G' we have N nilpotent. Thus to prove (i) we must show that N is a Hall subgroup of G (i.e., $(|N|, [G : N]) = 1$). Suppose this is not true; choose G in $\mathcal{Y}$ of minimal order among those groups for which (i) is false. We will show that such a group G cannot exist; this will follow from a series of steps. (Note that G is not nilpotent.)

(a) *N contains a unique minimal normal subgroup of G.* For suppose H_1 and H_2 are minimal normal subgroups of G lying in N with $H_1 \neq H_2$. The nilpotent residual of G/H_i is N/H_i since $(G/H_i)/(N/H_i) \cong G/N$, $i = 1$, 2. Therefore we have from the minimality of $|G|$ that N/H_i is a Hall subgroup of G/H_i. Let $|G/N| = m$, $|N| = n$, $|H_i| = p_i$. Then

$$(m, n/p_1) = (m, n/p_2) = 1.$$

If $p_1 \neq p_2$, then p_1 divides n/p_2 and p_2 divides n/p_1 so that neither divides m which implies $(m, n) = 1$. If $p_1 = p_2 = p$, then p^2 divides n and since p does not divide m we again have $(m, n) = 1$. So it would follow that N is a Hall subgroup of G contrary to our choice of G. Hence N contains a unique minimal normal subgroup of G which we will henceforth denote by H. Since G is supersolvable, H has order p, a prime.

(b) *$N = H$.* Since N is nilpotent it follows from (a) that N is a p-group, so suppose $|N| = p^k$ with $k > 1$. Then N/H is the nilpotent residual of G/H and $|N/H| = p^{k-1}$ is coprime to $|(G/H)/(N/H)| = |G/H|$. But this would imply that N is a Hall subgroup of G, contrary to our assumption. So $N = H$.

(c) *Each minimal normal subgroup of G has order p.* For suppose Q is a minimal normal subgroup of G with $|Q| = q \neq p$. Since $(G/Q)/(NQ/Q) \cong G/NQ$ is nilpotent, the nilpotent residual L/Q of G/Q lies in NQ/Q and so $|L/Q| = 1$ or p. If $|L/Q| = 1$, then G/Q is nilpotent which means that $G = G/N \cap Q$ is nilpotent which is contrary to the choice of N. If $|L/Q| = p$, then it follows that p is not a factor of $|G/NQ|$ which means that p is not a factor of $|G/N|$ and N is a Hall subgroup of G, contrary to the choice of G. Thus such a subgroup Q does not exist.

(d) *Fit(G) is a p-group containing N.* That Fit$(G) \geq N$ is immediate by definition of Fit(G). If Fit(G) contains a nontrivial q-subgroup, $q \neq p$, then

G would possess a minimal normal subgroup of order q, contrary to (c). Clearly Fit(G) is the Sylow p-subgroup of G.

(e) $N \cap \Phi(G) = \{1\}$. For otherwise $N \leq \Phi(G)$ and this implies that $G/\Phi(G)$, and hence G, is nilpotent.

(f) *N is the unique minimal normal subgroup of G.* Suppose $K \neq N$ is a minimal normal subgroup of G. By (c) we have $|K| = |N| = p$. Since G/N is nilpotent, it follows that $K \leq Z(G)$. Let $N = \langle x \rangle$, $K = \langle y \rangle$. Then xy is an element of order p. Let $r = |G/\text{Fit}(G)|$. Then r divides $[G:\langle xy \rangle]$ so $G \in \mathcal{Y}$ (in particular, G satisfies condition (3) above) implies existence of a subgroup R or order rp with $R \geq \langle xy \rangle$. $R \cap \text{Fit}(G) = \langle xy \rangle$ so that $\langle xy \rangle \lhd R$. Since R centralizes y and normalizes $\langle xy \rangle$, it follows easily that R centralizes x. Since x is centralized by Fit(G) we see that N lies in the center of $G = \text{Fit}(G)R$. But since G/N is nilpotent this means that G is nilpotent, a contradiction. This proves assertion (f).

Now we can show that an impossible situation exists. By (e) and (f), $\Phi(G) = \{1\}$. Apply Lemma 2.3 of Appendix C, keeping in mind that N is the only minimal normal subgroup of G, to conclude that Fit$(G) = N$. Since Fit(G) is the Sylow p-subgroup of G (as noted in (d)) then N is a Hall subgroup of G after all, the desired contradiction.

To prove (ii) consider a subgroup H of N. Let $m = |G/N|$. Since $G \in \mathcal{Y}$, there exists a subgroup L of G with $L \geq H$ and $[L:H] = m$. Then $L \cap N = H \lhd L$ so that $N_G(H) \geq L$. Since $NL = G$ we have $G = NN_G(H)$.

Remark Since the subgroup N of the theorem above is a Hall subgroup, it follows (by extended Sylow theory for solvable groups—see Appendix A) that N has a complement K in G and that all subgroups of G of order $|K|$ are conjugates of K. Then K normalizes at least one element from each class of conjugate (in G) subgroups of N. For if $H \leq N$, then $NN_G(H) = G$ implies that $N_G(H)$ contains a conjugate of K.

Continuing our program for proving Theorem 4.3, we consider the class $\mathcal{Y}'$ of groups G having the following property: G contains a normal Hall subgroup N such that (i) N and G/N are nilpotent, and (ii) for all $H \leq N$, $G = NN_G(H)$. That $\mathcal{Y} \subseteq \mathcal{Y}'$ is just the statement of Theorem 4.4.

LEMMA 4.5 *Let $G \in \mathcal{Y}'$ with defining normal Hall subgroup N. If $H \lhd N$, then $H \lhd G$ and $G/H \in \mathcal{Y}'$.*
Proof: From $G = NN_G(H)$ and $N_G(H) \geq N$ it follows that $H \lhd G$. Let $\bar{G} = G/H$ and $\bar{N} = N/H$. Since $\bar{N}$ is a factor of N and $\bar{G}/\bar{N} \cong G/N$ we see that $\bar{N}$ is a Hall subgroup of $\bar{G}$ with $\bar{N}$ and $\bar{G}/\bar{N}$ nilpotent. Let $\bar{K} = K/H$ be a subgroup of $\bar{N}$. Then $K \leq N$ is normalized by some complement L of N in G so that $HL/H = \bar{L} \leq N_{\bar{G}}(\bar{K})$. Therefore $\bar{G} = \bar{N}\bar{L} = \bar{N}N_{\bar{G}}(\bar{H})$ and the lemma is proved.

Note It follows easily from this lemma that the groups in $\mathcal{Y}'$ are supersolvable. For an element of prime order in $Z(N)$ for such a group G generates a normal subgroup of G.

THEOREM 4.6 $\mathcal{Y}' \subseteq \mathcal{Y}$.
Proof: Let $G \in \mathcal{Y}'$ with defining normal Hall subgroup N. Let K be a subgroup of G and let q be a prime divisor of $[G : K]$. We must show that there is a $K^* \leq G$ with $K^* \geq K$ and $[K^* : K] = q$, and we treat two cases, using induction in both.

Case 1. $q \,|\, |N|$. If $Z(N)$ contains a q-element not in K, then there is an $x \in Z(N) \backslash K$ such that $x^q \in K$. Since $\langle x \rangle \lhd N$, we have $\langle x \rangle \lhd G$ and we take $K^* = \langle x \rangle K$. If, however, the Sylow q-subgroup Q of $Z(N)$ lies in K, then we know from the lemma above that $G/Q \in \mathcal{Y}'$. Then by induction G/Q contains a subgroup K^*/Q with $K^*/Q \geq K/Q$ and $[K^*/Q : K/Q] = q$. Therefore $K^* \geq K$ and $[K^* : K] = q$.

Case 2. $q \nmid |N|$. Let $T = N \cap K$. Then $N_G(T)$ contains K and T is a Hall subgroup of K. So T has a complement S in K: $T \cap S = \{1\}$ and $TS = K$. Since $G \in \mathcal{Y}'$, we know that $N_G(T)$ contains a complement C of N in G, $|C| = |G/N|$. Then from Theorem 1 of Appendix A if follows that a conjugate in $N_G(T)$ of C contains S, so by adjusting notation we have $N_G(T) \geq C \geq S$. Therefore TC is a subgroup of G which contains K, and K/T is a subgroup of the nilpotent group $TC/T \cong C \cong G/N$ with $q \,|\, [CT/T : K/T]$. Since nilpotent groups are in $\mathcal{Y}$ we know there exists $K^*/T \leq TC/T$ with $[K^*/T : K/T] = q$. Then $G \geq K^* \geq K$ with $[K^* : K] = q$.

This provides one characterization of the class $\mathcal{Y}$; others will be given later in the book.

5. The Generalized Center and Central Series

In this section we present a characterization of supersolvability analagous with the characterization of nilpotency through the center and central series. The material is drawn primarily from the work of Ram K. Agrawal and O. H. Kegel.

Two subgroups H and K of a group G are said to **permute** or be **permutable** subgroups if $HK = KH$. It is easily seen that H and K permute if and only if the set HK is a subgroup of G.

DEFINITION An element x of group G is a **generalized central** element (g.c. element) of G provided the subgroup $\langle x \rangle$ permutes with every Sylow subgroup of G.

Clearly every element of a normal cyclic subgroup of G is a g.c. element of G, hence a supersolvable group contains nontrivial g.c. elements.

A g.c. element x of G is obviously a g.c. element of a subgroup $H \leq G$ if $x \in H$.

LEMMA 5.1   *Let x be a g.c. element of G and let θ be a homomorphism of G. Then x^θ is a g.c. element of G^θ.*
Proof: Each Sylow subgroup of G^θ is the image S^θ of a Sylow subgroup S of G. Since $\langle x \rangle S$ is a subgroup of G it follows that the set $\langle x \rangle^\theta S^\theta = \langle x^\theta \rangle S^\theta = (\langle x \rangle S)^\theta$ is a subgroup of G^θ. Hence x^θ is a g.c. element of G^θ.

LEMMA 5.2   *Let x be a g.c. element of G and $|x| = p^r m$ where p is a prime and $(p, m) = 1$. Then $x = yz = zy$ with $|y| = p^r$, $|z| = m$, y is a g.c. element of G, and $y \in O_p(G)$.*
Proof: The factorization $x = yz = zy$ is a well-known number-theoretic fact (y and z here are appropriate powers of x). Let P be a Sylow p-subgroup of G. Then since x is a g.c. element, $\langle x \rangle P$ is a subgroup of G of order $n = |\langle x \rangle P| = |x| |P| / |\langle x \rangle \cap P|$. Since the power of p which divides n is exactly $|P|$ we see that $|\langle x \rangle \cap P| = p^r$. This implies that $\langle y \rangle$, the Sylow p-subgroup of $\langle x \rangle$, lies in P. So $\langle y \rangle$ lies in every Sylow p-subgroup of G and hence $y \in O_p(G)$ (since $O_p(G)$ is the intersection of all the Sylow p-subgroups of G).

Now let Q be a Sylow q-subgroup of G, $q \neq p$, and consider the subgroup $\langle x \rangle Q$. Since $|\langle x \rangle Q| = |x| |Q| / |\langle x \rangle \cap Q|$, it is clear that $\langle y \rangle$ is a Sylow p-subgroup of $\langle x \rangle Q$. Since

$$(\langle x \rangle Q) \cap O_p(G) \geq \langle y \rangle$$

it follows that $(\langle x \rangle Q) \cap O_p(G) = \langle y \rangle$ and $\langle y \rangle \triangleleft \langle x \rangle Q$. Since $\langle y \rangle$ is normalized by Q we have $\langle y \rangle Q = Q \langle y \rangle$. Therefore y is a g.c. element of G.

DEFINITION   The **generalized center** of a group G is the subgroup

$$\mathrm{genz}(G) = \langle x : x \text{ is a g.c. element of } G \rangle.$$

THEOREM 5.3   *$\mathrm{genz}(G)$ is a nilpotent normal subgroup of G, and each of its Sylow subgroups is generated by g.c. elements of G.*
Proof: The normality of $\mathrm{genz}(G)$ follows from Lemma 5.1 and the nilpotency follows from Lemma 5.2 (for the Sylow p-subgroup of $\langle x \rangle$, when x is a g.c. element of G, lies in $O_p(G)$ which lies in $\mathrm{Fit}(G)$; hence $\langle x \rangle \leq \mathrm{Fit}(G)$).

Let $|G| = p_1^{e_1} \ldots p_n^{e_n}$ be the prime factorization of $|G|$. Let

$$P_i = \langle x_i : x_i \text{ is a g.c. element of } G \text{ and } |x_i| \mid p_i^{e_i} \rangle \qquad i = 1, \ldots, n.$$

Then it follows from Lemmas 5.1 and 5.2 that $P_i \triangleleft G$ and $P_i \leq O_p(G)$ for all i. Therefore $P = P_1 P_2 \ldots P_n$ is a normal nilpotent subgroup of G and clearly $P \leq \text{genz}(G)$. However, it follows from the factorization in Lemma 5.2 that P contains every g.c. element of G. Hence $P = \text{genz}(G)$.

The proof adjusts easily to provide a slightly more general result:

THEOREM 5.4 *Let the normal subgroup N of G be generated by a set of g.c. elements of G. Then N is nilpotent and each of its Sylow subgroups is generated by* g.c. *elements of G.*

DEFINITION A **generalized central series** of G is a set of subgroups $G_i \triangleleft G$ with
$$\langle 1 \rangle = G_0 \leq G_1 \leq G_2 \leq \ldots \leq G_r \leq G$$
such that G_{i+1}/G_i is generated by g.c. element of G/G_i for $i = 1, \ldots, r - 1$. G_r is the **terminal term** of this series.

For working with such series the following is useful.

LEMMA 5.5 *Let x be a* g.c. *p-element of G and let Q be a Sylow q-subgroup of G, $q \neq p$. Then*
(a) $Q \leq N_G(\langle x \rangle)$.
(b) *If, in addition* $(q, p(p - 1)) = 1$, *then* $[\langle x \rangle, Q] = \{1\}$.
Proof: In the proof of Lemma 5.2 we observed that $y \in O_p(G)$. In the present case, since x is a p-element, x is the element y of Lemma 5.2. So $x \in O_p(G)$. Then $(\langle x \rangle Q) \cap O_p(G)$ is clearly $\langle x \rangle$ and $\langle x \rangle \triangleleft \langle x \rangle Q$. This proves Part (a). As for Part (b), $N_G(\langle x \rangle)/C_G(\langle x \rangle)$ is isomorphic with a subgroup of the automorphism group of $\langle x \rangle$ whose order has no prime divisors not dividing $p(p - 1)$. Hence $Q \leq C_G(x)$.

LEMMA 5.6 *Let G_n be a term of a generalized central series of G. Then G_n satisfies the Sylow tower property.*
Proof: Note that the terms of the generalized central series which lie in G_n form a generalized central series of G_n with terminal term G_n. So it follows by induction that we need only treat the case $G = G_n$.

Let p be the largest prime divisor of $|G|$. Let G_i be the smallest term of the given generalized central series of G for which $p \mid |G_i|$ and let P be a Sylow p-subgroup of G_i. Then every prime divisor of $|G_{i-1}|$ is less than p, so from Lemma 5.5 we see that for $j < i$
$$[P, G_j] \leq G_{j-1}.$$
From this and the fact that $(|P|, |G_{i-1}|) = 1$ it follows easily that P is centralized by G_{i-1} (recall Lemma 1.11).

Now consider the subgroup PG_{i-1}. Since G_i/G_{i-1} is nilpotent by Theorem 5.4, we see that $PG_{i-1} \triangleleft G$. Therefore $P \triangleleft G$. The factor group G/P

has a generalized central series terminating in G/P (by Lemma 5.1) so by induction G/P satisfies the Sylow tower property. Hence G does likewise.

THEOREM 5.7 *Let G_n be a term of a generalized central series of G and let S be a supersolvable subgroup of G. Then SG_n is a supersolvable subgroup of G.*

Proof: Proceeding by induction on $|G|$ we need only treat the case $G = SG_n$. First we show that G satisfies the Sylow tower property. Let p be the largest prime divisor of $|G|$. If p divides $|G_n|$, then G_n satisfies the Sylow tower property by Lemma 5.6 and P_n, the Sylow p-subgroup of G_n, is a normal subgroup of G. Then G/P_n satisfies the induction hypothesis so that G/P_n, and hence G, satisfies the Sylow tower property. So suppose $p \nmid |G_n|$ and let P_S be the Sylow p-subgroup of S. As in the proof of Lemma 5.6 we see that $[P_S, G_n] = \{1\}$ so that $P_S \lhd G$. Then G/P_S satisfies the induction hypothesis and it follows that G satisfies the Sylow tower property.

Let P be the Sylow p-subgroup of G and consider $\Phi(G)$. If $\Phi(G) \neq \{1\}$, then $G/\Phi(G)$ satisfies the induction hypothesis and hence is supersolvable. By Corollary 3.2 we know that G is supersolvable if $G/\Phi(G)$ is. So we need only consider the case $\Phi(G) = \{1\}$. Then since $P \lhd G$, we have $\Phi(P) = \{1\}$ (Lemma 1.1 of Appendix C). Therefore P is an elementary Abelian p-group.

Let M be a maximal subgroup of G. We shall prove G supersolvable by showing that $[G:M]$ is a prime. If M contains a nontrivial subgroup $H \lhd G$, then G/H is supersolvable by induction so that $[G/H:M/H] = [G:M]$ is a prime. So we can assume that M contains no nontrivial normal subgroup of G. Then $G = MP$ and $[G:M] = p^a$. Let G_i be the smallest term of the given generalized central series of G with p not dividing $|G_i|$. If $i > 0$, then $G_i < M$. For $G_i \neq \{1\}$ and $G_i \not\lhd M$ imply that $G = MG_i$ holds and thus $[G:M]$ divides $|G_i|$ contrary to $[G:M] = p^a$. Consequently $i = 0$ and $G_1 \cap P = P_1$ is a nontrivial normal subgroup of G generated by g.c. elements of order p. Since $P_1 \not\lhd M$, there exists a g.c. element x of G lying in P_1 but not in M. Let Q be a Sylow q-subgroup of G. Then $\langle x \rangle Q$ is a subgroup of G and $(\langle x \rangle Q) \cap P = \langle x \rangle$. So $N_G(\langle x \rangle) \geq Q$. Certainly $N_G(\langle x \rangle) \geq P$ since P is Abelian. Therefore $\langle x \rangle \lhd G$. Then $\langle x \rangle M$ is a subgroup of G, and $\langle x \rangle M = G$ since M is maximal and $x \notin M$. This shows that $[G:M] = o(x) = p$, and the theorem is proved.

COROLLARY 5.8 *Each term of a generalized central series of G lies in every maximal supersolvable subgroup of G (and hence is supersolvable itself).*

A group possesses a canonical generalized central series, the **ascending generalized central series** defined by

$$G_0 = \{1\} \leq G_1 \leq G_2 \leq \ldots \leq G_r \leq G$$

with $G_{i+1}/G_i = \mathrm{genz}(G/G_i)$ for all i. The terminal term of this series is the **hyper-generalized-center** of G.

COROLLARY 5.9 *The hyper-generalized-center of G is a normal supersolvable subgroup of G contained in every maximal supersolvable subgroup of G.*

It is not yet known whether or not the intersection of the maximal supersolvable subgroups of G equals the hyper-generalized-center.

COROLLARY 5.10 *A group G is supersolvable if and only if it is equal to its hyper-generalized-center.*
Proof: Clearly the generalized center of a supersolvable group is nontrivial, hence the terminal term of the ascending generalized central series of a supersolvable group G is G.

6. Some Additional Conditions for Supersolvability

A group G is nilpotent if (and only if) every maximal subgroup of G is normal in G. Similarly it can be shown that a group is supersolvable if its "large" subgroups are "nearly" normal. These terms, of course, must be replaced by precise ones.

DEFINITION A subgroup H of the group G is **S-quasinormal** (Sqn) in G if H permutes with every Sylow subgroup of G.

Clearly this concept is closely related to that of generalized central elements in the previous section. It has been studied by Kegel, Deskins, Agrawal, and others.

THEOREM 6.1 *Let H be Sqn in G.*
(a) If θ is a homomorphism, then H^θ is Sqn in G^θ.
(b) If $G \geq K \geq H$, then H is Sqn in K.
Proof: From Sylow theory a Sylow subgroup of G^θ has the form S^θ for a Sylow subgroup S of G. Since $HS = SH$, we have $H^\theta S^\theta = S^\theta H^\theta$ and H^θ is Sqn in G^θ.
Let S_1 be a Sylow subgroup of K; then $S_1 = K \cap S$ for some Sylow subgroup S of G by the Sylow theory. Since HS is a subgroup, we have $K \cap SH = (K \cap S)H = S_1 H$ is a subgroup. Hence
$$HS_1 = S_1 H.$$

Now we shall show that an Sqn subgroup of G is close to being normal in G.

THEOREM 6.2 *If G is solvable and $H \leq G$ is Sqn in G, then $H/\mathrm{Core}_G(H)$ is nilpotent.*

Proof: Since $H/\mathrm{Core}_G(H)$ is Sqn in $G/\mathrm{Core}_G(H)$, it clearly suffices to prove the theorem when $\mathrm{Core}_G(H) = \{1\}$. Let $p_1, p_2, \ldots, p_k$ denote the prime divisors of $|G|$, let $\{S_1, S_2, \ldots, S_k\}$ be a Sylow basis (thus each S_i is a Sylow p_i-subgroup of G) and let $\{C_1, C_2, \ldots, C_k\}$ be the associated Sylow complements (i.e., $C_i = S_1 S_2 \ldots S_{i-1} S_{i+1} \ldots S_k$).

(a) *If R is a Sylow p_i-subgroup of H and $x \in C_i$, then $R^x \leq H$.* To see this consider $K = HC_i$. Clearly R^x is a Sylow p_i-subgroup of K since $|K| = |H||C_i|/|H \cap C_i| = |H|d$ where d is coprime to p_i. By the previous theorem H is Sqn in K, so HR^x is a subgroup of K of order $|H||R^x|/|H \cap R^x|$ and clearly this is $|H|$. Thus $R^x \leq H$.

(b) *For distinct primes p_i and p_j, $HS_i \cap HS_j = H$.* Indeed $[HS_i \cap HS_j : H]$ must divide both $[HS_i : H] = [S_i : S_i \cap H]$, which is a p_i-number and $[HS_j : H] = [S_j : S_j \cap H]$, which is a p_j-number. Thus $[HS_i \cap HS_j : H] = 1$.

(c) *For each prime p_i, $H \cap S_i$ is a Sylow p_i-subgroup of H.* This is because $H \cap S_i$ is a p_i-subgroup of H and $[H : H \cap S_i] = [HS_i : S_i]$, which is a p_i'-number.

To complete the proof we will show the Sylow subgroups of H are normal by showing that arbitrary elements $x \in H \cap S_i$ and $y \in H \cap S_j$ (where $p_i \neq p_j$) commute. Since we're assuming $\mathrm{Core}_G(H) = \{1\}$, it will suffice to show $(x, y) \in \mathrm{Core}_G(H)$. For any $g \in G$, $g \in C_i S_i$ so $g = \gamma \sigma$ where $\gamma \in C_i$, $\sigma \in S_i$. Then $g^{-1}(x, y)g = \sigma^{-1}(\gamma^{-1}x^{-1}\gamma)(\gamma^{-1}y^{-1}xy\gamma)\sigma$. Since $H \cap S_i$ and $y^{-1}(H \cap S_i)y$ are Sylow p_i-subgroups of H we get $\gamma^{-1}x^{-1}\gamma \in H$ and $\gamma^{-1}y^{-1}xy\gamma \in H$ by (a). Thus $g^{-1}(x, y)g \in S_i HHS_i = HS_i$. Similarly $g \in G = C_j S_j$ so $g^{-1}(x, y)g \in HS_j$. Thus $g^{-1}(x, y)g \in HS_i \cap HS_j$. Now apply (b) to get $g^{-1}(x, y)g \in H$; this for arbitrary $g \in G$. Hence $(x, y) \in \mathrm{Core}_G(H)$.

We note in passing that Theorem 6.2 (but not its proof) is also valid for nonsolvable groups (see [Deskins 1963]).

COROLLARY 6.3 *If G is a solvable group and H is Sqn in G, then H is subnormal in G.*

Proof: If $\mathrm{Core}_G(H)$ is nontrivial, then $H/\mathrm{Core}_G(H)$ is Sqn in $G/\mathrm{Core}_G(H)$ by Theorem 6.1 so by induction $H/\mathrm{Core}_G(H)$ is subnormal in $G/\mathrm{Core}_G(H)$ which implies H is subnormal in G. So we have left the case $\mathrm{Core}_G(H) = \{1\}$; then H is nilpotent by Theorem 6.2.

For S_i a Sylow p_i-subgroup of G note that $H \cap S_i$ is a Sylow p_i-subgroup of H and hence is the only such by nilpotency. This implies that the Sylow p_i-subgroup of H lies in every Sylow p_i-subgroup of G. Thus $H \cap S_i$

$\leq O_{p_i}(G)$ and it follows that $H \leq \mathrm{Fit}(G)$. Therefore H is subnormal in $\mathrm{Fit}(G)$ because all subgroups of a nilpotent group are subnormal. This and $\mathrm{Fit}(G) \lhd G$ imply H is subnormal in G.

As we noted for Theorem 6.2, the preceding corollary is also valid for nonsolvable groups.

COROLLARY 6.4 *Let H be* Sqn *in G, where G is solvable. If H is maximal in G, then $H \lhd G$.*
Proof: H Sqn G implies H is subnormal in G. Therefore $N_G(H) > H$.

Now we are able to prove a result due to Agrawal which gives a sufficient condition for supersolvability involving 2-maximal subgroups (definition: H is a **2-maximal subgroup** of the group G if H is a maximal subgroup of some maximal subgroup M of G).

THEOREM 6.5 *Let each 2-maximal subgroup of the solvable group G be* Sqn *in G. Then G is supersolvable. Moreover, if $|G|$ is divisible by 3 or more distinct primes, G is nilpotent.*
Proof: Suppose the prime factorization of $|G|$ is

$$|G| = p_1^{e_1} \dots p_n^{e_n} \quad \text{with } n \geq 3.$$

Let P_1 be a Sylow p_1-subgroup of G and let M_1 be a maximal subgroup of G with $M_1 \geq P_1$. Since G is solvable, $[G : M_1] = p_2^a$ (relabelling if necessary). Since $n \geq 3$, we see that P_1 lies in some maximal subgroup N of M_1, so that N is 2-maximal, hence Sqn, hence subnormal in G. Moreover P_1 is normal in N since M_1, hence N, is nilpotent. So P_1 is normal in N which is subnormal in G and hence P_1 is subnormal in G. Therefore when $n \geq 3$ all the Sylow subgroups of G are subnormal in G. From the Sylow theory we know that subnormal Sylow subgroups of G are normal in G, and so we have G nilpotent.

Now we have the case $|G| = p^a q^b$, $p \neq q$, remaining, and we shall prove G supersolvable by showing that each maximal subgroup of G has prime index. Let M be a maximal subgroup of G. Then $[G : M]$ is a prime power, say q^c. If $M \lhd G$, then $c = 1$ and we are finished, so suppose M is not normal in G. Let N be a maximal subgroup of M. Since N is Sqn in G, we have N is subnormal in G, and since M is not normal in G there exists $x \notin M$ which normalizes N. However, $N \lhd M$ since M is nilpotent, and so $N \lhd G$. Then if $N \neq \{1\}$, we have by induction that $[G/N : M/N] = q = [G : M]$. So we must consider the case when $N = \{1\}$.

When $N = \{1\}$, we have $|M| = p$ and $q^b = q^c$. Suppose $c > 1$, let Q be a Sylow q-subgroup of G, and let T be a maximal subgroup of Q. Since $[G : Q] = p$, Q is maximal in G. This and T maximal in Q yield T is 2-maximal, hence Sqn, in G. But this means that MT is a subgroup of G of index q, contradicting the maximality of M. Hence $c = 1$ and $[G : M] = q$. By Huppert's characterization, G is supersolvable.

A dualization of Wielandt's characterization of nilpotent groups as groups in which every maximal subgroup is normal, was developed by N. Ito. A key result is the following lemma which will *not* be proved here (reference: [Huppert 1967, p. 435]).

Ito's Lemma   *Let p be an odd prime. If each element of G of order p lies in $Z(G)$, then G possesses a normal p-complement.*

From this important result the following theorem is immediate.

THEOREM 6.6   *Let G be a group of odd order. If each subgroup of prime order lies in $Z(G)$, then G is nilpotent.*

Buckley generalized this to supersolvability.

THEOREM 6.7   *Let G be a group of odd order. If each subgroup of G of prime order is normal in G, then G is supersolvable.*
Proof: Let q be the smallest prime divisor of $|G|$. Since the automorphism group of a group of order q has order $q - 1$, we see that the elements of order q of G lie in $Z(G)$. So by Ito's lemma, G contains a normal q-complement N. Since each minimal subgroup of N is normal in G, hence in N, we have by induction that N is supersolvable. Let p be the largest prime divisor of $|N|$. Then P, the Sylow p-subgroup of N, is normal in N and hence in G, and

$$P_0 = \Omega(P) = \langle x \in P : |x| = p \rangle$$

is a normal subgroup of G lying in $Z(P)$ since each $\langle x \rangle \lhd P$. Let $C = C_G(P_0)$. Then $P \lhd C \lhd G$, and by Ito's lemma, C contains a normal p-complement M. Therefore C splits, $C = P \times M$.

Considering the action of G on P_0, since every subgroup of P_0 is normal in G, we see that G/C is Abelian of exponent dividing $p - 1$. So if H/K is a chief factor of G lying in P, we see that the action of G on H/K is effected by G/C and hence $|H/K| = p$. Since G/P is supersolvable by induction, we conclude that every chief factor of G has prime order, hence G is supersolvable.

Using the theory of formations, Yokoyama and Laue have extended the above results to the following:

THEOREM   *Let G be a solvable group of odd order and let $M = M(G)$ be the subgroup generated by the minimal normal subgroups of $\mathrm{Fit}(G)$. Then (a)*

G is nilpotent if and only if $M \leq Z^*(G)$, the hypercenter of G, (b) G is supersolvable if and only if each chief factor of G lying in M has prime order.

When normality is replaced by the weaker concept of S-quasinormality we obtain

COROLLARY 6.8 *Let G be a group of odd order. If each minimal subgroup of G is Sqn in G, then the nilpotent residual of G is supersolvable.*
Proof: Let $N = \gamma_\infty(G)$ be the nilpotent residual of G. Let $\langle x \rangle$ be a subgroup of N of order p. Thus x is a generalized central element and so $Q \leq N_G(\langle x \rangle)$ for all Sylow q-subgroups Q of G by Lemma 5.5. Hence $O^p(G) \leq N_G(\langle x \rangle)$ so that $N \leq N_G(\langle x \rangle)$ ($N = \bigcap\{O^p(G) : p \text{ divides } G\}$ by Theorem 5.2 of Appendix C). Conclude that every minimal subgroup of N is normal in N. Hence by Theorem 6.7, N is supersolvable.

The consequences of normality conditions on other small subgroups (such as 2-minimal subgroups) have been investigated by Sastry and Deskins.

Wielandt's characterization of nilpotency can be stated as: G is nilpotent if and only if G satisfies the **normalizer condition:** $H < N_G(H)$ for all $H < G$. By altering the concept of normalizer we obtain another characterization of odd order supersolvable groups.
DEFINITION Let $H \leq G$. The **permutizer** of H in G is the subgroup

$$P_G(H) = \langle x \in G : \langle x \rangle H = H \langle x \rangle \rangle.$$

Clearly $P_G(H) \geq N_G(H)$.
DEFINITION G is said to satisfy the **permutizer condition** if

$$H < P_G(H) \quad \text{for all } H < G.$$

LEMMA 6.9 *Let G be supersolvable and let M be a maximal subgroup of G. Then $P_G(M) = G$.*
Proof: Since G is supersolvable, $[G : M] = p$, a prime. Hence G contains a p-element x which is not in M. Then from the product formula for subgroups ($|HK| = |H| \cdot |K| / |H \cap K|$), we conclude that $\langle x \rangle M = G$ so that $\langle x \rangle M = M \langle x \rangle$. Therefore $P_G(M) = G$.

THEOREM 6.10 *Supersolvable groups satisfy the permutizer condition.*
Proof: Let G be supersolvable and let $H < G$. Choose any subgroup M of G such that H is a maximal subgroup of M. Then $M = P_M(H)$ by Lemma 6.9. Hence $H < M \leq P_G(H)$.

To prove the converse of this result for odd order groups we need the following.

LEMMA 6.11   *Let P be an odd order p-group, H a subgroup of index p^n, and X a normal complement of H which is elementary Abelian. If P contains an element y such that $P = \langle y \rangle H$, then $n = 1$.*

Proof: Assume the result is false and let P be a minimal counterexample. So $n > 1$. Since $X \triangleleft P$ we know $X \cap Z(P)$ contains $z \neq 1$. Clearly, for $H^* = H\langle z \rangle$ we have $P = XH^* = \langle y \rangle H^*$ and $[P : H^*] = p^{n-1}$. Let $\bar{P} = P/\langle z \rangle$, $\bar{X} = X/\langle z \rangle$, and $\overline{H^*} = H^*/\langle z \rangle$. Then $\bar{X}$ is a normal complement of $\overline{H^*}$ in $\bar{P}$ and $\bar{P} = \langle \bar{y} \rangle \overline{H^*}$, so by the minimality of $|P|$ we conclude $[\bar{P} : \overline{H^*}] = [P : H^*] = p^{n-1} = p$. Hence $n = 2$. Therefore $X = \langle x, z \rangle$ for some $x \in X$ and so

$$y = x^a z^b h$$

for some integers a and b and some $h \in H$. Then

$$y^p = ((x^a h) z^b)^p = (x^a h)^p (z^b)^p = (x^a h)^p.$$

Since $[P : H^*] = p$, H^* is a maximal subgroup of P and hence $H^* \triangleleft P$. Therefore the commutator (x^a, h) lies in both X and H^*, so lies in $X \cap H^* = \langle z \rangle$. In particular, (x^a, h) commutes with both x^a and h so an elementary result from commutator calculus ([Scott 1964, 3.4.4]) yields

$$y^p = (x^a h)^p = (x^a)^p h^p (h, x^a)^{p(p-1)/2} = h^p.$$

Therefore $y^p \in H$ and

$$|\langle y \rangle H| = p \cdot |H|$$

which is impossible since $\langle y \rangle H = P$.

The next example shows that the lemma is not necessarily true when $p = 2$.

EXAMPLE   Take P to be the dihedral group of order 8,

$$P = \langle a, b : a^4 = b^2 = 1, bab = a^3 \rangle.$$

Then for $H = \langle b \rangle$, $X = \langle a^2, ab \rangle$, and $y = a$ we have X is a normal complement of H and is elementary Abelian. Moreover, $P = \langle y \rangle H = \langle a, b \rangle$.

LEMMA 6.12   *Let G be a solvable group of odd order and let M be a maximal subgroup. If $P_G(M) = G$, then $[G : M]$ is a prime.*

Proof: Since G is solvable we have $[G : M] = p^n$. Since $P_G(M) = G$ we know there exists $y \in G$ such that $G = \langle y \rangle M$, and clearly y can be assumed to be a p-element of G. Suppose M contains a subgroup $K \neq \{1\}$, $K \triangleleft G$. Then $\bar{M} = M/K$ is maximal in $G/K = \bar{G}$ and $P_{\bar{G}}(\bar{M}) = \bar{G}$. So by induction $[\bar{G} : \bar{M}] = [G : M] = p$. So we need only treat the case when $\mathrm{Core}_G(M) = \{1\}$.

Let X be a minimal normal subgroup of G and note that $X \cap M = \{1\}$ since $\text{Core}_G(M) = \{1\}$. Then X is an elementary Abelian group of order p^n. By a basic result on factorizable groups (see [Scott 1964, 13.2.5]) there is a Sylow p-subgroup H of M such that $P = \langle y \rangle H$ is a Sylow p-subgroup of G. Then $H = P \cap M$ and consequently $P = XH$, $X \triangleleft P$, and $X \cap H = \{1\}$. Lemma 6.11 now applies to yield $|X| = p$. Hence $[G : M] = p$.

This result enables us to extend Theorem 6.10.

THEOREM 6.13 *Let G be a solvable group of odd order. Then G is supersolvable if and only if G satisfies the permutizer condition.*
Proof: The permutizer condition and the above lemma imply that each maximal subgroup of G has prime index in G. Thence, by Huppert's characterization (Theorem 3.1), G is supersolvable.

7. *Supersolvably Embedded Subgroups*

R. Baer initiated the study of normal subgroups which are "well-situated" in a group, and obtained a generalization of the hypercenter, $Z^*(G)$.

DEFINITION A normal subgroup H of G is **supersolvably embedded** (SE) in G if for all homomorphisms θ, $H^\theta \neq \{1\}$ implies there exists a nontrivial normal cyclic subgroup of G^θ lying in H^θ.

The following are easily verified.
(1) *If $H \triangleleft G$, then H is SE in G iff every chief factor of G which lies in H has prime order.*
(2) *If H, K are normal subgroups of G with $K \leq H$ and H SE in G, then H/K is SE in G/K.*
(3) *If H, K are normal subgroups of G with $K \leq H$ and H SE in G, then K is SE in G.* (This follows from (1).)
(4) *If H, K are normal subgroups of G with both H, K SE in G, then HK is SE in G.*

From this last property we see that a group G has a unique maximal supersolvably embedded subgroup, namely the product of all the SE normal subgroups of G. It will be denoted by $SE(G)$.

LEMMA 7.1

(i) $Z^*(G) \leq SE(G)$.

(ii) If $N \triangleleft G$ and $N \leq SE(G)$, then $SE(G/N) = SE(G)/N$.

Proof: To prove (i), let θ be a homomorphism such that $(Z^*(G))^\theta \neq \{1\}$. Now $(Z^*(G))^\theta \leq Z^*(G^\theta)$ (this follows from the fact that $(Z_n(G))^\theta \leq Z_n(G^\theta)$, which is easily seen by induction on n). This clearly implies nontriviality of $(Z^*(G))^\theta \cap Z(G^\theta)$ from which it follows that $Z^*(G)$ is SE in G, proving (i).

As for (ii), $SE(G)/N \leq SE(G/N)$ follows from (2) above while the reverse inclusion follows from (1).

A somewhat similar generalization of central elements was introduced by O. Ore.

DEFINITION An element x of a group G is **quasi-central** (q.c.) in G if $\langle x \rangle \langle y \rangle = \langle y \rangle \langle x \rangle$ for all $y \in G$.

Clearly if x is q.c. in G and θ is a homomorphism of G, then x^θ is q.c. in G^θ. The following useful characterization is easily verified.

LEMMA 7.2 *An element x of G is q.c. in G iff $\langle x \rangle \langle y \rangle = \langle y \rangle \langle x \rangle$ for all elements y of G of prime-power order.*

Of course a quasi-central element of G is also a generalized central element, so the results in Section 5 suggest (and are useful in the proof of) the following result.

LEMMA 7.3 *Let x be a q.c. element of G. Then $x = uv = vu$ where u and v are a p-element and p'-element respectively, and u is q.c. in G.*

Proof: The elements u and v are powers of x in the usual factorization of x, and we know from Section 5 that u is g.c. in G. Hence by Lemma 5.5, $Q \leq N_G(\langle u \rangle)$ for all Sylow q-subgroups Q of G where $q \neq p$. So we need only show that $\langle u \rangle$ permutes with $\langle y \rangle$ whenever y is a p-element of G. This is proved by induction on $|G|$ and obviously we need only treat the case $G = \langle x \rangle \langle y \rangle$. Then $|G| = |x| \cdot |y|/|\langle x \rangle \cap \langle y \rangle|$. Since $\langle u \rangle$ is the Sylow p-subgroup of $\langle x \rangle$, we have $\langle x \rangle \cap \langle y \rangle = \langle u \rangle \cap \langle y \rangle$. Therefore

$$|G| = |u| \cdot |v| \cdot |y|/|\langle u \rangle \cap \langle y \rangle|$$

and the p-factor of $|G|$ is clearly $|u| \cdot |y|/|\langle u \rangle \cap \langle y \rangle|$. Let P be a Sylow p-subgroup of G containing u. Then for some $w \in G$, $y \in P^w = wPw^{-1}$. Since $w = y^i x^j$ for some i, j, we have $\langle u, y \rangle \leq y^{-i}P^w y^i = x^j Px^{-j}$, a Sylow p-subgroup of G. Now

$$|\langle u \rangle \langle y \rangle| = |\langle u \rangle| \cdot |\langle y \rangle|/|\langle u \rangle \cap \langle y \rangle| = |P|$$

so that $x^j Px^{-j} = \langle u \rangle \langle y \rangle = \langle y \rangle \langle u \rangle$.

LEMMA 7.4   *Let u be a* q.c. *p-element of G. Then u^i is* q.c. *in G for all i.*
Proof: Since u is quasi-central, it is generalized central. Hence (Lemma 5.5) $Q \leq N_G(\langle u \rangle)$ for all Sylow q-subgroups Q where $q \neq p$. This implies $Q \leq N_G(\langle u^i \rangle)$ for all such Q. So we need only prove that $\langle u^i \rangle \langle y \rangle = \langle y \rangle \langle u^i \rangle$ for each p-element y of G. This is proved by induction on $|G|$, and we need only treat the case $G = \langle u \rangle \langle y \rangle$, a p-group. Then $Z(G)$ contains an element z of order p with $z \in \langle u \rangle$ and/or $z \in \langle y \rangle$. Induction applied to $G/\langle z \rangle$ leads readily to the conclusion $\langle u^i \rangle \langle y \rangle = \langle y \rangle \langle u^i \rangle$ for each i.

Note   In general the product of two quasi-central elements is not quasi-central, even when the two elements commute. As an example, let $G = \langle a, b, c : a^3 = b^3 = c^2 = 1, ab = ba, ac = ca, bc = cb^2 \rangle \cong S_3 \times \mathbb{Z}_3$. In this (supersolvable) group the elements a and b are both quasi-central. However ab is not since $\langle ab \rangle \langle c \rangle$ is not a subgroup of G.

An elegantly simple proof of Maier and Schmid enables us to show that a q.c. element of prime order in G is nicely situated in G.

THEOREM 7.5 (Maier–Schmid)   *Let x be* q.c. *in G of order p. Then either $x \in Z^*(G)$ or $\langle x \rangle \triangleleft G$.*
Proof   Assume the result false and let G be a group of minimal order having a quasi-central element x of prime order p with $x \notin Z^*(G)$ and $\langle x \rangle$ not normal in G. Let P be a Sylow p-subgroup of G. Then $x \in P$ since x is q.c. in G and has order p. Since $O^p(G) = \langle Q : Q$ is a Sylow q-subgroup of $G, q \neq p \rangle$ (see Section 5 of Appendix C) we have $G = PO^p(G)$ and (from Lemma 5.5) $N_G(\langle x \rangle) \geq O^p(G)$. Hence we conclude that $\langle x \rangle$ is not normal in P.

Suppose $C_G(x) \geq O^p(G)$; then for all $g \in G$, $C_G(x^g) \geq O^p(G)$. Since $x \in O_p(G)$ we see that $\langle x \rangle^G = \langle x^g : g \in G \rangle$ is a normal p-subgroup of G. Then from $C_G(x^g) \geq O^p(G)$ and $G = PO^p(G)$ we conclude that $\langle x \rangle^G$ lies in $Z^*(G)$ (Theorem 6.3 of Appendix C). Since this is contrary to the hypothesis of G we conclude that $C_G(x)$ does not contain $O^p(G)$. Hence there is a Sylow q-subgroup Q, $q \neq p$, such that $[x, Q] \neq \{1\}$.

Now form the subgroup $H = \langle x \rangle N_G(Q)$. We observe from the Frattini argument that H is contained in no proper normal subgroup of G since H contains the normalizer of a Sylow subgroup of G. This means that $HO^p(G) = G$. For if $HO^p(G) < G$, then $HO^p(G)/O^p(G)$ is a proper subgroup of the p-group $G/O^p(G)$ which would imply that H is contained in a proper normal subgroup of G. From $HO^p(G) = G$ we conclude that $H = G$. For if $H \neq G$, then from the minimality of G we have either $x \in Z^*(H)$ or $\langle x \rangle \triangleleft H$. But $x \in Z^*(H)$ implies that $C_H(x) \geq O^p(H) \geq Q$, contradicting $[x, Q] \neq \{1\}$. And $\langle x \rangle \triangleleft H$ implies $\langle x \rangle \triangleleft G$ since $N_G(\langle x \rangle) \geq O^p(G)$ and $G = HO^p(G)$. So we have $H = \langle x \rangle N_G(Q) = G$.

Consider Q^g for $g \in G$. Since $G = \langle x \rangle N_G(Q)$ we see that $Q^g = Q^{x'}$ for some $x' \in \langle x \rangle$. Hence the normal hull Q^G of Q lies in $\langle x \rangle Q$. Since $|\langle x \rangle Q| = p \cdot |Q| = pq^a$, $q \neq p$, we see that either $Q^G = Q$ or $Q^G = \langle x \rangle Q$. But $Q^G = Q$ implies that $[x, Q] = \{1\}$ since $N_G(\langle x \rangle) \geq O^p(G) \geq Q$, and $Q \triangleleft G$ means $[x, Q] \leq \langle x \rangle \cap Q = \{1\}$. On the other hand, $Q^G = \langle x \rangle Q$ implies that, for a Sylow p-subgroup P of G, $P \cap Q^G = \langle x \rangle \triangleleft P$, contradicting the earlier conclusion that $\langle x \rangle$ is not normal in P. Thus we have an impossible situation, and the theorem is proved.

Note Maier and Schmid actually proved that if H is a subgroup of G which permutes with every subgroup of G (such subgroups are called **quasinormal** in G), then $H^G/\text{Core}_G(H)$ lies in $Z^*(G/\text{Core}_G(H))$. Their proof can be reconstructed from the above, with some care.

This result enables us to characterize $SE(G)$ in a manner analogous to the usual characterization of $Z^*(G)$.

DEFINITIONS The **quasicenter**, $Q(G)$, of a group G is the subgroup generated by the set of all q.c. elements of G. The **hyperquasicenter**, $Q^*(G)$, is the largest term of the chain of subgroups

$$Q_0(G) = \{1\} \leq Q_1(G) = Q(G) \leq Q_2(G) \leq \cdots$$

where $Q_i(G)/Q_{i-1}(G) = Q(G/Q_{i-1}(G))$ for all $i > 0$.

Clearly $Q(G)$ is a characteristic subgroup of G and $Q(G) \geq Z(G)$. Then $Q^*(G)$ is a characteristic subgroup of G, and we shall see that $Q^*(G) \geq Z^*(G)$.

LEMMA 7.6 Let $M \triangleleft G$, $M < Q^*(G)$. Then $Q(G/M)$ is nontrivial.
Proof: Since $M < Q^*(G)$, there is some i such that M contains $Q_{i-1}(G)$ but does not contain $Q_i(G)$. Hence there exists x in $G \backslash M$ such that $Q_{i-1}(G)x$ is a q.c. element of $G/Q_{i-1}(G)$. Then Mx is a nontrivial q.c. element of G/M and we have $Q(G/M)$ nontrivial.

This enables us to prove the following basic result.

THEOREM 7.7 Let $N \triangleleft G$, $N \leq Q^*(G)$. Then $Q^*(G/N) = Q^*(G)/N$.
Proof: A straightforward induction argument establishes that $NQ_i(G)/N \leq Q_i(G/N)$ for all natural numbers i. Since $Q^*(G) = Q_k(G)$ for some k, it follows that $Q^*(G)/N \leq Q^*(G/N)$.

Now suppose $Q^*(G)/N < Q^*(G/N)$. Then applying the above lemma yields $Q((G/N)/(Q^*(G)/N))$ is nontrivial. This, however, is a contradiction since $(G/N)/(Q^*(G)/N) \cong G/Q^*(G)$ which has trivial quasicenter by definition of Q^*. Hence the result is proved.

COROLLARY 7.8 $Q^*(G) \geq Z^*(G)$.

Proof: If $Z(G)$ is trivial, then so is $Z^*(G)$ and the result holds, so we may assume $Z(G) \neq \{1\}$. We know $Z(G) \leq Q(G)$. Consider $G/Z(G)$. By induction, $Z^*(G/Z(G)) \leq Q^*(G/Z(G))$. Since

$$Z^*(G/Z(G)) = Z^*(G)/Z(G) \quad \text{and} \quad Q^*(G/Z(G)) = Q^*(G)/Z(G),$$

we see that $Z^*(G) \leq Q^*(G)$.

COROLLARY 7.9 $Q^*(G) \geq SE(G)$.

Proof: $SE(G)$ contains a subgroup $\langle x \rangle$ with $\langle x \rangle \lhd G$. Clearly x is a q.c. element of G, so that $\langle x \rangle \leq Q(G)$. Consider $G/\langle x \rangle$. By induction, $SE(G/\langle x \rangle) \leq Q^*(G/\langle x \rangle)$. Since

$$SE(G/\langle x \rangle) = SE(G)/\langle x \rangle \quad \text{and} \quad Q^*(G/\langle x \rangle) = Q^*(G)/\langle x \rangle$$

we conclude that $SE(G) \leq Q^*(G)$.

Of course the reason for introducing the quasicentral series was to obtain, if possible, a characterization of $SE(G)$. The following theorem of N. Mukherjee is the desired result.

THEOREM 7.10 (Mukherjee) $SE(G) = Q^*(G)$.

Proof: We proceed by induction on $|G|$. If $Z^*(G) \neq \{1\}$, then we consider $G/Z^*(G)$. By induction

$$SE(G/Z^*(G)) = Q^*(G/Z^*(G));$$

since

$$SE(G/Z^*(G)) = SE(G)/Z^*(G)$$

and

$$Q^*(G/Z^*(G)) = Q^*(G)/Z^*(G)$$

we have $SE(G) = Q^*(G)$.

If $Z^*(G) = \{1\}$, then by the theorem of Maier and Schmid, a q.c. element x of G of prime order generates a normal subgroup $\langle x \rangle$ of G and $\langle x \rangle$ clearly lies in both $SE(G)$ and $Q^*(G)$. So if $Q^*(G) \neq \{1\}$, then from Lemmas 7.3 and 7.4 we know that such an x exists. So we consider $G/\langle x \rangle$ and proceed as in the first case to obtain $SE(G) = Q^*(G)$. Finally, if $Q^*(G) = \{1\}$, then by Corollary 7.9 $SE(G) = \{1\}$ and we have $Q^*(G) = SE(G)$ in all cases.

It is easily verified that $Q^*(G)H$ is a supersolvable subgroup of G whenever H is supersolvable. Hence $Q^*(G)$ lies in every maximal supersolvable subgroup of G, just as the hyper-generalized-center of G does (as we saw in Section 5). In fact it can be shown, by proving an analog of

Theorem 7.7 for the hyper-generalized-center of G, genz*(G), that genz*$(G) \geq Q^*(G)$. The following example shows that this containment can be proper.

EXAMPLE   Let $G = S_3$ wr $\mathbb{Z}_3$. Then

$$Z(G) = Z^*(G) = Q(G) = Q^*(G) = \{1\}$$

while genz(G) has order 27 and genz*(G) has order 54.

We now present one further characterization of SE(G); this in terms of weakly central subgroups.

DEFINITIONS   Let H be a subgroup of the group G. The **weak centralizer** of H in G, $C_G^*(H)$, is defined by

$$C_G^*(H) = \bigcap \{ N_G(K) : K \leq H \}.$$

H is called **weakly central** in G whenever $G = C_G^*(H)$. The **weak center** of G is the product of all weakly central subgroups of G and is denoted by WZ(G).

WARNING   The weak center of a group G need not be weakly central in G.

It is relatively easy to verify that the weak center of a group G is precisely the product of all normal cyclic subgroups of G. We use this characterization of the weak center in the sequel.

THEOREM 7.11   *The weak center of a group G is nilpotent of class at most two.*
Proof: We may assume, with no loss of generality that $G = $ WZ(G), so that G itself is generated by cyclic normal subgroups. For all such cyclic normal subgroups $\langle x \rangle$, $G/C_G(x)$ is isomorphic with a subgroup of Aut$(\langle x \rangle)$ and hence is Abelian. Thus $G' \leq \bigcap \{ C_G(x) : \langle x \rangle \triangleleft G \} = Z(G)$ and the theorem follows.

DEFINITIONS   (1) A chain of normal subgroups $N_0 \leq N_1 \leq N_2 \leq \cdots$ $\leq N_k$ of the group G is called a **weakly central series** of G whenever N_{i+1}/N_i lies in WZ(G/N_i) for $i = 0, 1, 2, \ldots, k-1$.
(2) The **ascending weak central series** of the group G is the series

$$\{1\} - (\text{WZ})_0(G) \leq (\text{WZ})_1(G) \leq (\text{WZ})_2(G) \leq \cdots$$

where $(\text{WZ})_{i+1}(G)/(\text{WZ})_i(G) = \text{WZ}(G/(\text{WZ})_i(G))$ for $i \geq 0$.
(3) We set $(\text{WZ})^*(G) = \bigcup_i (\text{WZ})_i(G)$.

The next result shows that like the ascending central series, the ascending weak central series increases faster than any other series of its type.

THEOREM 7.12 *Let* $\{1\} = N_0 \leq N_1 \leq N_2 \leq \ldots$ *be a weakly central series of G. Then for* $i \geq 0$, $N_i \leq (WZ)_i(G)$.
Proof: We induce on i. For $i = 0$ the result is trivial, so induction begins. Assume now that $N_i \leq (WZ)_i(G)$. Let $W/N_i = WZ(G/N_i)$, so that $W \geq N_{i+1}$ and W/N_i is generated by cyclic normal subgroups of G/N_i. For all such subgroups $N_i\langle x\rangle/N_i$, it is easily seen that $(WZ)_i(G)\langle x\rangle$ is normal in G and hence that $(WZ)_i(G)\langle x\rangle/(WZ)_i(G)$ is a cyclic normal subgroup of $G/(WZ)_i(G)$. By the definition of $(WZ)_{i+1}(G)$ then, $x \in (WZ)_{i+1}(G)$. It now follows that $W \leq (WZ)_{i+1}(G)$ so that N_{i+1} is also in $(WZ)_{i+1}(G)$ and hence the theorem is valid.

As corollaries we have:

COROLLARY 7.13 *In the group G,* $(WZ)(G) = SE(G)$.
Proof: By the definition of the weak center, each term of the ascending weak central series is SE in G. Hence $SE(G) \geq (WZ)^*(G)$. Conversely, any G-chief series of $SE(G)$ is a weakly central series, and consequently $SE(G) \leq (WZ)^*(G)$ by Theorem 7.12.

COROLLARY 7.14 *The group G is supersolvable iff* $G = (WZ)^*(G)$.

It is rather easy to obtain a necessary condition for a normal subgroup of G to be SE in G.

THEOREM 7.15 *Let the normal subgroup N be SE in G. Then* $G/C_G(N)$ *is supersolvable (i.e., G induces a supersolvable group of automorphisms on N).*
Proof: This generalizes the familiar result, that the automorphism group of a group of prime order p is cyclic, and the proof is by induction on the length n of a G-chief series of N. When $n = 1$ we have $|N| = p$ and the conclusion follows. When $n \geq 2$, let M be a subgroup of G maximal with respect to $M \triangleleft G$ and $M < N$. Then N/M is a chief factor of G so that $|N/M| = p$ because N is SE in G. Since M is SE in G and has a G-chief series of length $n - 1$ we have by induction that $G/C_G(M)$ is supersolvable. Clearly $C_G(M) \geq C_G(N)$. Now let K be a minimal normal subgroup of G contained in M. Then K is cyclic of prime order. Also, N/K is SE in G/K and N/K has a chief series of length $n - 1$, so by induction the group $G/C_G(N/K)$ is supersolvable. Set $C_G(M) \cap C_G(N/K) = L$; then G/L is supersolvable.

Since N/M has order p, it is cyclic; $N/M = \langle Mx \rangle$. let $y \in L$; since $y \in C_G(N/K)$ we have

$$(y,x) = y^{-1}x^{-1}yx \in K.$$

For $u,v \in L$ we have, since $L \le C_G(M) \le C_G(K)$,

$$(u,x) \cdot (v,x) = u^{-1}x^{-1}uxv^{-1}x^{-1}vx = v^{-1}(u^{-1}x^{-1}ux)x^{-1}vx = (uv,x),$$

so that the mapping $\psi : L \to K$ given by

$$\psi(y) = (y,x) \quad \text{for all } y \in K,$$

is a homomorphism. Moreover, the kernel of ψ is clearly $C_G(N)$ since it consists of the elements from $C_G(M)$ in L which centralize x. Hence $L/C_G(N)$ is cyclic, and this means that $G/C_G(N)$ is supersolvable since G/L is supersolvable.

This necessary condition is not, however, sufficient as we see immediately by examining $G/C_G(V)$ for $G = S_4$ and V the normal subgroup of S_4 of order 4. In this example $G/C_G(V)$ is isomorphic with S_3 and hence is supersolvable, but V is not SE in G. However, R. Baer showed how automorphisms could be used to obtain sufficient conditions.

DEFINITION A group H and a subgroup $K \le \text{Aut}(H)$ form a **supersolvable pair** if every K-chief factor of H is cyclic. (A K-chief factor of H is a quotient group M/N where M and N are normal subgroups of H which are fixed by the elements of K, i.e., K-admissible normal subgroups or just K-subgroups, with $M > N$ and N a maximal normal K-subgroup of H in M.)

If (H,K) is a supersolvable pair, then it is clear that H is SE in the group $G = H[K]$ which consists of H extended by K. Moreover it follows from Theorem 7.15 that K is supersolvable, hence $G = H[K]$ is supersolvable. Conversely, if N is an SE normal subgroup of a group G, then $(N, G/C_G(N))$ is a supersolvable pair.

Now we derive a characterization of supersolvable pairs which extends an earlier characterization of supersolvability.

THEOREM 7.16 *Let K be a subgroup of $\text{Aut}(H)$ with $K \ge \text{Inn}(H)$. Then (H,K) is a supersolvable pair if and only if H satisfies the Sylow tower property and, for each prime p, K_p induces on each Sylow p-subgroup P of H a strictly p-closed group of automorphisms. (Note: K_p denotes the subgroup consisting of all elememts of K which fix P.)*
Proof: If (H,K) is a supersolvable pair then, as noted ebove, the extended group $G = H[K]$ is supersolvable. Then for a Sylow p-subgroup P of H, $G_1 = P[K_P]$ is a subgroup of G and hence is supersolvable. Then the chief

factors of G_1 lying in P all have order p, so that G_1, hence K_P, induces a cyclic group of automorphisms on each of these chief factors, of exponent dividing $p - 1$. So it follows, as in Theorem 1.12, that K_P induces on P a strictly p-closed group of automorphisms.

To prove the converse we note first that since $K \geq \text{Inn}(H)$ we have $N_H(P)/C_H(P)$ isomorphic with a subgroup of the group of automorphisms induced on P by K_P. Therefore H satisfies the Sylow tower property and, for every prime p, each Sylow p-subgroup P of H has the property: $N_H(P)/C_H(P)$ is strictly p-closed. So, by Theorem 1.12 H is supersolvable. And we know that, when p is the largest prime divisor of $|H|$, the Sylow p-subgroup H_p of H is characteristic in H. Let M be a minimal normal K-subgroup of H which is in H_p. Then M is an elementary Abelian p-group lying in $Z(H_p)$, and $K_M = K$ induces on H_p a strictly p-closed group K^* of automorphisms. Moreover K^* acts irreducibly on M because of the minimality of M, so it follows from Lemma 1.3 that $|M| = p$. Now we shall show that $(H/M, K_1)$ is a supersolvable pair, where K_1 is the group of automorphisms induced on H/M by K. To do this we note that H/M obviously has the Sylow tower property and show that $(K_1)_Q$ induces on each Sylow q-subgroup Q of H/M a strictly q-closed group of automorphisms, for every prime q.

Case 1: $q = p$. Then $Q = H_p/M$ and $(K_1)_Q = K_1$. So the group of automorphisms induced on Q by K_1 is a homomorphic image of the group of automorphisms induced on H_p by K, which is strictly p-closed. Hence K_1 induces a strictly p-closed group of automorphisms on Q.

Case 2: $q \neq p$. Then $Q = Q_1M/M$ where Q_1 is a Sylow q-subgroup of H. By hypothesis K_{Q_1} induces a strictly q-closed group of automorphisms on Q_1. Let $\alpha \in K_{Q_1M}$; then $Q_1M = (Q_1M)^\alpha = Q_1^\alpha M$ so that Q_1^α is a Sylow q-subgroup of Q_1M. Hence there is some $x \in M$ such that $Q_1^\alpha = Q_1^x$. But $K \geq \text{Inn}(H)$ so the inner automorphism π_x induced by x is in K, and $Q_1^\alpha = Q_1^{\pi_x}$. Then $\alpha \cdot \pi_x^{-1} \in K_{Q_1}$ and π_x^{-1} induces the identity automorphisms on H/M. This means the group of automorphisms induced on Q by $(K_1)_Q$ is isomorphic with the group of automorphisms induced on Q_1 by K_{Q_1}, hence both are strictly q-closed.

Thus in both cases $(K_1)_Q$ induces on each Sylow q-subgroup Q of H/M a strictly q-closed group of automorphisms, for every prime q. So, by induction we have $(H/M, K_1)$ is a supersolvable pair. Then it follows that (H, K) is a supersolvable pair.

Another characterization involving the Frattini subgroup can also be given. First some notation: If $K \leq \text{Aut}(H)$ and L is a normal K-subgroup of H, then K rel L will denote the automorphism group induced on H/L by K.

THEOREM 7.17 *Let K be a subgroup of $\mathrm{Aut}(H)$ with $\mathrm{Inn}(H) \le K$. Then (H, K) is a supersolvable pair if and only if $(H/\Phi(H), K$ rel $\Phi(H))$ is.*

Proof: If (H, K) is a supersolvable pair, then it is obvious that $(H/L, K$ rel $L)$ is a supersolvable pair when L is any normal K-subgroup of H. So suppose $(H/\Phi(H), K$ rel $\Phi(H))$ is a supersolvable pair. Then $H/\Phi(H)$ is supersolvable so that by Huppert's theorem, H is supersolvable. Let p be the largest prime divisor of $|H|$. Then P, the Sylow p-subgroup of H, is characteristic in H, and $\Phi(P) = P \cap \Phi(H)$. Also $P/\Phi(P) \cong P\Phi(H)/\Phi(H)$ is the Sylow p-subgroup of $H/\Phi(H)$ so that K rel $\Phi(H)$ induces a strictly p-closed group of automorphisms in $P\Phi(H)/\Phi(H)$ by the previous result. Now the automorphism group K induces on P the group K^* of automorphisms, and if K^{**} denotes the subgroup of K^* of elements which induce the identity map on $P/\Phi(P)$, we see that K^*/K^{**} is basically the group of automorphisms induced on $P/\Phi(P)$ by K. Since the group of automorphisms induced on $P/\Phi(P)$ by K is basically the same as the group of automorphisms induced on $P\Phi(H)/\Phi(H)$ by K rel $\Phi(H)$, we see that K^*/K^{**} is a strictly p-closed group. For if $\alpha \in K$ and $x \in P$ the induced automorphisms are:

$$\alpha_1 : x \to x^\alpha, \qquad \alpha_2 : x\Phi(H) \to x^\alpha\Phi(H),$$

$$\alpha_3 : x\Phi(P) \to x^\alpha\Phi(P).$$

Then K^{**} is the kernel of the homomorphism θ given by $\theta(\alpha_1) = \alpha_2$ and of the homomorphism ψ given by $\psi(\alpha_1) = \alpha_3$.

Now by a result of Burnside (given below as Lemma 7.18) an automorphism of a p-group P which induces the identity map on $P/\Phi(P)$ is of order a power of p. Hence K^{**} is a p-group and K^* is a p-group extended by a strictly p-closed group. Hence K^* is a strictly p-closed group. This means that a minimal normal K-subgroup M of H lying in P has $|M| = p$.

Consider the pair $(H/M, K$ rel $M)$. Since $(H/M)/\Phi(H/M)$ is a homomorphic image of $H/\Phi(H)$, we know that $((H/M)/\Phi(H/M), (K$ rel $M)$ rel $\Phi(H/M))$ is a supersolvable pair. Then by induction $(H/M, K$ rel $M)$ is a supersolvable pair and we conclude that (H, K) is also.

LEMMA 7.18 *Let α be an automorphism of p-group P such that α induces the identity map on $P/\Phi(P)$. Then the order of α is a power of p.*

Proof: Let $|P/\Phi(P)| = p^n$. Then each minimal generating set of P has n elements. Pick a minimal generating set $\{x_1, \ldots, x_n\}$ for P and form $\Omega = \{(x_1 y_1, x_2 y_2, \ldots, x_n y_n) : \text{for all } y_i \in \Phi(P)\}$. Thus $|\Omega| = p^{mn}$, where $|\Phi(P)| = p^m$, α acts as a permutation on Ω since α is trivial on $P/\Phi(P)$, and for each $\lambda \in \Omega$, the orbit of λ under α has length equal to the order of α since λ is an ordered generating set of P. Hence the order of α divides p^{mn}.

The two results above on supersolvable pairs can be restated in terms of supersolvably embedded normal subgroups.

THEOREM 7.19 (R. Baer)   *The following properties of the normal subgroup H of group G are equivalent*:
(i) *H is SE in G.*
(ii) *H satisfies the Sylow tower property and for each Sylow p-subgroup P of H, $N_G(P)/C_G(P)$ is strictly p-closed, for all primes p.*
(iii) *$H/\Phi(H)$ is SE in $G/\Phi(H)$.*

EXERCISE   Let G be a solvable group. If $\text{Fit}(G)/\Phi(G)$ is SE in $G/\Phi(G)$, then G is supersolvable. (Hint: Take $\Phi(G) = \{1\}$ and recall that, since G is solvable, $C_G(\text{Fit}(G)) \leq \text{Fit}(G)$.)

8.   Weak Normality

In preceding sections subgroup permutability has been used to generalize the center and hypercenter of finite groups. In this section we develop these ideas further with a concept of "normality" based on the permuting of certain subgroups. This concept leads to a description of supersolvable groups similar to well-known descriptions of nilpotent groups. A knowledge of P. Hall's generalization of the Sylow theorems is crucial for this section. The reader is referred to Appendix A for the necessary concepts, notation, and results.

DEFINITIONS   Let H be a subgroup of the solvable group G and Σ_H a Sylow system of H.
   (1) The element $x \in G$ is called Σ_H-**quasinormal** whenever $\langle x \rangle A = A \langle x \rangle$ for all $A \in \Sigma_H$.
   (2) The **weak system normalizer** of Σ_H in G is the subgroup of G generated by all Σ_H-quasinormal elements of G; it is denoted by $N_G^*(\Sigma_H)$. If $G = N_G^*(\Sigma_H)$, then Σ_H is said to be **weakly normal** in G.
   (3) The **weak normalizer** of H in G is the subgroup $N_G^*(H) = HN_G^*(\Sigma_H)$. We say H is **weakly normal** in G whenever $G = N_G^*(H)$.

LEMMA 8.1   *The weak normalizer of a subgroup H in the group G is independent of the choice of Sylow system.*
Proof: Let Σ_H and $\Sigma_{\widetilde{H}}$ be two Sylow systems of H. By Theorem 4 of Appendix A, Σ_H and $\Sigma_{\widetilde{H}}$ are conjugate in H, hence there is an element $h \in H$ with $\Sigma_{\widetilde{H}} = \{A^h : A \in \Sigma_H\}$. If x is Σ_H-quasinormal in G, then $\langle x^h \rangle A^h = (\langle x \rangle A)^h = (A \langle x \rangle)^h = A^h \langle x^h \rangle$ for all $A \in \Sigma_H$ and hence x^h is $\Sigma_{\widetilde{H}}$-quasinormal in G. Similarly if x is $\Sigma_{\widetilde{H}}$-quasinormal $x^{h^{-1}}$ is Σ_H-quasinormal and it follows that $[N_G^*(\Sigma_H)]^h = N_G^*(\Sigma_{\widetilde{H}})$. Therefore $N_G^*(H) = HN_G^*(\Sigma_H) = H(N_G^*(\Sigma_H))^h = HN_G^*(\Sigma_{\widetilde{H}})$.

LEMMA 8.2 *Let H be a subgroup of the group G and Σ_H a Sylow system of H. The element $x \in G$ is Σ_H-quasinormal if*
(1) $\langle x \rangle$ permutes with each Sylow subgroup of Σ_H;
(2) $\langle x \rangle$ permutes with each Sylow complement of Σ_H.

Proof: Let π be a set of primes and H_π the Hall π-subgroup of H in Σ_H. Since H_π is the product of Sylow p-subgroups from Σ_H, it follows that, if $\langle x \rangle$ permutes with each Sylow subgroup, it must also permute with H_π and thus is Σ_H-quasinormal.

We assume now that $\langle x \rangle$ permutes with the Sylow p-complements of H in Σ_H. We need only show $\langle x \rangle$ permutes with each Hall π-complement H^π in Σ_H and we do this by induction on the number of primes in π. Thus we assume that $\langle x \rangle$ permutes with H^π. Let p be a prime with $p \notin \pi$. Then $\langle x \rangle H^\pi \cap \langle x \rangle H^p$ is a subgroup of G. By use of the Dedekind identity ($A \leq C$ implies $AB \cap C = A(B \cap C)$), $\langle x \rangle H^\pi \cap \langle x \rangle H^p = \langle x \rangle [H^\pi \cap \langle x \rangle H^p] = \langle x \rangle [H^\pi \cap H \cap H^p \langle x \rangle] = \langle x \rangle [H^\pi \cap H^p(H \cap \langle x \rangle)]$. Since $[H : H^p] = [H : H^p(H \cap \langle x \rangle)][H^p(H \cap \langle x \rangle) : H^p] = p^\alpha$, it follows that $H^p(H \cap \langle x \rangle) = \langle y \rangle H^p = H^p \langle y \rangle$, where $\langle y \rangle$ is the Sylow p-subgroup of $H \cap \langle x \rangle$. Moreover, since $(H \cap \langle x \rangle)H^\pi$ is a subgroup of H and $[H : H^\pi] = [H : (H \cap \langle x \rangle)H^\pi][(H \cap \langle x \rangle)H^\pi : H^\pi]$ is a π-number, it follows that $\langle y \rangle \leq H^\pi$. Therefore, $\langle x \rangle H^\pi \cap \langle x \rangle H^p = \langle x \rangle [H^\pi \cap H^p(H \cap \langle x \rangle)] = \langle x \rangle [H^\pi \cap \langle y \rangle H^p] = \langle x \rangle [\langle y \rangle (H^\pi \cap H^p)] = \langle x \rangle H^{\pi \cup \{p\}}$ proving $\langle x \rangle H^{\pi \cup \{p\}}$ is a subgroup so that $\langle x \rangle$ and $H^{\pi \cup \{p\}}$ permute as desired.

LEMMA 8.3 *Let Σ be a Sylow system of the group G and let z be a p-element of G. Then z is Σ-quasinormal if and only if $z \in G_p$ and $\langle z \rangle G^p = G^p \langle z \rangle$, where G_p and G^p are the Sylow p-subgroup and Sylow p-complement of G in Σ.*

Proof: If z is Σ-quasinormal, then $\langle z \rangle$ permutes with G^p and G_p. Since $\langle z \rangle G_p$ is a p-group, it follows that $z \in G_p$. Conversely suppose $z \in G_p$ and $\langle z \rangle$ permutes with G^p. Let G^q be the Sylow q-complement of G in Σ for some prime q. If $q \neq p$, then $z \in G_p \leq G^q$ so that $\langle z \rangle G^q = G^q = G^q \langle z \rangle$. If $q = p$, then $\langle z \rangle$ permutes with G^p by hypothesis. Thus $\langle z \rangle$ permutes with each Sylow q-complement of G so z is Σ-quasinormal by Lemma 8.2.

LEMMA 8.4 *Let Σ be a Sylow system of the group G and x a Σ-quasinormal element of G. Let $\langle y \rangle$ be the Sylow p-subgroup of $\langle x \rangle$, p a prime. Then y is Σ-quasinormal.*

Proof: Let G_p and G^p be the Sylow p-subgroup and Sylow p-complement of G in Σ. Then $\langle x \rangle G_p$ and $\langle x \rangle G^p$ are both subgroups of G. Since $([\langle x \rangle G_p : G_p], p) = 1$, it follows that $\langle x \rangle \cap G_p = \langle y \rangle$. Since $[\langle x \rangle G^p : G^p] = p^a =$ the order of y, it follows that $\langle x \rangle G^p = \langle y \rangle G^p$. Lemma 8.3 can now be applied to show that y is Σ-quasinormal in G.

LEMMA 8.5 *Let Σ be a Sylow system of the group G. Then*

(1) *$N_G^*(\Sigma)$ is generated by Σ-quasinormal elements of prime-power order.*

(2) *$N_G^*(\Sigma) \cap G_p$ is a Sylow p-subgroup of $N_G^*(\Sigma)$ and $N_G^*(\Sigma) \cap G_p = \langle x : x$ is a Σ-quasinormal p-element of $G \rangle$, where G_p is the Sylow p-subgroup of G in Σ.*

Proof: If x is Σ-quasinormal of order $p_1^{a_1} p_2^{a_2} \ldots p_r^{a_r}$, then $x = y_1 y_2 \ldots y_r$ where the order of y_i is $p_i^{a_i}$ and y_i generates the Sylow p_i-subgroup of $\langle x \rangle$. By Lemma 8.4 y_i is Σ-quasinormal and, since x may be replaced by y_1, $y_2, \ldots, y_r$ in any generating set, statement (1) is valid.

For the prime q, let S_q be the subgroup of $N_G^*(\Sigma)$ generated by Σ-quasinormal elements of order a power of q. By Lemma 8.3, S_q is a subgroup of the Sylow q-subgroup G_q in Σ, moreover, S_q permutes with each member of Σ. If $q \neq p$, then $S_p G_q \cap S_q G_p = S_q(S_p G_q \cap G_p) = S_q(S_p(G_q \cap G_p)) = S_q S_p$ and hence the subgroups S_p must permute with one another. By (1), $N_G^*(\Sigma)$ is generated by the subgroups S_p. These last two facts imply that each S_p is a Sylow p-subgroup of $N_G^*(\Sigma)$. Then, since $G_p \cap N_G^*(\Sigma) \geq S_p$, we conclude that $S_p = G_p \cap N_G^*(\Sigma)$ and (2) is valid.

LEMMA 8.6 *Let $G = PQ$ be a solvable group where P is a p-group, Q is a cyclic q-group, p and q are primes. If M is a maximal subgroup of G with $M \geq P$, then $[G : M] = q$.*

Proof: If $p = q$, the result is trivial; hence we assume $p \neq q$. Let H be a minimal normal subgroup of G. If $M \geq H$, then we apply induction to conclude that $[G/H : M/H] = [G : M] = q$. We therefore assume M does not contain H. In this situation $G = MH$ and hence H is a q-group. Since Q is a Sylow q-subgroup of G, $Q \geq H$, and H is cyclic. Hence $|H| = q$ and $[G : M] = |H| = q$.

THEOREM 8.7 *For a finite solvable group G the following are equivalent:*

(a) *G is supersolvable.*

(b) *Each Sylow system of G is weakly normal in G.*

(c) *Each maximal subgroup of G is weakly normal in G.*

Proof:

(a)$\Rightarrow$(b) Let p denote the largest prime dividing $|G|$ and let Σ denote a Sylow system of G. We also let G_p and G^p denote, respectively, the Sylow p-subgroup and Sylow p-complement of G in Σ. The Sylow system Σ reduces into G^p, so that $\Sigma \cap G^p$ is a Sylow system of G^p. By induction we may assume that $\Sigma \cap G^p$ is weakly normal in G^p and since G_p is a normal subgroup of G (Theorem 1.8), it is easily seen that G^p is generated by Σ-quasinormal elements. We now direct our attention to G_p. Since G_p is a normal subgroup of G, $\Phi(G_p)$ is also and $G_p/\Phi(G_p)$ is an elementary Abelian normal subgroup of G. We may view $G_p/\Phi(G_p)$ as a vector space over $\mathrm{GF}(p)$ and $\mathrm{Aut}(G_p/\Phi(G_p))$ as a group of linear transformations acting on $G_p/\Phi(G_p)$. As $(p, |\mathrm{Aut}_G(G_p/\Phi(G_p))|) = 1$, we may apply the theorem of

Maschke (see [Scott 1964, 12.1.3]) to conclude that $G_p/\Phi(G_p)$ is the direct sum of minimal normal subgroups of $G/\Phi(G_p)$. Moreover, since G is supersolvable, each of these minimal normal subgroups must have order p. Let $\langle x \rangle \Phi(G_p)/\Phi(G_p)$ be any one of the mimimal normal subgroups of $G/\Phi(G_p)$ contained in $G_p/\Phi(G_p)$.

If $\langle x \rangle \Phi(G_p) = G_p$, then $G_p = \langle x \rangle$ and since G_p is a normal subgroup of G, x is Σ-quasinormal. It now follows that $N_G{}^*(\Sigma) \geq G_p G^p = G$, so that Σ is weakly normal in G.

If $\langle x \rangle \Phi(G_p) < G_p$, then $T = \langle x \rangle \Phi(G_p) G^p$ is a proper subgroup of G. Σ reduces into T and we may apply induction to conclude that $\Sigma \cap T$ is a weakly normal Sylow system of T. By Lemma 8.5, $\langle x \rangle \Phi(G_p)$ is generated by $(\Sigma \cap T)$-quasinormal elements and, by Lemma 8.3, these $(\Sigma \cap T)$-quasinormal p-elements are also Σ-quasinormal. Thus $\langle x \rangle \Phi(G_p)$ is generated by Σ-quasinormal elements.

G_p is generated by the subgroups $\langle x \rangle \Phi(G_p)$, where $\langle x \rangle \Phi(G_p)$ is a minimal normal subgroup of $G/\Phi(G_p)$ contained in $G_p/\Phi(G_p)$. Therefore G_p is generated by Σ-quasinormal elements and $N_G{}^*(\Sigma) \geq G_p G^p = G$. It now follows that Σ is weakly normal in G.

(b) $\Rightarrow$ (c) Let p be the largest prime dividing $|G|$. We will first show that G has a normal Sylow p-subgroup. Let Σ be a fixed Sylow system of G and G_p the Sylow p-subgroup of G in Σ. Let G_q be a Sylow q-subgroup of G in Σ, where q is a prime and $q \neq p$. G_q is generated by Σ-quasinormal elements; let x be one of these generators. In the group $T = \langle x \rangle G_p$, $\langle x \rangle$ is a Sylow q-subgroup. $N_T(\langle x \rangle)/C_T(\langle x \rangle)$ is isomorphic to a p-subgroup of $\mathrm{Aut}(\langle x \rangle)$. If the order of x is q^{a+1}, then $|\mathrm{Aut}(\langle x \rangle)| = q^a(q-1)$ and, since $p > q$, it follows that $N_T(\langle x \rangle)/C_T(\langle x \rangle)$ trivial. By Burnside's theorem ([Scott 1964, 6.2.9]), we know that G_p is normal in T and it is now apparent that $N_G(G_p) \geq G_q$ for any Sylow q-subgroup of G in Σ. Therefore $N_G(G_p) = G$.

Consider now a maximal subgroup M of G with $[G : M] = q^a$ for some prime q. Let Σ be a Sylow system of G which reduces into M. If G^q is the Sylow q-complement of G in Σ, then G^q is contained in M. We first assume that $q = p$.

In this case, $G = MG_p$, where G_p is the Sylow p-subgroup of G. $M \cap G_p$ is normal in M and $N_{G_p}(M \cap G_p) > M \cap G_p$ hence $N_G(M \cap G_p) > M$ and $M \cap G_p$ is normal in G. G_p is generated by Σ-quasinormal elements and not all of these lie in M; say x is a Σ-quasinormal element of G_p with $x \notin M$. Since $M \cap G_p$ is normal in G, $\langle x \rangle$ permutes with $M \cap G_p$. If M_r is the Sylow r-subgroup of M in $\Sigma \cap M$ with $r \neq p$, then $M_r = G_r$ is the Sylow r-subgroup of G in Σ. Hence $\langle x \rangle$ permutes with M_r and $x \in N_G{}^*(\Sigma \cap M)$. It follows that $G = MN_G{}^*(\Sigma \cap M)$ and that M is weakly normal in G.

Assume now that $q \neq p$, so that $M \geq G_p$. Let G^p denote the Sylow p-complement of G in Σ. $G = G_p G^p$ and hence by the Dedekind identity $M = M \cap G = M \cap G_p G^p = G_p(M \cap G^p)$. Σ reduces into G^p and $\Sigma \cap G^p$ is weakly normal in G^p. Since $M \cap G^p$ must be a maximal subgroup of G^p, we may apply induction to conclude that $M \cap G^p$ is weakly normal in

G^p. As each Sylow subgroup of $M \cap G^p$ is also a Sylow subgroup of M, it is apparent from the normality of G_p that any element of G^p which permutes with a Sylow system of $M \cap G^p$ will also permute with a Sylow system of M. Hence $N_G{}^*(M) \geq N_{G^p}{}^*(M \cap G^p) = G^p$. It now follows that $N_G{}^*(M) = G$ and in this case we also have M weakly normal in G.

(c)$\Rightarrow$(a) Let M be a maximal subgroup of G with $[G : M] = p^a$ for some prime p. We wish to show that $[G : M] = p$. If H is a nontrivial normal subgroup of G, then one easily sees that the maximal subgroups of G/H are weakly normal in G/H. By induction we may assume that G/H is supersolvable and, if M contains H, we may apply Theorem 1.7 to conclude that $[G/H : M/H] = [G : M] = p$. We therefore assume that M contains no nontrivial normal subgroups of G. Let Σ be a Sylow system of M, since M is weakly normal in G there are Σ-quasinormal elements of G which do not lie in M. Among these choose x so that the order of x is as small as possible. As $\langle x \rangle$ permutes with each subgroup in Σ, $\langle x \rangle$ permutes with M and by the maximality of M we must have $G = \langle x \rangle M$. Then $(\langle x \rangle \cap M)^G = (\langle x \rangle \cap M)^{\langle x \rangle M} = (\langle x \rangle \cap M)^M \leq M$. Since M does not contain a nontrivial normal subgroup, it follows that $\langle x \rangle \cap M = \{1\}$ and hence the order of x is equal to $[G : M] = p^a$. Let M_q be the Sylow q-subgroup of M in Σ. Since x is Σ-quasinormal, $\langle x \rangle M_q$ is a subgroup of G. Let R be a maximal subgroup of $\langle x \rangle M_q$ with $R \geq M_q$. By Lemma 8.6 $[\langle x \rangle M_q : R] = p$, moreover, $R = R \cap M_q \langle x \rangle = M_q(R \cap \langle x \rangle)$ and hence $p = [\langle x \rangle M_q : R] = [\langle x \rangle : R \cap \langle x \rangle]$. It now follows that $R \cap \langle x \rangle = \langle x^p \rangle$ and that $R = M_q \langle x^p \rangle$. Since $\langle x^p \rangle$ permutes with each Sylow subgroup of Σ, we may apply Lemma 8.2(1) to conclude that x^p is Σ-quasinormal. Then, by the choice of x, we must have $x^p \in M \cap \langle x \rangle = \{1\}$. Therefore x has order p and $[G : M] = p$. The supersolvability of G now follows from Theorem 3.1.

DEFINITION   A subgroup H of the group G is called **weakly subnormal** in G whenever there is a chain of subgroups, $G = H_n \geq H_{n-1} \geq \ldots \geq H_1 \geq H_0 = H$ such that H_{i-1} is a weakly normal subgroup of H_i for $i = 1, 2, \ldots, n$.

As a corollary to Theorem 8.7 we have

COROLLARY 8.8   *For the solvable group G the following are equivalent.*
(1) *G is supersolvable.*
(2) *Each proper subgroup of G is properly contained in its weak normalizer.*
(3) *Each subgroup of G is weakly subnormal in G.*

M-groups

B. M. Puttaswamaiah

The class of (finite) M-groups is a proper subclass of the class of finite solvable groups and in turn it properly contains the class of finite nilpotent groups. An M-group is defined in terms of induced representations. The importance of an M-group lies in its property that its irreducible representations can be constructed from certain homomorphisms from its subgroups into the group of all nonzero elements of the complex field. Therefore a familiarity with elementary results from representation theory is essential for a study of M-groups. The relevant material can be found in Chapter 12 of [Scott 1964] or else the reader can work out the necessary results from the brief description given in the following section (together with the help of any book on representation theory).

1. Results from Representation Theory

Let G be a group of finite order g and $\mathbb{C}$ the complex field. By a **representation** T of G of degree n (≥ 1)(over $\mathbb{C}$), we mean a function from G into the general linear group $\mathrm{GL}(V)$ of all invertible linear transformations of a vector space V of dimension n over $\mathbb{C}$, such that

(1) $$T(xy) = T(x)T(y)$$

for all x, y in G. In this case we say that V is a **G-space** or that V **affords** T. As the general linear group $GL(V)$ of degree n is isomorphic to the group GL_n of all n-by-n nonsingular matrices, a representation T of G of degree n can also be considered as a homomorphism from G into GL_n, in which case T will be called a **matrix representation** of G. For any v in V and x in G, vx denotes the image of v under $T(x)$. For any nonempty subset W of V and x in G, let $Wx = \{wx : w \in W\}$. A **G-subspace** W of V is a subspace of V such that $W \neq \{0\}$ and $Wx = W$ for all x in G. In this case, W is clearly a G-space which affords a representation. If V has no G-subspace other than V, then V and T are said to be **irreducible**. Otherwise they are said to be **reducible**. A well-known result of Maschke asserts that (since $\mathbb{C}$ has characteristic 0) any G-space V can be written as a direct sum $V = W_1 \oplus W_2 \oplus \ldots \oplus W_k$ of irreducible G-subspaces W_i. Each W_i affords an irreducible representation T^i of G. In this case we say that T^i is an **irreducible constituent** of T and that W_i is an **irreducible direct summand** of V.

Let U and V be G-spaces. A **G-map** θ from U to V is a linear transformation such that $\theta(ux) = (\theta(u))x$ for all x in G and all u in U. If θ is a bijective G-map, then U and V can be considered as essentially the same G-space and in this case we say that the representations afforded by U and V are **equivalent**. A basic property of irreducible G-spaces is given by Schur's lemma: *If U and V are irreducible G-spaces, then a G-map from U to V is either 0 or a bijection.*

Let V be a G-space of dimension n which affords the representation T. The **character** ζ of degree n afforded by V (or T) is the unique function from G into $\mathbb{C}$ given by $\zeta(x) = \mathrm{tr}\, T(x)$, x in G, where tr denotes the trace. Since the trace is independent of the basis of V, all equivalent representations have the same character. In addition, characters are class functions in the sense that $\zeta(x) = \zeta(y)$ whenever x and y lie in the same conjugacy class of G. A character ζ of G afforded by V is **irreducible** or **reducible** according as V is irreducible or reducible respectively.

An elementary but fundamental result in the theory of characters is that if h is the number of conjugacy classes of G, then G has exactly h equivalence classes of irreducible representations and hence h irreducible characters. As a class function, a character of G corresponds to an h-tuple of complex numbers (its values on the classes). The h-by-h matrix whose h rows are those h-tuples corresponding to the irreducible characters of G is the **character table** of G. By Maschke's theorem, any character can be expressed as a sum of irreducible characters. Moreover, a character of G determines the corresponding class of representations. It is easily verified that the sum and product of characters of G are also characters afforded by the direct sum and (inner) tensor product of G-spaces.

A representation T (and the character afforded by it) is said to be **linear** if it has degree 1. For example, every group G has the linear representation (called the trivial representation) in which each element of G is mapped on the identity linear transformation of a one-dimensional G-space. The trivial character takes the value 1 at every element of G. For any group G, G/G' is Abelian and hence it has $|G/G'|$ linear characters (by Schur's lemma) each of which gives a linear character of G. These are the only linear characters of G.

The set $\mathrm{Cl}(G)$ of all class functions on G is an h-dimensional vector space over $\mathbb{C}$. An inner product in $\mathrm{Cl}(G)$ is defined by

$$(2) \qquad (\theta, \varphi)_G = \frac{1}{|G|} \sum_{x \in G} \theta(x) \overline{\varphi(x)},$$

where the bar denotes the complex conjugate. This is an hermitian inner product on $\mathrm{Cl}(G)$. The irreducible characters $\zeta^1, \zeta^2, \ldots, \zeta^h$ of G form an orthonormal basis relative to this inner product. From the orthogonality of the irreducible characters, it follows that a character θ of G is irreducible if and only if

$$(3) \qquad (\theta, \theta)_G = 1.$$

There is an equivalent way in which the orthogonality of the irreducible characters of G can be expressed. For any a in G,

$$(4) \qquad \sum_{i=1}^{h} \zeta^i(a) \overline{\zeta^i(x)} = \begin{cases} |C_G(a)| & \text{if } x \text{ is conjugate to } a, \\ 0 & \text{otherwise.} \end{cases}$$

Since each element of G has finite order, the eigenvalues of the linear transformation $T(x)$ are roots of unity. Therefore the values of any character ζ of G are algebraic integers, i.e., roots of a monic polynomial with rational integer coefficients. Also $|\zeta(x)| \leq \zeta(1)$ with equality if and only if $\zeta(x) = \zeta(1)\omega(x)$ for $\omega(x)$ some g-th root of unity. If K is a conjugacy class of G, then it can be shown that $|K|\zeta(x)/\zeta(1)$, $x \in K$, is an algebraic integer and this leads to a result of Schur which asserts that $\zeta(1)$ is a factor of $|G|$ for every irreducible character ζ of G. From (4), it follows that

$$\sum_{i=1}^{h} \zeta^i(1) \overline{\zeta^i(1)} = |G|.$$

This relation will be referred to as the degree relation of G.

Let $a_1, a_2, \ldots, a_g$ be the elements of G and V be a g-dimensional vector space. Let $\{v_{a_1}, v_{a_2}, \ldots, v_{a_g}\}$ be a basis of V indexed by elements of G. Each element a of G determines a unique linear transformation T_a of V which sends v_{a_i} to $v_{a_i a}$, for $i = 1, 2, \ldots, g$. Then the function T from G into $GL(V)$ defined by $T(a) = T_a$ is a representation of G of degree g, since $T(ab) = T_{ab} = T_a T_b = T(a)T(b)$ for all a, b in G. This representation is called the **regular** representation of G. The character Φ afforded by the regular representation of G is called the **regular** character of G. Obviously $\Phi(1) = g$ and $\Phi(x) = 0$ for $x \neq 1$. Since $(\Phi, \zeta)_G \neq 0$ for any irreducible character ζ of G, every irreducible representation of G is a constituent of the regular representation. If m_i is the multiplicity of the irreducible character ζ^i in Φ, then $m_i = (\Phi, \zeta^i)_G = \zeta^i(1)$.

If H is a subgroup of G, it is obvious that the restriction T_H (or V_H or ζ_H) of a representation T (or G-space V or character ζ) of G to H is a representation (or H-space or character) of H. If S is a matrix representation of H of degree n, then we can construct a representation S^* of G as follows: Choose a set of coset representatives $a_1, a_2, \ldots, a_m$ of H in G. For x in G, let $A_{ij}(x) = \delta_{ij} S(a_i^{-1} x a_j)$, where δ_{ij} is 1 or 0 according as $Ha_i = Ha_j$ or $Ha_i \neq Ha_j$. Obviously $A_{ij}(x) = 0$ for all x in $G \backslash H$. Then S^* defined by $S^*(x) = (A_{ij}(x))$, $i, j = 1, 2, \ldots, m$, is a representation of G of degree mn. Indeed if $x, y \in G$, then

$$\sum_{k=1}^{m} A_{ik}(x) A_{kj}(y) = \sum_{k=1}^{m} \delta_{ik} S(a_i^{-1} x a_k) \delta_{kj} S(a_k^{-1} y a_j)$$

$$= \delta_{ij} S(a_i^{-1} x y a_j)$$

$$= A_{ij}(xy),$$

so that $S^*(xy) = S^*(x) S^*(y)$. The representation S^* of G is said to be **induced** by S. The representation S^* depends on the set of coset representatives. However, it is easily verified that an alternate choice of coset representatives gives an equivalent representation. If S affords the character χ of H, then S^* affords the character χ^* of G given by

$$\chi^*(x) = \sum_{i=1}^{m} \chi(a_i^{-1} x a_i) = \frac{1}{|H|} \sum_{a \in G} \chi(a^{-1} x a),$$

where $\chi(h) = 0$ for h in $G \backslash H$.

Let $\zeta^1, \zeta^2, \ldots, \zeta^h$ be the irreducible characters of G and $\chi^1, \chi^2, \ldots, \chi^k$ be the irreducible characters of H. Then there exists a unique h-by-k matrix (a_{ij}) with nonnegative integer coefficients a_{ij} such that

$$\zeta_H^i = \sum_{j=1}^{k} a_{ij} \chi^j.$$

Then it is easily checked that

$$(\chi^j)^* = \sum_{i=1}^{h} a_{ij} \zeta^i$$

so that we have the Frobenius reciprocity formula

(5) $$\left(\zeta_H^i, \chi^j\right)_H = \left(\zeta^i, (\chi^j)^*\right)_G$$

which is a very useful result in applications.

Suppose N is a normal subgroup of G and S is a representation of N which affords the character χ. Then for any fixed a of G, S^a defined by $S^a(x) = S(a^{-1} x a)$, $x \in N$, is a representation of N which affords the character χ^a where $\chi^a(x) = \chi(a^{-1} x a)$, $x \in N$. The representations S and

S^a, and the characters χ and χ^a are said to be **conjugate**. Obviously χ is irreducible if and only if χ^a is irreducible. We denote by $I(\chi)$ the subgroup $\{a : a \in G \text{ and } \chi^a = \chi\}$ and call $I(\chi)$ the **inertial group** of χ in G.

THEOREM 1.1 (Clifford) *Let N be a normal subgroup of a group G, and ζ be an irreducible character of G. Then there exists an irreducible character θ of N and a positive integer e such that*

$$\zeta_N = e \sum_{i=1}^{r} \theta^{a_i},$$

where $\{a_1 = 1, a_2, \ldots, a_r\}$ is a set of coset representatives of $I(\theta)$ in G. If χ is an irreducible constituent of $\zeta_{I(\theta)}$ such that θ is an irreducible constituent of χ_N, then $\zeta = \chi^*$.

Proof: Let V be a G-space which affords the irreducible character ζ and W be an N-subspace of V_N which affords an irreducible character θ of N. Since θ is a constituent of ζ_N, it follows by the Frobenius reciprocity formula (5) that

$$0 \neq (\theta, \zeta_N)_N = (\theta^*, \zeta)_G$$

so that ζ is a constituent of θ^*. For any a in N,

$$\theta^*(a) = \frac{1}{|N|} \sum_{x \in G} \theta(x^{-1}ax) = \frac{1}{|N|} \sum_{x \in G} \theta^x(a)$$

with $\theta^x = \theta^y$ if and only if $xy^{-1} \in I(\theta)$ so that

$$\theta^* = [I(\theta) : N] \sum_{i=1}^{r} \theta^{a_i},$$

where $\{a_1 = 1, a_2, \ldots, a_r\}$ is a set of coset representatives of $I(\theta)$ in G. Since ζ is a constituent of θ^* such that θ is a constituent of ζ_N, it follows that

$$\zeta_N = e \sum_{i=1}^{r} \theta^{a_i}$$

for some integer e which depends on N and ζ.

Let $H = I(\theta)$ and let χ be an irreducible constituent of ζ_H such that θ is a constituent of χ_N. Then $\chi_N = f\theta$ for some positive integer f. Clearly ζ is an irreducible constituent of χ^*. Since χ is a constituent of ζ_H, we have $f \leq e$. Therefore

$$e r \theta(1) = \zeta(1) \leq \chi^*(1) = r\chi(1) = rf\theta(1) \leq er\theta(1),$$

and hence $\zeta(1) = \chi^*(1)$ and $e = f$. In particular $\zeta = \chi^*$.

Remark If $I(\theta) = G$, then Clifford's theorem does not throw much light on ζ. A special case of this nature arises in this chapter.

Let G and G_1 be two groups, $\zeta^1, \zeta^2, \ldots, \zeta^h$ be the irreducible characters of G and $\chi^1, \chi^2, \ldots, \chi^k$ be the irreducible characters of G_1. Then $\zeta^i \chi^j$ given by $\zeta^i \chi^j(a,b) = \zeta^i(a)\chi^j(b)$ is a character of the direct product $G_0 = G \times G_1$. Since

$$\sum_{a \in G} \sum_{b \in G_1} \zeta^i \chi^j(a,b) \overline{\zeta^i \chi^j(a,b)} = |G_0|,$$

it follows that $\zeta^i \chi^j$ is an irreducible character of G_0. As

$$\sum_{i=1}^{h} \sum_{j=1}^{k} \zeta^i(1)\chi^j(1)\zeta^i\chi^j$$

is the regular character of G_0, it follows that $\{\zeta^i \chi^j : 1 \leq i \leq h, 1 \leq j \leq k\}$ is the complete set of irreducible characters of G_0.

The Groups A_4, S_4, $2A_4$, and $2S_4$

We conclude this section by determining the character tables of the alternating group A_4, the symmetric group S_4, and the double groups $2A_4$ and $2S_4$ (to be defined presently). These groups have orders 12, 24, 24, and 48 respectively. These examples serve to illustrate the salient points of the theory of M-groups. Hence the reader is urged to carry out the calculations in detail.

Let U_2 denote the group of all 2-by-2 (complex) unitary matrices of determinant 1 and O_3 denote the group of all 3-by-3 (real) orthogonal matrices of determinant 1. The mapping θ from U_2 to O_3 defined by

$$(6) \qquad \theta\begin{pmatrix} a & \beta \\ -\overline{\beta} & \overline{a} \end{pmatrix} =$$

$$\begin{pmatrix} (a^2 - \beta^2 + \overline{a}^2 - \overline{\beta}^2)/2 & i(a^2 + \beta^2 - \overline{a}^2 - \overline{\beta}^2)/2 & -a\beta - \overline{a}\overline{\beta} \\ i(-a^2 + \beta^2 + \overline{a}^2 - \overline{\beta}^2)/2 & (a^2 + \beta^2 + \overline{a}^2 + \overline{\beta}^2)/2 & -i(a\beta - \overline{a}\overline{\beta}) \\ a\overline{\beta} + \overline{a}\beta & -i(a\overline{\beta} - \overline{a}\beta) & a\overline{a} - \beta\overline{\beta} \end{pmatrix}$$

is a (group) epimorphism. Clearly the kernel of θ is just the center $Z(U_2) = \{\begin{pmatrix} 1 & 0 \\ 0 & 1 \end{pmatrix}, \begin{pmatrix} -1 & 0 \\ 0 & -1 \end{pmatrix}\}$. Let

$$u = \frac{1}{\sqrt{2}}\begin{pmatrix} 1 & i \\ i & 1 \end{pmatrix} \quad \text{and} \quad v = \frac{1}{2}\begin{pmatrix} 1+i & 1-i \\ -1-i & 1-i \end{pmatrix}$$

Then their images are

$$\theta(u) = \begin{pmatrix} 1 & 0 & 0 \\ 0 & 0 & 1 \\ 0 & -1 & 0 \end{pmatrix} \quad \text{and} \quad \theta(v) = \begin{pmatrix} 0 & 0 & 1 \\ 1 & 0 & 0 \\ 0 & 1 & 0 \end{pmatrix}.$$

The orders of the elements $u, v, \theta(u)$, and $\theta(v)$ are $8, 6, 4$, and 3 respectively. The subgroup $K = \langle \theta(u^2), \theta(vu^2v^{-1}) \rangle$ has order 4; $A_4 = K\langle \theta(v) \rangle = \langle \theta(v) \rangle K$ has order 12; and $S_4 = \langle \theta(u) \rangle \langle \theta(u^2), \theta(v) \rangle$ has order 24. The subgroup K is the commutator subgroup of A_4 and A_4 is the commutator subgroup of S_4. Hence A_4 has three linear characters and S_4 has two linear characters. The group A_4 has 4 conjugacy classes with representatives $\theta(1)$, $\theta(u^2)$, $\theta(v)$, and $\theta(v^2)$ having $1, 3, 4$, and 4 elements respectively. Hence A_4 has 4 irreducible characters of which three have degree 1 and one has degree 3. The character of degree 3 is induced by a nontrivial linear character of $\langle \theta(u^2), \theta(vu^2v^{-1}) \rangle$. The three linear characters of A_4 are determined by the linear characters of A_4/K.

Character Table of A_4

$$
\begin{array}{cccc}
1 & 1 & 1 & 1 \\
1 & 1 & \omega & \omega^2 \\
1 & 1 & \omega^2 & \omega \\
3 & -1 & 0 & 0
\end{array}
\qquad \omega^3 = 1, \omega \neq 1
$$

The group S_4 has five conjugacy classes with representatives $\theta(1), \theta(u^2)$, $\theta(v), \theta(v^2u)$, and $\theta(u)$ having $1, 3, 8, 6$, and 6 elements respectively. We have already determined two linear characters. The character of S_4 induced by the character $\langle 1, 1, \omega, \omega^2 \rangle$ of A_4 is easily verified to be irreducible. Of course the defining matrices $\theta(u)$ and $\theta(v)$ determine an irreducible character of degree 3. Hence the remaining irreducible character has degree 3. This information together with the orthogonality relations give the following character table of S_4:

Character Table of S_4

$$
\begin{array}{ccccc}
1 & 1 & 1 & 1 & 1 \\
1 & 1 & 1 & -1 & -1 \\
2 & 2 & -1 & 0 & 0 \\
3 & -1 & 0 & -1 & 1 \\
3 & -1 & 0 & 1 & -1
\end{array}
$$

It may be observed that the characters ζ^4 and ζ^5 are the characters of S_4 induced by linear characters of a subgroup of order 8 of S_4.

The preimage $2A_4$ and the preimage $2S_4$ of A_4 and S_4 in U_2 are called the double groups of A_4 and S_4 respectively, and have orders 24 and 48.

Since $2A_4/Z(2A_4) \cong A_4$, each irreducible character of A_4 can be extended to $2A_4$. The group $2A_4$ has seven conjugacy classes with representatives $1, u^2, v,$ $v^{-1}, v^3, v^2,$ and v^4 having $1, 6, 4, 4, 1, 4,$ and 4 elements respectively. Hence $2A_4$ has seven irreducible characters. Four of these characters can be obtained from those of A_4. The defining matrices u^2 and v give an irreducible character of degree 2. The remaining two irreducible characters have degree 2 each. The orthogonality relations give the following character table of $2A_4$:

Character Table of $2A_4$

1	1	1	1	1	1	1	
1	1	ω	ω^2	1	ω	ω^2	
1	1	ω^2	ω	1	ω^2	ω	
3	-1	0	0	3	0	0	$\omega^3 = 1, \omega \neq 1$
2	0	1	-1	-2	-1	1	
2	0	ω	ω^2	-2	$-\omega$	$-\omega^2$	
2	0	ω^2	ω	-2	$-\omega^2$	$-\omega$	

The double group $2A_4$ has no subgroup of order 12. Hence no irreducible character of degree 2 can be induced by a linear character of a subgroup. However the three-dimensional character is induced by a nontrivial linear character of the subgroup $\langle \left(\begin{smallmatrix} 0 & i \\ i & 0 \end{smallmatrix}\right), \left(\begin{smallmatrix} -i & 0 \\ 0 & i \end{smallmatrix}\right) \rangle$.

The double group $2S_4$ has eight conjugacy classes with representatives $1, u^2, v, uv^2, u^5, v^3, v^2,$ and u having $1, 6, 8, 12, 6, 1, 8,$ and 6 elements respectively and so $2S_4$ has exactly 8 irreducible characters. We have already determined five irreducible characters, namely the ones obtained from those of S_4. The above matrices of u and v give a two-dimensional representation which is easily verified to be irreducible. If we replace $\sqrt{2}$ by $-\sqrt{2}$ we get another irreducible two-dimensional representation. Hence the remaining irreducible representation has degree 4. The above information together with the orthogonality relations yield the character table of $2S_4$.

Character Table of $2S_4$

1	1	1	1	1	1	1	1
1	1	1	−1	−1	1	1	−1
2	2	−1	0	0	2	−1	0
3	−1	0	1	−1	3	0	−1
3	−1	0	−1	1	3	0	1
2	0	1	0	$\sqrt{2}$	−2	−1	$-\sqrt{2}$
2	0	1	0	$-\sqrt{2}$	−2	−1	$\sqrt{2}$
4	0	−1	0	0	−4	1	0

The group $2S_4$ has subgroups of orders $1, 2, 3, 4, 6, 8, 12, 16, 24$, and 48. The subgroup $2A_4$ is a normal subgroup of $2S_4$.

2. Monomial Representations

The problem of writing down the actual matrices of a representation of a group G is a rather difficult one. Indeed there is no known simple algorithm which enables this to be done in the general case. Some of the representations can be constructed as induced representations. An induced representation of G is constructed with the aid of a representation of a subgroup. In general an induced representation is reducible. This section is concerned with representations of a group G induced by representations of degree 1 of subgroups of G. The concept of a monomial representation of a group G will be presented in more general form than it is actually needed for the theory of M-groups. This generalization has been successfully used to study the structure of finite groups. In particular our version includes the study of wreath products, transfers, and induced representations.

Let A be an arbitrary group, P be a permutation group on a set Ω, A^Ω be the collection of all functions from Ω to A, and $A \wr P$ be the cartesian product of the set A^Ω with P. In $A \wr P$, define a multiplication by

$$(f, a)(g, b) = (h, ab), \quad \text{where } h(i) = f(i) g(i^a) \text{ for } i \in \Omega.$$

This multiplication is easily verified to be associative. Moreover, an identity element and inverses exist in $A \wr P$. The identity element is $(1, 1)$, where the first coordinate is the function which maps each $i \in \Omega$ to the identity of A and the second coordinate is the identity of P. Inverses are given by $(f, a)^{-1} = (g, a^{-1})$, where $g(i) = f(i^{a^{-1}})^{-1}$. Hence $A \wr P$ is a group.

DEFINITION The group $A \wr P$ is called the **permutational wreath product** of A by P.

The permutational wreath product $A \wr P$ has a simple structure. Indeed, it is a semi-direct product of a direct product of $|\Omega|$ copies of A by a copy of P. If $E = \{1\}$ is the trivial subgroup of P, then $A^\Omega \times E$ is a subgroup of $A \wr P$, which will be denoted by $A \wr E$ and called the **base group** of $A \wr P$. Then $A \wr E$ is isomorphic to the direct product of $|\Omega|$ copies of A. The mapping θ from $A \wr P$ to P defined by $(f, a)^\theta = a$ is an epimorphism with kernel $A \wr E$. Since $(f, a) = (f, 1)(1, a)$, the subgroup $P_0 = \{(1, a) : a \in P\}$ is isomorphic to P and $(A \wr E) \cap P_0 = \{(1, 1)\}$. Hence $A \wr P = (A \wr E)P_0$ and so $A \wr P$ is a semi-direct product of $A \wr E$ by P_0.

Now let P be an arbitrary group. Each element p of P determines a unique permutation σ_p of P given by $\sigma_p(x) = xp$, $x \in P$, called a regular permutation of P. The set of all regular permutations of P forms a group which is isomorphic to P. If A is any group, then the permutational wreath product of A by the group of regular permutations of P is called the **standard wreath product** of A by P and is denoted $A \operatorname{wr} P$.

Let P be a permutation group on Ω and P_1 be a subgroup of P. Then P_1 is a permutation group on Ω. Then obviously $A_1 \wr P_1$ can be considered as a subgroup of $A \wr P$ for any subgroup A_1 of A. The following result is similar but lies a little deeper.

LEMMA 2.1 *Let A_1 be a subgroup of a group A and P_1 be a subgroup of P. Then there is a monomorphism from $A_1 \operatorname{wr} P_1$ into $A \operatorname{wr} P$.*
Proof: For each function f from P_1 to A_1, let $\bar{f}$ be the function from P to A given by

$$\bar{f}(p) = \begin{cases} f(p) & \text{if } p \in P_1, \\ 1 & \text{if } p \in P \setminus P_1. \end{cases}$$

Now define δ from $A_1 \operatorname{wr} P_1$ into $A \operatorname{wr} P$ by $(f, p)^\delta = (\bar{f}, p)$. If (f, p) and (g, p') are any two elements of $A_1 \operatorname{wr} P_1$, let $(f, p)(g, p') = (h, pp')$ (so that h is described by $h(z) = f(z)g(zp)$ for $z \in P_1$). Then $[(f, p)(g, p')]^\delta = (h, pp')^\delta = (\bar{h}, pp')$, while $(f, p)^\delta(g, p')^\delta = (\bar{f}, p)(\bar{g}, p') = (k, pp')$ with k described by $k(z) = \bar{f}(z)\bar{g}(zp)$ for all $z \in P$. Now note that $z \in P_1$ if and only if $zp \in P_1$. Thus

$$k(z) = \begin{cases} f(z)g(zp) = h(z) & \text{if } z \in P_1 \\ 1 & \text{if } z \notin P_1 \end{cases}$$

$$= \bar{h}(z)$$

so that $k = \bar{h}$ proving δ is a homomorphism. If $(f, p)^\delta = (1, 1)$, then $\bar{f} = 1$ and $p = 1$, so that δ is a monomorphism.

THEOREM 2.2 (Universal embedding theorem) *Let N be a normal subgroup of the group G. Then there is a monomorphism from G into the group $N \operatorname{wr} G/N$.*

Proof: Let R be a transversal for N in G. Thus for every $y \in G$ there is a unique $n(y) \in N$ and a unique $r(y) \in R$ such that $y = n(y)r(y)$. We define a mapping δ from G into $N \operatorname{wr} G/N$ by $\delta(y) = (f, Ny)$ where f is given by $f(Nz) = r(z)yr(zy)^{-1}$ for all $Nz \in G/N$. To show δ is a homomorphism, let $\delta(y) = (f, Ny)$, $\delta(w) = (g, Nw)$, and $\delta(yw) = (h, Nyw)$. By definition of multiplication in the wreath product we have $(f, Ny)(g, Nw) = (k, NyNw)$ where k is given by $k(Nz) = f(Nz)g(NzNy)$. Thus $k(Nz) = r(z)yr(zy)^{-1}r(zy) \cdot wr(zyw)^{-1} = r(z)ywr(zyw)^{-1} = h(Nz)$ for all $Nz \in G/N$. Consequently $k = h$ and so δ is a homomorphism. If y belongs to $\ker \delta$, then $y \in N$ and $r(z)yr(zy)^{-1} = 1$ for all $z \in G$. In particular, for $z = 1$ we get $y = r(y)$. As $y \in N$, this forces $y = 1$. Hence δ is a monomorphism completing the proof.

In order to achieve a clearer insight into permutational wreath products, we express them in matrix form as follows. Let $|\Omega| = n$. The elements of P can be naturally represented as n-by-n permutation matrices. This group of matrices will be denoted by $\tilde{P}$ and it is obviously isomorphic to P. For example, the permutation group $\{(1), (1,2)\}$ can be represented as the matrix group $\{(\begin{smallmatrix} 1 & 0 \\ 0 & 1 \end{smallmatrix}), (\begin{smallmatrix} 0 & 1 \\ 1 & 0 \end{smallmatrix})\}$. Replace the 1's in the identity matrix of $\tilde{P}$ by elements of A in all possible ways. These diagonal matrices over A form a group D with respect to "matrix multiplication," and this group is isomorphic to $A \wr E$. Then the semi-direct product $D\tilde{P}$ is isomorphic to $A \wr P$. We write $A \wr^* P$ for $D\tilde{P}$. An element of $A \wr^* P$ is called a **monomial matrix** of degree n over A.

DEFINITION Let A and G be two groups. A **monomial representation** T of G of degree n over A is a homomorphism of G into $A \wr^* S_n$ where S_n is the symmetric group.

EXAMPLE Let A be a cyclic group of order m, n be a multiple of m, and $G = \langle a, b : a^n = b^2 = (ab)^2 = 1 \rangle$ be the dihedral group of order $2n$. Then the wreath product $A \wr^* S_2$ has order $2m^2$. If $A = \langle c \rangle$, define T from G into $A \wr^* S_2$ by

$$T(a) = \begin{pmatrix} c & 0 \\ 0 & c^{-1} \end{pmatrix} \quad \text{and} \quad T(b) = \begin{pmatrix} 0 & 1 \\ 1 & 0 \end{pmatrix}$$

Then T is a monomial representation of G of degree 2 over A.

Next let $*$ be a homomorphism of A into the group $\langle \omega \rangle$ of the mth roots of unity given by $c^* = \omega$. Then $*$ determines a representation T^* of G

$$T^*(a) = \begin{pmatrix} c^* & 0 \\ 0 & c^{*-1} \end{pmatrix} \quad \text{and} \quad T^*(b) = \begin{pmatrix} 0 & 1^* \\ 1^* & 0 \end{pmatrix} = \begin{pmatrix} 0 & 1 \\ 1 & 0 \end{pmatrix} .$$

The representation T^* is not obtained as a representation of G induced by a representation of a subgroup of G. Thus we see that the concept of a monomial representation of a group generalizes the concept of an induced representation.

Frobenius has shown how to construct, for a given group G with a subgroup A of index n, a monomial representation T of G of degree n over A. Let $r_1, r_2, \ldots, r_n$ be a set of coset representatives of A in G. For any $x \in G$, let $\underline{r}x$ be the n-by-1 matrix with $(i, 1)$th entry equal to $r_i x$ (we'll just use $\underline{r}$ to denote $\underline{r}1$). Then for any $x \in G$, each $r_i x$ belongs to some $A r_k$ and so can be uniquely decomposed in the form $r_i x = a_i r_k$, $a_i \in A$, where a_i and r_k depend on i. Moreover $i \to k$ is a permutation of $\{1, 2, \ldots, n\}$. Hence we obtain $\underline{r}x = T(x)\underline{r}$ for a unique matrix $T(x)$ in $A \wr^* S_n$.

THEOREM 2.3 *With notation as in the preceding paragraph, the mapping $T: G \to A \wr^* S_n$ is a homomorphism.*
Proof: Let $x, y \in G$. Let V be the 1-by-1 matrix (y). Then $\underline{r}(xy) = \underline{r}x V$ $= (T(x)\underline{r})V = T(x)(\underline{r}V) = T(x)\underline{r}y = T(x)T(y)\underline{r}$ so that $T(x)T(y) = T(xy)$ by uniqueness of the matrix $T(xy)$. Hence T is a homomorphism.

If $*$ is a representation of A, we can let $x^* = 0$ if $x \in G \backslash A$ and replace each element a in $T(x)$ by the corresponding a^* to obtain $T^*(x)$. This gives a representation T^* of G, which is equivalent to the representation of G induced by the representation $*$ of the subgroup A.

In the following pages, a representation T (or a character ζ) of a group G will be called a **monomial representation (or character)** of G if G has a subgroup H and H has a linear representation S (or a linear character χ) such that S^* is equivalent to T (or $\zeta = \chi^*$, the induced character).

If T is a monomial representation of G induced by a linear representation S of a subgroup H, then the degree of T is equal to the index of H in G, and hence a necessary condition for a representation T to be monomial is that G must have a subgroup H such that $[G:H]$ equals the degree of T. This condition is by no means sufficient. Examples 2 and 3 below give an illustration of (i) a case where a subgroup of appropriate index does not exist and (ii) a case where a subgroup of appropriate index exists, but the subgroup does not have a representation of degree 1 which induces the given representation.

EXAMPLE 1 Every linear character of a group G is monomial. In particular the trivial character of G is monomial.

EXAMPLE 2 The double group $2A_4$ has 3 linear characters, one irreducible character of degree 3 and three irreducible characters of degree 2 each.

The irreducible character of degree 3 is a monomial character. Since $2A_4$ has no subgroup of order 12, none of the characters of degree 2 is monomial.

EXAMPLE 3 The double group $2S_4$ has two linear characters, three irreducible characters of degree 2, two irreducible characters of degree 3 and one irreducible character of degree 4. The group $2A_4$ is a subgroup of $2S_4$ of index 2. Since the trivial character of a proper subgroup of a group can not induce an irreducible character, it follows that not all irreducible characters of degree 2 are monomial, although $2S_4$ has a subgroup of order 24.

Our concern is to study the irreducible monomial representations of a group. For this purpose, we give a simple criterion for a monomial representation to be irreducible. Let G be a group, H be a subgroup of G, and θ be a class function on H. For each a in G, define a class function θ^a on $H^a = a^{-1}Ha$ by $\theta^a(x) = \theta(axa^{-1})$ for all x in H^a. Then $(\theta^a)^b = \theta^{ab}$ for all a, b in G. If θ is a character of H, then θ^a is a character of H^a. Moreover θ is irreducible if and only if θ^a is irreducible.

THEOREM 2.4 *Let H be a subgroup of a group G and θ be a linear character of H. The induced character θ^* is irreducible if and only if for each a in $G \backslash H$, θ^a and θ are distinct linear characters of the subgroup $H \cap H^a$.*
Proof: Let $G = Ha_1 H \cup \ldots \cup Ha_n H$ be the double coset decomposition of G with respect to H, where a_1 belongs to H, and let $H_i = a_i^{-1}Ha_i \cap H$. Then by definition

$$(\theta^*, \theta^*)_G = \frac{1}{|G|} \sum_{x \in G} \theta^*(x) \overline{\theta^*(x)}$$

$$= \frac{1}{|G|} \frac{1}{|H|^2} \sum_{a, b, x \in G} \theta(axa^{-1}) \overline{\theta(bxb^{-1})}$$

$$= \frac{1}{|G||H|^2} \sum_{a, b, x \in G} \theta^{ab^{-1}}(x) \overline{\theta(x)} \,.$$

Let $ab^{-1} = ha_i k$ for h, k in H. Then $\theta^{ab^{-1}} = \theta^{a_i k}$ and

$$\sum_{x \in G} \theta^{ab^{-1}}(x) \overline{\theta(x)} = \sum_{x \in G} \theta^{a_i k}(x) \overline{\theta(x)}$$

$$= \sum_{x \in G} \theta^{a_i}(x) \overline{\theta^{k^{-1}}(x)}$$

$$= \sum_{x \in G} \theta^{a_i}(x) \overline{\theta(x)}$$

$$= \sum_{x \in H_i} \theta^{a_i}(x) \overline{\theta(x)}, \quad \text{since } \theta = 0 \text{ on } G \backslash H$$

$$= |H_i| (\theta^{a_i}, \theta)_{H_i}.$$

Thus

$$(\theta^*, \theta^*)_G = \frac{1}{|G||H|^2} \sum_{i=1}^{n} \sum_{a,b \in G;\ ab^{-1} \in Ha_iH} |H_i|(\theta^{a_i}, \theta)_{H_i}$$

$$= \frac{1}{|G||H|^2} \sum_{i=1}^{n} \sum_{a \in G} |Ha_iH||H_i|(\theta^{a_i}, \theta)_{H_i}$$

$$= \sum_{i=1}^{n} \frac{1}{|H|^2} |Ha_iH||H_i|(\theta^{a_i}, \theta)_{H_i}$$

$$= \sum_{i=1}^{n} (\theta^{a_i}, \theta)_{H_i}, \quad \text{since } |Ha_iH||H_i| = |H|^2.$$

If for each a in $G \backslash H$, θ^a and θ are distinct linear characters of $H \cap H^a$, then $(\theta^a, \theta)_{H \cap H^a} = 0$. In particular $(\theta^{a_i}, \theta)_{H_i} = 0$ for $i = 2, 3, \ldots, n$ and $(\theta, \theta)_H = 1$. From the above equation it follows that $(\theta^*, \theta^*)_G = 1$ and hence θ^* is irreducible.

Conversely, let θ^* be irreducible, so that $(\theta^*, \theta^*)_G = 1$. Again the above equation yields $\sum_{i=2}^{n} (\theta^{a_i}, \theta)_{H_i} = 0$. Since $(\theta^{a_i}, \theta)_{H_i} \geq 0$, it follows that $(\theta^{a_i}, \theta)_{H_i} = 0$. Hence $(\theta^a, \theta)_{H^a \cap H} = 0$ for all a in $G \backslash H$, and θ^a and θ are distinct linear characters of $H \cap H^a$. This completes the proof of the result.

COROLLARY 2.5 *Let H be a normal subgroup of a group G and θ be a linear character of H. Then the induced character θ^* is irreducible if and only if θ and θ^a are distinct linear characters of H for every a in $G \backslash H$.*

EXAMPLE 4 As was seen in Section 1, the subgroup A_4 of S_4 has two nontrivial characters $\theta = (1, 1, \omega, \omega^2)$ and $\varphi = (1, 1, \omega^2, \omega)$. For any a in $S_4 \backslash A_4$, $\theta^a = \varphi$ and hence $\theta^a \neq \theta$. Thus θ^* is an irreducible character of S_4.

The next result answers the question: When is an irreducible representation of G monomial?

THEOREM 2.6 *Let T be an irreducible representation of G afforded by a G-space V of dimension n. Then T is monomial if and only if V has a basis $\{v_1, v_2, \ldots, v_n\}$ such that*

$$v_i x = \alpha_i(x) v_{\sigma_x(i)}; \qquad i = 1, 2, \ldots, n \text{ and } x \in G,$$

where σ_x is a permutation of $\{1, 2, 3, \ldots, n\}$ and $\alpha_i(x) \neq 0$ is a complex number.

Proof: Let $T = S^*$ where S is a representation of degree 1 of a subgroup H of G. By definition of induced representation, each row and each column of the matrix $S^*(x)$ has exactly one nonzero entry. Hence there is a basis $\{v_1, v_2, \ldots, v_n\}$ such that $v_i x = \alpha_i(x) v_{\sigma_x(i)}$, for $i = 1, 2, \ldots, n$ and x in G.

Conversely, assume that V has such a basis $\{v_1, v_2, \ldots, v_n\}$. Since T is irreducible, $P = \{\sigma_x : x \in G\}$ is a transitive permutation group on $\{1, 2, \ldots, n\}$. Let $H = \{x : x \in G \text{ and } \sigma_x(1) = 1\}$ and $[G:H] = m$. For all i, $1 \leq i \leq m$, let a_i be any element of G such that σ_{a_i} takes 1 to i. Then H is a subgroup of G and $G = \bigcup_{i=1}^{m} Ha_i$, since $\{a_1, a_2, \ldots, a_m\}$ is a set of coset representatives of H in G.

Let $\langle v \rangle$ be the linear H-subspace of V_H given by $vh = \alpha_1(h)v$ for all h in H. Since $\langle v \rangle$ is linear, it is an irreducible H-space. The induced G-space $\langle v \rangle^*$ has $\{va_1, va_2, \ldots, va_m\}$ as a basis. For any x in G, $a_i x = h_i a_{i'}$, for some h_i in H, so that $v(a_i x) = v(h_i a_{i'}) = (vh_i)a_{i'} = \alpha_1(h_i)va_{i'}$. But V is irreducible and hence the elements $w_i = v_1 a_i = \alpha_1(a_i)v_i$, $i = 1, 2, \ldots, m$ span all of V. Thus $n = m$. We have

$$w_i x = v_1 a_i x = v_1 (h_i a_{i'}) = \alpha_1(h_i)v_1 a_{i'} = \alpha_1(h_i)w_{i'}.$$

The mapping θ from $\langle v \rangle^*$ into V defined by

$$(v_i a_i)\theta = v_1 a_i = w_i$$

is a G-homomorphism. Hence $\langle v \rangle^* \cong V$ and T is a monomial representation of G.

3. M-groups

DEFINITION A group G is called a **monomial group** or in short an **M-group** if every irreducible representation of G is induced by a representation of degree 1 of a subgroup of G.

Note The term M-group is also used on occasion (e.g., in [Scott 1964]) in an entirely different context; namely to mean an (infinite) group possessing a finite normal series whose factor groups are either finite or infinite cyclic.

Example 5 The group A_4 is a monomial group. On the other hand, the double group $2A_4$ has no subgroup of index 2, but it has an irreducible representation of degree 2. Clearly this representation is not induced by a representation of degree 1 of a subgroup. The group S_4 is a monomial group. The double group $2S_4$ has the unique subgroup $2A_4$ of index 2, but it has irreducible representations of degree 2 which are not induced by representations of degree 1 of $2A_4$.

Example 6 Let G be the group generated by the matrices

$$a = \begin{pmatrix} i & 0 & 0 & 0 \\ 0 & i & 0 & 0 \\ 0 & 0 & -i & 0 \\ 0 & 0 & 0 & -i \end{pmatrix}, \qquad c = \begin{pmatrix} i & 0 & 0 & 0 \\ 0 & -i & 0 & 0 \\ 0 & 0 & i & 0 \\ 0 & 0 & 0 & -i \end{pmatrix}$$

$$\text{and} \qquad u = \frac{-i}{2} \begin{pmatrix} 1 & i & i & -1 \\ -1 & i & -i & -1 \\ -1 & -i & i & -1 \\ 1 & -i & -i & -1 \end{pmatrix}$$

Show that G is an M-group of order 96 and that G has subgroups which are not M-groups.

Proof: Let $u^{-1}au = b$ and $u^{-1}cu = d$. Then the elements a, b, c, d, and u satisfy the relations

$$a^4 = u^3 = 1, \qquad a^2 = b^2 = c^2 = d^2, \qquad ac = ca,$$
$$ad = da, \qquad bc = cb, \qquad bd = db, \qquad ab = ba^3,$$
$$cd = dc^3, \qquad u^{-1}au = b, \qquad u^{-1}cu = d,$$
$$u^{-1}du = cd, \quad \text{and} \quad u^{-1}bu = ab.$$

So G has center of order 2 generated by a^2. Each of the subgroups $\langle a,b \rangle$ and $\langle c,d \rangle$ is isomorphic to the quaternion group of order 8, so that $H = \langle a,b,c,d \rangle$ is a subgroup of order 32. Moreover, H is the derived subgroup of G. Since $u^{-1}\langle a,b \rangle u = \langle a,b \rangle$, $u^{-1}\langle c,d \rangle u = \langle c,d \rangle$, and $c^3 = 1$, it follows that $|G| = 96$. As $|G/H| = 3$, G has exactly three linear characters. The subgroup H has 17 conjugacy classes with representatives 1, a^2, a, b, c, d, ab, bc, bd, ad, ac, cd, abc, abd, acd, bcd, and $abcd$. The first two classes have one element in each while the remaining classes have two elements in each. The character of degree 4 of H (afforded by the defining matrices of H)

$$(4, -4, 0, 0, 0, 0, 0, 0, 0, 0, 0, 0, 0, 0, 0, 0, 0)$$

is easily verified to be irreducible. Hence H has 16 linear characters. The following are five of them.

$$\chi_4 = (1, 1, -1, 1, -1, 1, -1, 1, -1, 1, -1, 1, -1, 1, -1, 1 - 1),$$
$$\chi_5 = (1, 1, 1, 1, 1, -1, -1, -1, -1, 1, 1, 1, 1, -1, -1, -1, -1),$$
$$\chi_6 = (1, 1, -1, 1, 1, -1, -1, 1, -1, 1, 1, 1, -1, -1, -1, -1, 1),$$
$$\chi_7 = (1, 1, 1, -1, -1, -1, -1, 1, 1, 1, 1, -1, -1, -1, -1, 1, 1)$$
$$\chi_8 = (1, 1, -1, -1, 1, -1, 1, 1, -1, 1, -1, 1, 1, -1, 1, 1, -1).$$

By Theorem 2.4 $\zeta^4 = \chi_4{}^*$, $\zeta^5 = \chi_5{}^*$, $\zeta^6 = \chi_6{}^*$, $\zeta^7 = \chi_7{}^*$, and $\zeta^8 = \chi_8{}^*$ are irreducible characters of degree 3 each.

Let $K = \langle ac, bd, a^2, u \rangle$. Clearly $\langle ac, bd \rangle$ is a subgroup of order 4. Since $u^{-1}acu = bd$ and $u^{-1}bdu = abcd$, it follows that K is isomorphic to the direct product of the alternating group A_4 and a cyclic group of order 2. Hence H has 8 irreducible characters, of which 6 are linear characters. The following are three of the irreducible characters of K:

$$\theta_9 = (1, -1, 1, -1, 1, -1, 1, -1)$$
$$\theta_{10} = (1, -1, 1, -1, \omega, -\omega, \omega^2, -\omega^2)$$
$$\theta_{11} = (1, -1, 1, -1, \omega^2, -\omega^2, \omega, -\omega)$$

where $\omega^3 = 1$ and $\omega \neq 1$. Again by Theorem 2.4, $\zeta^9 = \theta_9{}^*$, $\zeta^{10} = \theta_{10}{}^*$, and $\zeta^{11} = \theta_{11}{}^*$ are all irreducible characters of G. Moreover, they are all distinct characters of degree 4 each. By the degree relation for G, it follows that ζ^1, $\zeta^2, \zeta^3, \zeta^4, \zeta^5, \zeta^6, \zeta^7, \zeta^8, \zeta^9, \zeta^{10}$, and ζ^{11} are all the irreducible characters of G. Hence G is an M-group.

Character Table of G

1	1	1	1	1	1	1	1	1	1	1
1	1	1	1	1	1	1	ω	ω^2	ω	ω^2
1	1	1	1	1	1	1	ω^2	ω	ω^2	ω
3	3	-1	3	-1	-1	-1	0	0	0	0
3	3	3	-1	-1	-1	-1	0	0	0	0
3	3	-1	-1	-1	3	-1	0	0	0	0
3	3	-1	-1	-1	-1	3	0	0	0	0
3	3	-1	-1	3	-1	-1	0	0	0	0
4	-4	0	0	0	0	0	1	-1	1	-1
4	-4	0	0	0	0	0	ω	$-\omega$	ω^2	$-\omega^2$
4	-4	0	0	0	0	0	ω^2	$-\omega^2$	ω	$-\omega$

Since $u^{-1}au = b$, $u^{-1}bu = ab$, $u^{-1}cu = d$, and $u^{-1}du = cd$, it follows that $\langle a, b, u \rangle$ and $\langle c, d, u \rangle$ are subgroups of order 24. Each of these subgroups is isomorphic to $2A_4$, which is not an M-group, although G is an M-group.

The following theorem restricts the nature of an M-group.

LEMMA 3.1 *Let G be a group, H be a subgroup, and θ a character of H. Then $\ker \theta^* = \bigcap_{y \in G}(\ker \theta)^y$.*

Proof: We can extend θ to G by letting $\theta(y) = 0$ for y in $G \backslash H$. By definition, $\theta^*(x) = (1/|H|)\sum_{y \in G}\theta(y^{-1}xy)$. An element $x \in G$ belongs to $\ker\theta^*$ if and only if $\theta^*(1) = [G:H]\theta(1) = (1/|H|)\sum_{y \in G}\theta(y^{-1}xy)$, so that $\sum_{y \in G}\theta(y^{-1}xy) = |G|\theta(1)$. Since $|\theta(y^{-1}xy)| \leq \theta(1)$, it follows that $\theta(y^{-1}xy) = \theta(1)$ for all y in G and for all x in $\ker\theta$. Hence the lemma is proved.

THEOREM 3.2 *Let G be an M-group and let $1 = z_1 < z_2 < \ldots < z_r$ be the distinct degrees of the irreducible characters of G. If ζ^i is an irreducible character of degree z_i, then the ith derived group $G^{(i)}$ is contained in $\ker\zeta^i$.*
Proof: The result will be proved by induction on r. For $r = 1$, ζ is a linear character, so that $G^{(1)} = [G,G] \leq \ker\zeta$. Assume $r > 1$ and that the result holds for all $G^{(j)}$ with $j < r$.

Since G is an M-group, we may choose a subgroup $H < G$ and a linear character χ of H such that $\zeta = \chi^*$ with $\zeta(1) = z_r$. As $(1_H^G, 1_G)_G = (1_H, 1_H)_H = 1$ (where 1_H denotes the trivial character of H), the induced character 1_H^G is reducible. Let ψ be an irreducible constituent of $(1_H)^G$. Then $\psi(1) < [G : H] = \chi^*(1) = \zeta(1) = z_r$ and thus $G^{(r-1)} \leq \ker\psi$ by the induction hypothesis. So $G^{(r-1)} \leq \ker 1_H^* \leq H$ by Lemma 3.1.

Now $G^{(r)} \leq [H,H] \leq \ker\chi$ and since $G^{(r)}$ is normal in G, we have $G^{(r)} \leq \bigcap_{x \in G}(\ker\chi)^x = \ker\zeta$, which completes the proof.

COROLLARY 3.4 (Taketa) *An M-group is solvable.*

THEOREM 3.4 *Let $\mathfrak{M}$ be a class of finite groups with the properties:*
(i)*If $G \in \mathfrak{M}$, then each subgroup of G is in $\mathfrak{M}$.*
(ii)*If $G \in \mathfrak{M}$, then each quotient group of G is in $\mathfrak{M}$.*
(iii)*If $G \in \mathfrak{M}$, then G is either Abelian of else G has a noncentral normal Abelian subgroup.*
Then every group in $\mathfrak{M}$ is an M-group.
Proof: Since all finite Abelian groups are M-groups, it suffices to prove the result for non-Abelian groups in $\mathfrak{M}$. The proof will be by induction on the order of the groups. Assume that G is a non-Abelian group in $\mathfrak{M}$ and that the result holds for all groups in $\mathfrak{M}$ of order less than $|G|$.

Let T be an irreducible representation of G. If T is not faithful, then T can be considered as an irreducible representation of the factor group $G/\ker T$, which by the induction hypothesis is monomial. So we may assume that $\ker T = \{1\}$. By condition (iii), G has a noncentral normal Abelian subgroup N. By Clifford's theorem, there is a subgroup K of G and an irreducible representation S of K such that $N \leq K \leq G$ and $T = S^*$.

First we shall show that $K \neq G$. Suppose that $K = G$. Let V be a G-space which affords the representation T. By Clifford's theorem,

$$V_N = W_1 \oplus W_2 \oplus \ldots \oplus W_t$$

where each $W_i = \langle w_i \rangle$ is a one-dimensional N-subspace of V_N such that W_i is isomorphic to W_1 for all i. Let a be an element of $N \backslash Z(G)$. Set

$w_1 a = \lambda w_1$ for some complex number λ. Let θ_i be the N-isomorphism from W_1 to W_i given by $\theta_i(w_1) = \gamma_i w_i$ for some nonzero element γ_i in the complex field. Then

$$w_i a = \left(\theta_i(w_1)\gamma_i^{-1}\right)a = \theta_i(w_1)a\gamma_i^{-1} = \theta_i(w_1 a)\gamma_i^{-1}$$

$$= \theta_i(\lambda w_1)\gamma_i^{-1} = \lambda\theta_i(w_1)\gamma_i^{-1} = \lambda\gamma_i w_i\gamma_i^{-1} = \lambda w_i$$

so that $va = \lambda v$ for all v in V, which in turn implies that $a \in Z(G)$, contrary to the choice of a. Hence $K \neq G$.

Since $|K| < |G|$, it follows by the induction hypothesis that S is a monomial representation of K, say $S = U^*$ for some one-dimensional representation U of a subgroup L of K. Then $T = S^* = (U^*)^*$ is monomial. Since T is an arbitrary irreducible representation of G, G is an M-group. This completes the proof of the result.

COROLLARY 3.5 *Supersolvable groups (and hence nilpotent groups) are M-groups.*
Proof: Theorem 3.4 can be applied once it is shown that a supersolvable group is either Abelian or has a noncentral normal Abelian subgroup. Let G be a non-Abelian supersolvable group. Let $A/Z(G)$ be a minimal normal subgroup of $G/Z(G)$. By supersolvability, $A/Z(G)$ is cyclic, say $\langle Z(G)a \rangle$. The subgroup A is clearly noncentral and normal. To show A is Abelian, let $x, y \in A$. Then $Z(G)x, Z(G)y \in \langle Z(G)a \rangle$ so $x = z_1 a^n$ and $y = z_2 a^m$ for some $z_1, z_2 \in Z(G)$ and some integers n, m. Then $xy = z_1 a^n z_2 a^m = z_1 z_2 a^{n+m} = yx$.

The following result will be used to exhibit another class of groups which satisfies the conditions (i)–(iii) of Theorem 3.4.

THEOREM 3.6 *Let G be a solvable group and let K/L be a chief factor of G. Let χ be an irreducible character of K such that $\chi^x = \chi$ for all x in G. Then one of the following holds:*
(i) χ_L *is irreducible.*

(ii) $\chi_L = \sqrt{[K:L]}\ \varphi$, φ *irreducible.*

(iii) $\chi_L = \sum_{i=1}^{t}\varphi_i$, φ_i *are distinct irreducible conjugates.*
Proof: Let φ be an irreducible constituent of χ_L and let $N = \{a : a \in G$ and $\varphi^a = \varphi\}$. Then the number of G-conjugates of φ is $[G:N]$. Since $\chi^x = \chi$ for all x in G, the G-conjugates of φ are also K-conjugates of φ, so that $[G:N] = [K:N \cap K]$. Thus $G = NK$. Since $L \leq N \cap K$ and K/L is Abelian, $N \cap K$ is normal in both N and K. Thus $N \cap K$ is a normal subgroup of $G = NK$ and so $N \cap K = L$ or $N \cap K = K$ since K/L is a chief factor.

If $N \cap K = L$, then $\chi_L = e\sum_{i=1}^{t}\varphi_i$ where $\varphi_1 = \varphi$, $t = [K:L]$, and the φ_i are distinct. So $\chi(1) = e[K:L]\varphi(1)$. The character χ is a constituent of φ^*, and $\chi(1) \leq [K:L]\varphi(1)$. Thus $e = 1$ which is the situation in (iii).

Next assume that $N \cap K = K$. Then $\varphi^a = \varphi$ for all a in K and $\chi_L = e\varphi$ for some e. If λ is an irreducible character of K/L, then $\lambda\chi$ is an irreducible character of K such that $(\lambda\chi)_L = \chi_L = e\varphi$. Suppose that all the characters $\lambda\chi$ are distinct as λ ranges over all the irreducible characters of K/L. Each of these $[K:L]$ characters is an irreducible constituent of φ^* with multiplicity e and we have

$$e[K:L]\chi(1) \leq \varphi^*(1) = [K:L]\varphi(1).$$

Therefore $e^2\varphi(1) \leq \varphi(1)$ and $e = 1$. This is the situation in (i).

Suppose now that the characters $\lambda\chi$ are not all distinct. Let $\lambda\chi = \mu\chi$ where $\lambda \neq \mu$. If $U = \ker(\lambda\bar{\mu})$, then $L \leq U \leq K$ and χ vanishes on $K\backslash U$. Since $\chi^a = \chi$ for all a in G, it follows that χ vanishes on $K\backslash U^a$ for all a in G. As there is no normal subgroup of G between L and K, $\bigcap_{a\in G}U^a = L$ and so χ vanishes on $K\backslash L$. The ramification index (i.e., the number of irreducible constituents of χ_L) is $[K:L]$. Then

$$e^2 = (e\varphi, e\varphi)_L = \frac{1}{|L|}\sum_{x\in L}\chi(x)\overline{\chi(x)}$$

$$= \frac{1}{|L|}\sum_{x\in K}\chi(x)\overline{\chi(x)} = [K:L].$$

Thus $\chi_L = \sqrt{[K:L]}\,\varphi$ which is the situation in (ii). This completes the proof.

COROLLARY 3.7 *Let N be a normal subgroup of G of prime index p for some prime. If ζ is an irreducible character of G, then either ζ_N is irreducible or ζ_N is a sum of p distinct (conjugate) irreducible characters.*
Proof: Since p is a prime, the case (ii) can not arise.

COROLLARY 3.8 *Let N be a normal subgroup of G of prime index p. If θ is an irreducible character of N such that $\theta^a = \theta$ for all a in G, then G has an irreducible character ζ such that $\zeta_N = \theta$.*
Proof: Let ζ be an irreducible constituent of θ^*. Then $\zeta_N = e\theta$ for some number $e > 0$. From Corollary 3.7, $e = 1$. Hence the result follows from the Frobenius reciprocity formula.

THEOREM 3.9 *Let G be a solvable group in which every chief factor of every subgroup has nonsquare order. Then G is an M-group.*
Proof: Let ζ be an irreducible character of G. We shall show that ζ is monomial. Let K be minimal among the (not necessarily proper) normal subgroups W of G such that ζ_W is irreducible. Choose a normal subgroup L

of G such that $L \leq K$ and K/L is a chief factor. Then by hypothesis, K/L is an Abelian normal subgroup of nonsquare order.

Now we apply Theorem 3.6 to the chief factor K/L and the character ζ_K of K. Obviously $\zeta^a = \zeta$ for all a in G. As ζ_L is reducible and $[K:L]$ is nonsquare, we have $\zeta_L = \sum_{i=1}^{t}\varphi_i$, where φ_i are the distinct irreducible (conjugate) characters of L and $t = [K:L] > 1$.

Now we let $N = \{a : a \in G$ and $\varphi_1^a = \varphi_1\}$, so that $N \geq L$. By Clifford's theorem, $\zeta = \psi^*$ for some irreducible character ψ of N. But $N < G$, so by induction on $|G|$, we conclude that N is an M-group and that $\psi = \theta^*$ for some linear character θ of a subgroup H of N. Then $\zeta = \psi^* = (\theta^*)^*$, so that ζ is monomial. Hence G is an M-group.

Note that this result implies Corollary 3.5. Finally we conclude this section with a result of Huppert.

THEOREM 3.10 (Huppert) *Let G be a group with a normal solvable subgroup N such that G/N is supersolvable. If the Sylow subgroups of N are all Abelian, then G is an M-group.*

Proof: Let $\mathfrak{M}$ denote the class of groups satisfying the hypothesis of the theorem. We will show that $\mathfrak{M}$ satisfies the three conditions of Theorem 3.4.

If $G \in \mathfrak{M}$, $H \leq G$, and $N \triangleleft G$, then it is easily seen that $H \in \mathfrak{M}$ and $G/N \in \mathfrak{M}$.

Finally we let $G \in \mathfrak{M}$, with normal subgroup N with Abelian Sylow subgroups and supersolvable associated quotient group, and seek to produce a noncentral Abelian normal subgroup of G. First note that if $N \leq Z(G)$, then G is supersolvable (this follows easily from Theorem 1.2 of Chapter 1). Hence, as was shown in the proof of Corollary 3.5, G has a noncentral Abelian normal subgroup. We may therefore assume N is noncentral. The subgroup $\text{Fit}(N)$ is Abelian because it is nilpotent and all its Sylow subgroups are Abelian. Since $\text{Fit}(N)$ is characteristic in N which is normal in G, $\text{Fit}(N) \triangleleft G$. Thus $\text{Fit}(N)Z(G)$ is an Abelian normal subgroup of G. If $\text{Fit}(N)Z(G)$ is noncentral, we are done. Otherwise, $\text{Fit}(N) \leq Z(G)$ which implies $\text{Fit}(N) \leq Z(N)$ and so $N \leq C_N(\text{Fit}(N))$. Thus $N \leq \text{Fit}(N)$ by Theorem 2.6 of Appendix C. Hence $N = \text{Fit}(N)$ is Abelian. This and N noncentral yields N as the desired noncentral Abelian normal subgroup.

An appeal to Theorem 3.4 now completes the proof.

COROLLARY 3.11 *Let G be a group with an Abelian normal subgroup N such that G/N is nilpotent. Then G is an M-group.*

4. Subgroups of M-groups

A factor group of an M-group is always an M-group. However, a subgroup of an M-group need not be an M-group. The example $2A_4$ shows that a solvable group is not an M-group in general. We shall prove in this section a result of Dade which asserts that every solvable group can be embedded in an M-group. First we give a generalization of Corollary 3.8.

LEMMA 4.1 *Let N be a normal subgroup of a group G such that G/N is cyclic. If χ is an irreducible character of N such that $\chi^a = \chi$ for all a in G, then there exists an irreducible character ζ of G such that $\zeta_N = \chi$.*

Proof: Let S be an irreducible matrix representation of N which affords χ. For any x in G, S^x defined by $S^x(a) = S(x^{-1}ax)$, $a \in N$ is also a representation of N. By hypothesis, S and S^x are equivalent, and so there is a nonsingular matrix A_x such that $S^x(a) = A_x^{-1}S(a)A_x$ for all a in N. In particular, choose an x such that $G/N = \langle xN \rangle$, and let $A_x = A_0$ and $|G/N| = n$. Then $x^n \in N$, so that $A_0^{-n}S(a)A_0^n = S^{x^n}(a) = S(x^{-n})S(a) \cdot S(x^n)$ for all a in N. By Schur's lemma $S(x^n)A_0^{-n} = \alpha_0 \cdot 1$ for some complex number α_0. Choose α such that $\alpha^n = \alpha_0$. Then $S(x^n) = \alpha^n A_0^n = (\alpha A_0)^n = A^n$ where $A = \alpha A_0$. Now define T on G by $T(ax^i) = S(a)A^i$ for all a in N and $i = 0, 1, \ldots, n-1$. We shall show that T is a representation of G. Indeed, let $a, b \in N$ and $0 \le i, j < n$. Then

$$
\begin{aligned}
T(ax^i)T(bx^j) &= S(a)A^iS(b)A^j \\
&= S(a)A^iS(b)A^{-i}A^{i+j} \\
&= S(a)A_0^iS(b)A_0^{-i}A^{i+j} \\
&= S(a)S(x^ibx^{-i})A^{i+j} \\
&= S(ax^ibx^{-i})A^{i+j}.
\end{aligned}
$$

If $0 \le i + j < n$, then $S(ax^ibx^{-i})A^{i+j} = T(ax^ibx^{-i}x^{i+j}) = T(ax^ibx^j)$. If $n \le i + j < 2n$, then $S(ax^ibx^{-i})A^{i+j} = S(ax^ibx^{-i})S(x^n)A^{i+j-n} = S(ax^ibx^{-i}x^n)A^{i+j-n} = T(ax^ibx^{-i}x^{i+j-n}x^n) = T(ax^ibx^j)$. Hence T is a representation of G. The restriction of T to N equals the irreducible representation S and so T is irreducible. If ζ is the character of G afforded by T, then $\zeta_N = \chi$ as required.

COROLLARY 4.2 *Under the hypothesis of the theorem, there are at most $|G/N|$ irreducible characters ζ of G such that $\zeta_N = \chi$.*

Proof: This follows from the observation that there are exactly n nth roots of α_0. (Indeed each of these roots gives such a character provided $\operatorname{tr}A_0 \neq 0$.)

THEOREM 4.3 *Let N be a normal subgroup of a group G of prime index p and let C_p be a subgroup of order p with $G = NC_p$. Let χ be an irreducible character of N induced by a linear character ψ of a subgroup H, such that $\chi^a = \chi$ for all a in G. If $HC_p = C_pH$, then there are exactly p irreducible monomial characters ζ of G such that $\zeta_N = \chi$.*

Proof: Let $\{a_1 = 1, a_2, \ldots, a_m\}$ be a set of coset representatives of H in N. Then it is also a set of coset representatives of HC_p in $NC_p = G$. If λ is any linear character of C_p, then $\lambda\psi$ defined by $(\lambda\psi)(ch) = \lambda(c)\psi(h)$ for all $c \in C_p$, $h \in H$, is a linear character of HC_p. If λ and μ are linear characters of C_p, such that $\lambda\psi = \mu\psi$, then $(\lambda\psi)(ch) = \lambda(c)\psi(h) = \mu(c)\psi(h)$ for all c in C_p and h in H, so that $\lambda = \mu$. Hence if $\lambda_1, \lambda_2, \ldots, \lambda_p$ are the linear characters of C_p, then $\lambda_1\psi, \lambda_2\psi, \ldots, \lambda_p\psi$ are p distinct linear characters of HC_p, whose restriction to H equals ψ. Since $(\lambda_i\psi)^*(a) = \sum_{j=1}^m \lambda_i\psi(a_j^{-1}aa_j)$ for any a in G, we get $(\lambda_i\psi)^*(a) = \sum_{j=1}^m \lambda_i\psi(a_j^{-1}aa_j) = \psi^*(a)$ for all a in N. Hence $(\lambda_1\psi)^*, (\lambda_2\psi)^*, \ldots, (\lambda_p\psi)^*$ are all irreducible characters of G whose restriction to N equals $\psi^* = \chi$. If $(\lambda_i\psi)^* = (\lambda_j\psi)^*$, then for any c in C_p, $(\lambda_i\psi)^*(c) = \sum_{k=1}^m \lambda_i\psi(a_k^{-1}ca_k) = \sum_{k=1}^m \lambda_i(c)\psi(c^{-1}a_kca_k) = \sum_{k=1}^m \lambda_i(c) \cdot \psi(c^{-1}a_kc)\psi(a_k) = \lambda_i(c)\psi^*(1) = \lambda_j(c)\psi^*(1)$, so $i = j$. Hence $(\lambda_1\psi)^*$, $(\lambda_2\psi)^*, \ldots, (\lambda_p\psi)^*$ are the only irreducible characters of G whose restriction to N coincide with χ. Moreover, these are all monomial. The proof is complete.

LEMMA 4.4 *Let G_1 and G_2 be groups and let ζ_i be an irreducible monomial character of G_i for $i = 1, 2$. Then $\zeta_1\zeta_2$ is an irreducible monomial character of $G_1 \times G_2$.*

Proof: By Section 1, $\zeta_1\zeta_2$ is an irreducible character of $G_1 \times G_2$. Let $\zeta_i = \psi_i^*$, where ψ_i is a linear character of the subgroup H_i of G_i for $i = 1, 2$. Then

$$\zeta_1\zeta_2(a,b) = \zeta_1(a)\zeta_2(b) = \frac{1}{|H_1|}\sum_{x \in G_1}\psi_1(x^{-1}ax)\frac{1}{|H_2|}\sum_{y \in G_2}\psi_2(y^{-1}by)$$

$$= \frac{1}{|H_1 \times H_2|}\sum_{(x,y) \in G_1 \times G_2}\psi_1\psi_2\big((x,y)^{-1}(a,b)(x,y)\big)$$

$$= (\psi_1\psi_2)^*(a,b)$$

for all (a,b) in $G_1 \times G_2$. Hence $\zeta_1\zeta_2 = (\psi_1\psi_2)^*$ is monomial.

THEOREM 4.5 *Let G be an M-group, p a prime, and $\mathbb{Z}_p$ a group of order p. Then $G \operatorname{wr} \mathbb{Z}_p$ is an M-group.*

Proof: Let N be the base group of $G \operatorname{wr} \mathbb{Z}_p$. Then $|N| = |G|^p$ and N is isomorphic to a direct product of p copies of G. Since G is monomial, by Lemma 4.4, N is monomial. Every irreducible character of N has the form $\zeta_1\zeta_2 \cdots \zeta_p$ where ζ_i is an irreducible character of G which is induced by a linear character ψ_i of H_i of G.

If Φ is an irreducible character of $G \operatorname{wr} \mathbb{Z}_p$, then Φ_N is either irreducible or else Φ_N is a sum of p conjugate characters φ_i of N, in which case $\Phi = \varphi_1^* = \varphi_2^* = \ldots = \varphi_p^*$. In the latter case Φ is monomial, since φ_i is monomial. If Φ_N is irreducible, then by Theorem 4.3, there are p irreducible monomial characters $\Phi_1 = \Phi$, Φ_2, ..., Φ_p whose restriction to N coincides with Φ_N. Thus every irreducible character of $G \operatorname{wr} \mathbb{Z}_p$ is monomial.

THEOREM 4.6 (Dade) *Every solvable group can be embedded in an M-group.*

Proof: Let G be a solvable group. The result will be proved by induction on the order of G. The result is obvious for a group of order 1, so assume that $|G| > 1$ and the result holds for all solvable groups of order less than $|G|$. Since G is solvable, it has a normal subgroup N such that G/N is of prime order p. Since N is solvable and $|G/N| < |G|$, by the induction hypothesis there is an M-group U and a monomorphism from N into U. Theorem 2.2 shows that there is a monomorphism from G into $N \operatorname{wr} G/N$. By Lemma 2.1, there is a monomorphism from $N \operatorname{wr} G/N$ into $U \operatorname{wr} G/N$. Hence G can be embedded in $U \operatorname{wr} G/N$ which, by Theorem 4.5, is an M-group.

We close this chapter by raising two questions about M-groups: (i) Is there a purely group-theoretical characterization (i.e., a characterization not involving representation theory) of M-groups? (ii) Is a normal subgroup of an M-group an M-group?

The second second question has been answered by Dade and Dornhoff. Dornhoff has shown that a normal Hall subgroup of an M-group is an M-group. An outline of this result is given in Exercise 18. On the other hand Dade has constructed examples to show that in general an arbitrary normal subgroup of an M-group is not an M-group.

The first question is still open.

Exercises

1. Show that the group Q generated by the matrices

$$\begin{pmatrix} 1 & 0 & 0 \\ 0 & 0 & -1 \\ 0 & 1 & 0 \end{pmatrix}, \quad \begin{pmatrix} 1 & 0 & 0 \\ 0 & 0 & i \\ 0 & i & 0 \end{pmatrix}$$

is a group of order 8. Determine the character table of Q and hence show that every irreducible character is induced by a linear character of a subgroup of Q.

2. Determine a set of generators of the double group $2Q$ of Q. Construct the character table of $2Q$. Show that every irreducible character of $2Q$ is monominal.

3. Let G be a metabelian group (i.e., $G'' = \{1\}$) and N be a maximal Abelian normal subgroup of G with $G' \leqslant N$. If $n \neq 1$ is the degree of an irreducible representation of G, then show that $n = [G : N]$.

4. Let G, A, n, and T be as in Theorem 2.3. Let $\{r_1, r_2, \ldots, r_n\}$ be a set of coset representatives of A in G, and let D be the diagonal matrix in $G \wr^* S_n$ whose diagonal entries are $r_1, r_2, \ldots, r_n$. Then show that $D^{-1}T(x)D = \Pi_x \Gamma$ where Π_x is a permutation matrix and Γ is obtained from Π_x by replacing each 1 by x.

5. Let P be a permutation matrix group of degree n and let A and B be abstract groups. Show that any homomorphism from A into B induces a natural homomorphism π from $A \wr^* P$ into $B \wr^* P$.

6. In Exercise 5, let B be an Abelian group. For u in $B \wr^* P$, let u^δ be the product of the nonzero coefficients of u. Show that δ is a homomorphism from $B \wr^* P$ into B. (The homomorphism $\pi\delta$ is called a transfer.)

7. Show that $\mathbb{Z}_2 \wr^* S_4$ is an M-group.

8. Show that the semi-direct product of S_4 extended by S_3 (S_3 considered as a group of automorphisms of S_4) is an M-group.

9. Show that the group generated by the permutations $(1,2)$; $(2,3)$; and $(1,4)(2,5)(3,6)$ is an M-group.

10. Show that the group generated by the permutations $(1,2)$; $(1,3,4)$; and $(1,5)(2,6)(3,7)(4,8)$ is an M-group.

11. Show that the group generated by the permutations $(1,2,3)$; $(2,3,4)$; and $(1,5)(2,6)(3,7)(4,8)$ is an M-group.

12. Let $G = \langle a, b : a^{52} = b^6 = 1, b^{-1}ab = a^3 \rangle$. Determine all the irreducible matrix representations of G. Determine the character table of G.

13. Let V be the vector space $F \times F$ where F is a 3-element field. Show that the automorphism group $SL(2,3)$ of all linear transformations of V of determinant 1 has order 24, and thus that the semi-direct product of V by $SL(2,3)$ has order 216. Is this semi-direct product an M-group? Why? (Note: $SL(2,3)$ is isomorphic to $2A_4$.)

14. Show that there is an M-group of order 1536 in which $SL(2,3)$ is embedded.

15. Find the order of the group G generated by the matrices

$$\frac{1}{\sqrt{2}} \begin{pmatrix} 1 & i \\ i & 1 \end{pmatrix}, \quad \frac{1}{2} \begin{pmatrix} 1+i & -1-i \\ 1-i & 1-i \end{pmatrix}, \quad \text{and} \quad \begin{pmatrix} 0 & 1 \\ 1 & 0 \end{pmatrix}.$$

Show that G is not an M-group. Determine the smallest order of an M-group in which G can be embedded.

16. Let ζ be an irreducible monomial character of a group G and K a subgroup of G such that ζ_K is irreducible. Show that ζ_K is a monomial character of K.

17. Let N be a normal subgroup of the group G. Let ζ be an irreducible character of G and θ an irreducible constituent of ζ_N. Show that $\zeta(i)/\theta(1)$ divides $[G : N]$.

18. Let G be an M-group and N a normal Hall subgroup of G. Show that N is an M-group. [Hint: Let θ be an irreducible character of N. Then $(\zeta_N, \theta)_N \neq 0$ for some irreducible constituent ζ of θ^G. Use the assumption that ζ is monomial to obtain a linear character ψ of a subgroup H of G such that $\zeta = \psi^G$. Use Exercise 17 to conclude that $\psi^{NH}(1) = \theta(1)$. Finally conclude that the restriction of a monomial character of G to a subgroup is monomial.]

CLT and non-CLT Groups

Henry G. Bray

0. Definitions, Comments on Notation and Terminology, and a Lemma

DEFINITION The finite group G is a **CLT group** iff it satisfies the following converse to Lagrange's theorem: for every divisor d of $|G|$, G has a subgroup of order d.

The class of CLT groups will be denoted by $\mathcal{CLT}$.

We shall adopt the definitions, terminology, notation, and viewpoint in [Mac Lane and Birkhoff 1968] with regard to functions and their composition. In particular, to put it loosely, $f \circ g$ means g first, then f. To be consistent with this, we must compose permutations (whenever it is possible to do so) from right to left: e.g., $(1,2) \circ (1,3) = (1,3,2)$ (*not* $(1,2,3)$).

For positive integers a and n with $(a,n) = 1$, the smallest positive integer k such that $a^k \equiv 1 \bmod n$ is called the **exponent** to which a belongs $\bmod\, n$. It is denoted $k = \exp(a, n)$.

In a great deal of our work, we will be involved with the notion of the semi-direct product of an ordered pair of groups relative to an appropriate homomorphism. When dealing with this concept, it is possible for one to get into some nasty logical difficulties if one identifies certain isomorphic groups. The material which follows is put down so as to attempt to avoid the logical difficulties (at least those of which we are aware).

We shall be using finite elementary Abelian p-groups a great deal, and we wish to be very specific about precisely which such groups we will use; we shall also be making a great deal of use of fields of order p, and we wish to be very specific about these also. For p a prime and k an integer, let $[k]_p$ denote the residue class $[k]_p = \{x : x \in \mathbb{Z} \text{ and } x \equiv k \pmod{p}\}$, let B_p denote the set of all such residue classes, and let $+$ and $\cdot$ denote, respectively, the usual mod p addition and multiplication on B_p (i.e., $[x]_p + [y]_p = [x+y]_p$ and $[x]_p \cdot [y]_p = [xy]_p$). Now, one can "add and multiply across congruences," and it is a classical fact that this shows that $+$ and $\cdot$ are both binary operations on B_p. In view of this, we shall interpret $\mathbb{Z}_p$ and $\mathrm{GF}(p)$ as follows: $\mathbb{Z}_p = (B_p, +)$ and $\mathrm{GF}(p) = (B_p, +, \cdot)$. It is a classical fact that $\mathbb{Z}_p$ is a cyclic group and $\mathrm{GF}(p)$ is a field, both of order p.

Now, continuing to let p denote a prime, for positive integers a, we let B_p^a denote the usual Cartesian product $B_p^a = B_p \times B_p \times \ldots \times B_p$ with a factors, and let addition be defined coordinatewise on B_p^a. It is a classical fact that this addition (which we will also denote by $+$) is a binary operation on B_p^a; in view of this, we shall interpret $\mathbb{Z}_p^a$ as follows: $\mathbb{Z}_p^a = (B_p^a, +)$. It is a classical fact that $\mathbb{Z}_p^a$ is an elementary Abelian p-group of order p^a.

One of the reasons that we have been so fussy is that we shall use some machinery from linear algebra. We shall indicate below (in Lemma 2.8) that there is a way of making $\mathbb{Z}_p^a$ into an a-dimensional vector space over $\mathrm{GF}(p)$ in such a way that every automorphism of $\mathbb{Z}_p^a$ is, in fact, a linear automorphism of the vector space; we shall also show that there is a way of interpreting $\mathrm{GL}(a, p)$ so that we may write $\mathrm{Aut}(\mathbb{Z}_p^a) = \mathrm{GL}(a, p)$ rather than just $\mathrm{Aut}(\mathbb{Z}_p^a) \cong \mathrm{GL}(a, p)$.

By extension of our notation above, when n is a positive integer (not necessarily a prime), we will denote by B_n the set of all residue classes mod n and will let $\mathbb{Z}_n = (B_n, +)$. It is a classical fact that $\mathbb{Z}_n$ is a cyclic group of order n.

We shall not need to be nearly so fussy about finite fields other than $\mathrm{GF}(p)$, and so we shall use $\mathrm{GF}(p^a)$ to denote any finite field of order p^a, where p is a prime and a is a positive integer, $a \geq 2$. If F is any field, we shall use F^* to denote the multiplicative group of F, i.e., $(F \setminus \{0\}, \cdot)$.

The alternating group A_4 of all even permutations on $\{1, 2, 3, 4\}$ will play a key role in many of our future discussions; it happens that A_4 is the world's smallest non-CLT group. There are many different ways of showing that A_4 is not CLT, but we happen to like the argument which we present below, as it uses nothing heavier than Lagrange's theorem and the fact that any noncyclic group of order 6 must be isomorphic to S_3.

LEMMA 0.1 $|A_4| = 12$ and A_4 has no subgroups of order 6.
Argument: Suppose, to the contrary, that A_4 has a subgroup H of order 6.

Now A_4 consists of 8 three-cycles (all of which have order 3) together with $V = \{1, (1,2)(3,4), (1,3)(2,4), (1,4)(2,3)\}$. Letting $S = \{x : x \in A_4$ and x is a three-cycle$\}$, we thus have $A_4 = S \cup V$ and $S \cap V = \emptyset$, and we may conclude two things:

(i) $|A_4| = |S| + |V| = 8 + 4 = 12$.

(ii) A_4 has no elements of order 6.

Now, (ii) shows that H is not isomorphic to $\mathbb{Z}_6$, and hence we must have $H \cong S_3$. Since S_3 contains precisely three elements of order 2, the same must be true of H; also, we must have $1 \in H$ (since H is a subgroup). From above, we see that A_4 contains precisely three elements of order 2, and that all of these elements are members of V; it follows that $V \subseteq H$.

Now, a little computation will show that V is a subgroup of A_4. It follows that $V \leq H$. Lagrange's theorem now serves to banish H out of existence.

We wish to add a small postscript to Lemma 0.1. With the notation as in this lemma, we claim that $V \triangleleft A_4$. In the proof of Lemma 0.1, we obtained that $V \leq A_4$ and $V \cup S = A_4$ and $V \cap S = \emptyset$. Now if we pick any $g \in A_4$ and any $v \in V$, then $g \circ v \circ g^{-1} \notin S$ (conjugation preserves cycle structure, v is not a three-cycle, and all of the members of S *are* three-cycles). It follows that $g \circ v \circ g^{-1} \in V$ for all $g \in A_4$ and $v \in V$; and we are done.

1. Some Structural Results

We first indicate, in Theorem 1.3 below, the relative position of the class of all CLT groups in the nilpotent-to-solvable hierarchy.

THEOREM 1.1 *If G is a finite group, then G is supersolvable iff every subgroup of G (including G itself) is* CLT. (This is just a restatement of Theorem 4.1 of Chapter 1.)

THEOREM 1.2 (P. Hall) *If G is a finite group, then G is solvable iff G has Hall subgroups of all possible orders.*

A proof that solvable groups have Hall subgroups is given as part of Theorem 1 in Appendix A. For a proof of the converse, see [Hall 1959, Theorem 9.3.3].

THEOREM 1.3 $\mathfrak{N} \subset \mathfrak{SS} \subset \mathcal{CLT} \subset \mathfrak{S}$ *and all inclusions are proper.*

Argument: The fact that $\mathfrak{N} \subset \mathfrak{SS}$ follows easily from the material in Chapter 1 (e.g., Chapter 1, Theorem 3.1). It follows immediately from Theorem 1.1 that $\mathfrak{SS} \subset \mathcal{CLT}$, and it follows immediately from Theorem 1.2 that $\mathcal{CLT} \subset \mathfrak{S}$. To (more than) justify that all all of the above inclusions are proper, we have:

(i) If H is any finite group of odd order, then $A_4 \times H$ is solvable and not CLT.

(ii) If G is any CLT group, then $(A_4 \times \mathbb{Z}_2) \times G$ is CLT and not supersolvable.

(iii) If K is any finite noncyclic group of square-free order (e.g., S_3), then K is supersolvable and not nilpotent.

For proofs of (i) and (ii), see [Bray 1968, Lemmas 3, 4, and 5]. We shall show later that any finite group of square-free order is supersolvable, and we offer a quick argument now to show that any finite nilpotent group of square-free order must be cyclic. For such a group G ($\neq \{1\}$), all Sylow subgroups of G have prime order hence are cyclic. Since G is nilpotent, it is isomorphic to the direct product of its Sylow subgroups. Thus G is isomorphic to a direct product of finite cyclic groups of pairwise coprime orders. It follows that G must be cyclic.

We now investigate the questions of closure of the class $\mathcal{CLT}$ under subgroups, quotient groups, extensions, and finite direct products.

It is well-known that the classes cyclic, Abelian, nilpotent, supersolvable, and solvable are all closed under subgroups and quotient groups. Unfortunately, the class $\mathcal{CLT}$ is not so obliging.

THEOREM 1.4 *The class $\mathcal{CLT}$ is neither subgroup-closed nor quotient-closed.*

Proof: The direct product $A_4 \times \mathbb{Z}_2$ is CLT, and said direct product has a copy of A_4 as a subgroup, and also has a copy of A_4 as a quotient group; Lemma 0.1 shows that A_4 is not CLT.

Despite the above bad news, the class $\mathcal{CLT}$ *is* closed under Hall subgroups (for a proof, see [Baskaran 1973]).

It is well-known that the class $\mathcal{S}$ is closed under extensions. The analogue for the class $\mathcal{CLT}$, however, is false.

THEOREM 1.5 *The class $\mathcal{CLT}$ is not extension-closed.*

Proof: Let $G = A_4$ and let $V = \{1, (1,2)(3,4), (1,3)(2,4), (1,4)(2,3)\}$. Then $V \triangleleft G$; $V \cong \mathbb{Z}_2^2$; and $G/V \cong \mathbb{Z}_3$. Thus V and G/V are both CLT, but G is not.

Finally, something pleasant occurs for a change.

THEOREM 1.6 *The class $\mathcal{CLT}$ is closed under finite direct products.*

For a proof, see Chapter 6, Theorem 1.1 or [McCarthy 1971, page 589]. (In passing, we wish to commend this last reference. It contains (among other things) structural results about groups satisfying various different strong converses of Lagrange's theorem; some history about the subject; and a good list of references.)

2. SBP Groups, BNCLT Groups, and McCarthy Groups; Some Numerical Results

In this section we shall do four things:

(1) Characterize those finite groups for which Sylow's first theorem is best possible as a partial converse to Lagrange's theorem.

(2) Characterize those finite groups which fail to be CLT as badly as possible.

(3) Indicate how certain semi-direct products of pairs of finite groups give a very natural and very useful way of constructing certain kinds of non-CLT groups.

(4) Describe a certain class of semi-direct products which generalizes the class described in (2), and which will give us some examples of groups which are not CLT but which do not fail to be CLT as badly as possible.

Sylow's first theorem tells us that for all finite groups G: If d is a prime-power dividing $|G|$, then G has at least one subgroup of order d. Thus Sylow's first theorem may be regarded as a partial converse to Lagrange's theorem. In [McCarthy 1970], McCarthy shows that this partial converse is "sharp" by proving the following:

THEOREM 2.1 *Let d be any positive integer which is not a prime-power. Then there is a finite group G such that d divides $|G|$ and G has no subgroups of order d.*

One may interpret Theorem 2.1 as saying that Sylow's first theorem is "best possible" as a partial converse to Lagrange's theorem. Our interpretation is different, (although related to McCarthy's, as we shall see). We illustrate our viewpoint with our old friend A_4.

The order of A_4 is $12 = 2^2 \cdot 3$, and the set of positive integer divisors of 12 is $\{1, 2, 3, 4, 6, 12\}$. Of course A_4 must have subgroups of orders 1 and 12, and Sylow's theorem says that A_4 must have subgroups of orders 2, 3, and 4. Since A_4 has no subgroups of order 6 (Lemma 0.1), Lagrange's theorem shows that if H is any nontrivial proper subgroup of A_4, then $|H|$ is 2, 3, or 4; so that A_4 does not have any nontrivial proper subgroups of any orders other than those that it must have by virtue of Sylow's theorem. We consider then that A_4 fails to be CLT as badly as possible, and that Sylow's first theorem is best possible (as a partial converse to Lagrange's theorem) for A_4.

DEFINITION The finite group G is **SBP (Sylow Best Possible)** iff every proper subgroup H of G has prime-power order (the prime depending on H).

DEFINITION The finite group G is **BNCLT (Badly Non-CLT)** iff G is not CLT and is SBP.

We caution the reader of two things: (1) a group (S_3, $\mathbb{Z}_6$, or any p-group, for example) can be both CLT and SBP; (2) BNCLT does *not* stand for "Bacon, Cheese, Lettuce, and Tomato."

We characterize the classes of SBP and BNCLT groups in the next two theorems.

THEOREM 2.2 *Let $G \neq \{1\}$ be an* SBP *group. Then precisely one of the following holds*:
(I) G *is a q-group, some prime q.*
(II) $|G| = pq$, *where p and q are distinct primes.*
(III) $|G| = p^a q$, *where p and q are distinct primes, $a = \exp(p,q)$, and $a \geq 2$. Moreover, $G \cong \mathbb{Z}_p^a \times_\theta \mathbb{Z}_q$ for some monomorphism $\theta : \mathbb{Z}_q \to \mathrm{Aut}(\mathbb{Z}_p^a)$. Furthermore, if G is* BNCLT, *then* (III) *must hold.*

THEOREM 2.3 *Let p and q be distinct primes such that $2 \leq a = \exp(p,q)$. Then*
(1) *There exists a monomorphism $\theta : \mathbb{Z}_q \to \mathrm{Aut}(\mathbb{Z}_p^a)$.*
(2) *For all monomorphisms $\lambda : \mathbb{Z}_q \to \mathrm{Aut}(\mathbb{Z}_p^a)$, the semi-direct product $\mathbb{Z}_p^a \times_\lambda \mathbb{Z}_q$ is* BNCLT.
(3) *For any pair of monomorphisms $\theta_1, \theta_2 : \mathbb{Z}_q \to \mathrm{Aut}(\mathbb{Z}_p^a)$, $\mathrm{Im}(\theta_1)$ and $\mathrm{Im}(\theta_2)$ are conjugate in $\mathrm{Aut}(\mathbb{Z}_p^a)$, and hence $\mathbb{Z}_p^a \times_{\theta_1} \mathbb{Z}_q \cong \mathbb{Z}_p^a \times_{\theta_2} \mathbb{Z}_q$.*
(4) *Any two* BNCLT *groups of the same order are isomorphic.*

Before getting to the proofs of the above theorems, we give an example to illustrate the last conclusion of Theorem 2.2.

We have seen that A_4 is BNCLT. Consider the subgroups V of Lemma 0.1 and $Q = \langle (1,2,3) \rangle$. Then $V \triangleleft A_4$, Q is not normal, $V \cap Q = \{1\}$, and $VQ = A_4$, $V \cong \mathbb{Z}_2^2$, and $Q \cong \mathbb{Z}_3$. This is sufficient to show there exists a monomorphism $\theta : \mathbb{Z}_3 \to \mathrm{Aut}(\mathbb{Z}_2^2)$ such that $G \cong \mathbb{Z}_2^2 \times_\theta \mathbb{Z}_3$. (We will put in more detail in the proof of Theorem 2.2.) The above shows that A_4 is isomorphic to a *split* extension of $\mathbb{Z}_2^2$ by $\mathbb{Z}_3$. (Note that $2 = \exp(2,3)$.)

The proofs of the two theorems are long and involved; we will do them by a sequence of lemmas. (Some of the information contained in these lemmas will also be used in our work after the proof of Theorem 2.3.)

First, some notation. If H and K are any two groups and $\theta : K \to \mathrm{Aut}(H)$ is any homomorphism, we use $*_\theta$ to denote the binary operation in the semi-direct product $H \times_\theta K$. Most of the details about semi-direct products which we omit may be found in the careful treatment of semi-direct products on pages 7–17 of [Weinstein 1977].

LEMMA 2.4 *Let H, M, K be any three groups such that $M \leq H$ and let $\theta : K \to \mathrm{Aut}(H)$ be any homomorphism. Then, in $H \times_\theta K$, the following are equivalent*:
(i) $M \times K$ *is closed under $*_\theta$.*
(ii) M *is invariant under $\mathrm{Im}(\theta)$.*
(iii) $M \times K$ *is a subgroup of $H \times_\theta K$.*

Proof: (i)$\Rightarrow$(ii) Pick any $m \in M$ and $k \in K$ and seek to show $(\theta(k))(m)$ $\in M$. Since $(1, k)$, $(m, 1) \in M \times K$, by (i) we have $(1, k)*_\theta (m, 1) = ((\theta(k))$ $(m), k) \in M \times K$. Hence $(\theta(k))(m) \in M$ as required.

(ii)$\Rightarrow$(iii) To show product closure, let (m, k), $(m_1, k_1) \in M \times K$. Then $(m, k)*_\theta (m_1, k_1) = (m \cdot (\theta(k))(m_1), k \cdot k_1)$ which is in $M \times K$ because $(\theta(k))$ $(m_1) \in M$ by (ii). To show closure under inverses, let $(m, k) \in M \times K$. Then $(m, k)^{-1} = (\theta(k^{-1})(m^{-1}), k^{-1})$ which again belongs to $M \times K$ since $\theta(k^{-1})(m^{-1}) \in M$ by (ii).

(iii)$\Rightarrow$(i) This is immediate.

DEFINITION If f is an automorphism of the group G, we say that f is **invariant-subgroup free (i.s.f.)** iff no nontrivial subgroup of G is invariant under f.

DEFINITION If V is a vector space and $T: V \rightarrow V$ is a linear map, we say that T is **invariant-subspace free (i.s.f.)** iff no nontrivial subspace of V is invariant under T. (In the parlance of linear algebra, we have "T is irreducible on V" or just plain "T is irreducible.")

NOTATION If V is a vector space over the field F and $T: V \rightarrow V$ is a linear map, we use m_T to denote the minimum polynomial (in $F[x]$) of T.

LEMMA 2.5 *If V is a nontrivial finite-dimensional vector space over the field F and $T: V \rightarrow V$ is a linear map, the following are equivalent:*
(1) T is i.s.f.
(2) m_T is irreducible in $F[x]$ and $\deg(m_T) = \dim(V)$.
    For a proof, see [Jacobson 1953, vol II, Theorem 2, p. 69 and Theorem 3, p. 128].

NOTATION For n a positive integer and p a prime, we use F_n^p to denote a specific minimal splitting field (over $\mathrm{GF}(p)$) for the polynomial $x^n - [1]_p$ $\in \mathrm{GF}(p)[x]$.

We need some information about certain cyclotomic polynomials. For convenience, we gather what we need and put it in the form of two lemmas and a definition. See [Rédei 1967, pages 508–513] for proofs of these two lemmas.

LEMMA 2.6 *For all primes p and positive integers n such that $p \nmid n$, we have:*
(1) If $H(n)$ denotes the set of all roots of $x^n - [1]_p$ in F_n^p, then $H(n)$ is a subgroup of $(F_n^p)^*$ and $H(n) \cong \mathbb{Z}_n$.
(2) There exists a primitive nth root of unity $u \in (F_n^p)^*$.
(3) There exists a unique polynomial $f \in \mathrm{GF}(p)[x]$ satisfying

(i) $f \mid x^n - [1]_p$ in $\mathrm{GF}(p)[x]$, and f is monic.
(ii) $\deg(f) = \varphi(n)$ (*Euler's phi function*).
(iii) F_n^p is a minimal splitting field for f over $\mathrm{GF}(p)$.
(iv) *For all* t, t *is a root of* f (*in any field containing* $\mathrm{GF}(p)$) *iff* $t \in F_n^p$ *and* t *is a primitive* nth *root of unity.*

DEFINITION   For all primes p and positive integers n such that $p \nmid n$, we denote the unique polynomial $f \in \mathrm{GF}(p)[x]$ which appears in (3) of Lemma 2.6 by Φ_n^p, and call it the **nth cyclotomic polynomial** in $\mathrm{GF}(p)[x]$.

LEMMA 2.7   *For all primes* p *and positive integers* n *such that* $p \nmid n$, *we have*: $x^n - [1]_p = \prod_{d \mid n} \Phi_d^p$ *and there exist* $f_1, f_2, \ldots, f_s \in \mathrm{GF}(p)[x]$ *such that*
(i) $s \cdot (\exp(p, n)) = \varphi(n)$ (*Euler's phi function*)
(ii) $f_i \neq f_j$ *unless* $i = j$
(iii) *each* f_i *is monic irreducible in* $\mathrm{GF}(p)[x]$
(iv) $\deg(f_i) = \exp(p, n)$ *for all* i
(v) $\Phi_n^p = \prod_{i=1}^{s} f_i$. *Furthermore*, $F_n^p \cong \mathrm{GF}(p^e)$ *where* e *denotes* $\exp(p, n)$.

The information contained in the next lemma will be vital to us for most of the rest of our work.

LEMMA 2.8   *For all primes* p *and positive integers* a *define* $*$ *on* $\mathrm{GF}(p) \times \mathbb{Z}_p^a \to \mathbb{Z}_p^a$ *by* $[s]_p * ([x_1]_p, [x_2]_p, \ldots, [x_a]_p) = ([sx_1]_p, [sx_2]_p, \ldots, [sx_a]_p)$. *Then*
(1) $(\mathbb{Z}_p^a, *)$ *is a vector space over* $\mathrm{GF}(p)$, *which we will henceforth denote as* V_p^a.
(2) *For all* $w \in V_p^a$ *and for all* $[s]_p \in \mathrm{GF}(p)$ *with* $1 \leq s \leq p$,
$$[s]_p * w = w + w + \ldots + w \ (s \ times).$$
(3) *Every subgroup of* $\mathbb{Z}_p^a$ *is a subspace of* V_p^a.
(4) *Every automorphism of* $\mathbb{Z}_p^a$ *is a linear automorphism of* V_p^a (*i.e., preserves* $*$).
(5) $\dim(V_p^a) = a$.
(6) *There is a way of interpreting* $\mathrm{GL}(a, p)$ *so that* $\mathrm{Aut}(\mathbb{Z}_p^a) = \mathrm{Aut}(V_p^a)$ $= \mathrm{GL}(a, p)$.
Proof: We leave the details of (1) to the reader noting that the proof rests on the fact that we can "multiply across congruences" (i.e., $x \equiv y \pmod{p}$ implies $sx \equiv sy \pmod{p}$).

The proof of (2) is immediate from the definition of the scalar multiplication $*$.

To prove (3), let H be a subgroup of $\mathbb{Z}_p^a$, let $h \in H$, and let $[s]_p$ be a scalar. By (2) then, $[s]_p * h = h + h + \cdots + h$ (s times) which belongs to H. Thus H is closed under scalar multiplication, hence is a subspace.

To show (4), let $T \in \mathrm{Aut}(\mathbb{Z}_p^a)$, let $w \in V_p^a$, and   let $[s]_p \in \mathrm{GF}(p)$. Then $T([s]_p * w) = T(w + w + \ldots + w \; (s \text{ times}))$ by (2) which is $T(w) + T(w) + \ldots + T(w) \; (s \text{ times})$ since $T \in \mathrm{Aut}(\mathbb{Z}_p^a)$. But this, by another application of (2), is $[s]_p * T(w)$. Conclude $T([s]_p * w) = [s]_p * T(w)$ which proves (4).

To show (5), we define, for $1 \le i \le a$, $b_i \in V_p^a$ to be the ordered a-tuple having $[1]_p$ as ith coordinate with all other coordinates equal to $[0]_p$. The reader may verify that $\{b_1, b_2, \ldots, b_a\}$ is a basis of V_p^a. Hence $\dim(V_p^a) = a$.

Finally we let $\mathrm{GL}(a, p)$ denote the group whose underlying set is $\{T : T$ is a linear automorphism of $V_p^a\}$ and whose binary operation is composition of functions. Then (6) holds.

LEMMA 2.9 *For all primes p, all positive integers a, and all positive integers $s \le a$, the following are equivalent:*
(1) $H \le \mathbb{Z}_p^a$ *and* $|H| = p^s$.
(2) H *is a subspace of* V_p^a *of dimension* s.
Proof: Suppose H is a subspace of dimension s. That $H \le \mathbb{Z}_p^a$ is immediate. Let $\{b_1, b_2, \ldots, b_s\}$ be a basis for H. Define a function c on $\mathrm{GF}(p) \times \mathrm{GF}(p) \times \ldots \times \mathrm{GF}(p) \; (s \text{ times})$ into H by $c(t_1, t_2, \ldots, t_s) = t_1 * b_1 + t_2 * b_2 + \ldots + t_s * b_s$. The fact that each $h \in H$ can be written uniquely as a linear combination of the basis elements shows that c is a bijection. Hence $|H| = |\mathrm{GF}(p) \times \mathrm{GF}(p) \times \ldots \times \mathrm{GF}(p) \; (s \text{ times})| = |\mathrm{GF}(p)|^s = p^s$.

Conversely, suppose H is a subgroup of $\mathbb{Z}_p^a$ of order p^s. H is a subspace of V_p^a by Lemma 2.8, part (3). Let k be the dimension of H. An argument analogous to the one above can be used to concoct a bijection f from $\mathrm{GF}(p) \times \mathrm{GF}(p) \times \ldots \times \mathrm{GF}(p) \; (k \text{ times})$ to H. Conclude $|H| = p^k$. Thus $k = s$.

We are finally ready to attack Theorem 2.2.

LEMMA 2.10 *Let G be an SBP group, $|G| \ge 2$. Then*
(i) *G is solvable.*
(ii) *G has a normal subgroup P, with $[G : P] = q$, a prime.*
(iii) *P is a p-group for some prime p.*
(iv) *$|G| = p^a q$ for some integer $a \ge 0$.*
(v) *If $p = q$ or $a = 0$, then I of Theorem 2.2 holds and G is CLT.*
(vi) *If $p \ne q$ and $a = 1$, then II of Theorem 2.2 holds and G is CLT.*
(vii) *If $p \ne q$ and $a \ge 2$, then G is BNCLT and III of Theorem 2.2 holds.*
Proof:
(i) If G is nilpotent, it is certainly solvable. If G is not nilpotent, then since all proper subgroups of G are nilpotent, solvability follows from [Scott 1964, 6.5.7].

(ii) By solvability $G' < G$. Letting P be a maximal subgroup of G which contains G', we get $[G : P] = q$, a prime and $P \lhd G$.

Now (iii) holds by definition of SBP and (iv) follows immediately.

(v) This is also immediate.

(vi) If $p \neq q$ and $a = 1$, we get $|G| = pq$. Here either Cauchy's theorem or Sylow's theorem will apply to show G is CLT.

(vii) We now assume $p \neq q$ and $a \geq 2$. Since $1 < pq < |G|$, and G has no subgroups of order pq by definition of SBP, we conclude G is not CLT, hence *is* BNCLT.

Now let Q be any Sylow q-subgroup of G. Then $|Q| = q$, $P \cap Q = \{1\}$ and (by a cardinality consideration) $G = PQ$. The last two equations together with $P \lhd G$ imply there is a homomorphism $\omega : Q \to \mathrm{Aut}(P)$ such that $G \cong P \times_\omega Q$ (see [MacLane and Birkhoff 1968, Proposition 5, page 462]). If ω is trivial, then $P \times_\omega Q = P \times Q$. But $P \times Q$ is nilpotent (both P and Q are), hence CLT. This would imply G is CLT, a contradiction. Thus ω is not trivial. Then $\{1\} \leq \ker(\omega) < Q$ and $|Q| = q$ forces $\ker(\omega) = \{1\}$, i.e., ω is a monomorphism.

We now claim that P is a minimal normal subgroup of G, for suppose to the contrary that $L \lhd G$, $\{1\} < L < P$. Then $|L| = p^s$ where $1 \leq s \leq a - 1$ and so $|LQ| = p^s q$. In this case LQ is a proper subgroup of non-prime-power order, a contradiction to G being SBP. This establishes the claim.

Thus P is elementary Abelian, $P \cong \mathbb{Z}_p^a$. We may now conclude that there is some monomorphism $\theta : \mathbb{Z}_q \to \mathrm{Aut}(\mathbb{Z}_p^a)$ so that $G \cong \mathbb{Z}_p^a \times_\theta \mathbb{Z}_q$.

It remains to show that $a = \exp(p, q)$. Let $T = \theta(g)$ where g is a generator of $\mathbb{Z}_q$. Note that T has order q in $\mathrm{Aut}(\mathbb{Z}_p^a)$.

We claim that T is i.s.f. Suppose to the contrary that there exists $M \leq \mathbb{Z}_p^a$ such that M is invariant under T, $M \neq \{1\}$ and $M \neq \mathbb{Z}_p^a$. Then M is invariant under $\mathrm{Im}(\theta)$ and Lemma 2.4 applies to show that $M \times \mathbb{Z}_q \leq \mathbb{Z}_p^a \times_\theta \mathbb{Z}_q$. Since $M \neq \{1\}$, $\mathbb{Z}_p^a$, we have $|M| = p^s$ where $1 \leq s \leq a - 1$. Hence $|M \times \mathbb{Z}_q| = p^s q$. Thus $M \times \mathbb{Z}_q$ is a proper subgroup of $\mathbb{Z}_p^a \times_\theta \mathbb{Z}_q$. Then $|M \times \mathbb{Z}_q| = p^s q$ contradicts the fact that $\mathbb{Z}_p^a \times_\theta \mathbb{Z}_q \cong G$ is an SBP group. Conclude T is i.s.f. as claimed.

Since V_p^a is a vector space over $\mathrm{GF}(p)$ of dimension a, we can now apply Lemma 2.5 to conclude that m_T is irreducible in $\mathrm{GF}(p)[x]$ and that m_T has degree a.

Now let id denote the identity map of V_p^a and let Θ be the zero map (i.e., $\Theta(x) = 0$ for all $x \in V_p^a$). Then $T^q - \mathrm{id} = \Theta$. Letting h denote $x^q - [1]_p \in \mathrm{GF}(p)[x]$, we have that h annihilates T. Since m_T is the minimum polynomial of T, we may conclude $m_T | h$. But $h = (x - [1]_p)\Phi_q^p$, so that $m_T | (x - [1]_p)\Phi_q^p$. This and m_T irreducible imply either $m_T | x - [1]_p$ or $m_T | \Phi_q^p$. But if the first case holds, then $x - [1]_p$ would annihilate T, which would mean $T = \mathrm{id}$, a contradiction. Conclude then that $m_T | \Phi_q^p$. Since m_T is monic irreducible, we may now apply Lemma 2.7 to get $\deg(m_T) = \exp(p, q)$.

We now have $\exp(p, q) = \deg(m_T) = a$ which completes the proof.

LEMMA 2.11 *If G is a* BNCLT *group, then* III *of Theorem* 2.2 *must hold.*
Proof: We have shown in Lemma 2.10 that one of I, II, or III must hold; but if either I or II holds, then G is CLT.

Lemmas 2.10 and 2.11 complete the proof of Theorem 2.2.
The following will be used in the proof of Theorem 2.3.

THEOREM 2.12 *Let H and K be two groups and let* $\theta : K \to \mathrm{Aut}(H)$ *be a homomorphism. Then*
(1) $\{1_H\} \times K \leq H \times_\theta K.$
(2) *The following are equivalent*:
  (i) $\{1_H\} \times K \vartriangleleft H \times_\theta K.$
  (ii) $(h, 1_K)^{-1} *_\theta (1_H, k) *_\theta (h, 1_K) \in \{1_H\} \times K$ *for all* $h \in H$, $k \in K$.
  (iii) θ *is trivial.*
  (iv) $H \times_\theta K = H \times K$; *i.e.*, $(h, k) *_\theta (h_1, k_1) = (hh_1, kk_1)$ *for all* $h, h_1 \in H$, $k, k_1 \in K$.
Proof: The proof of (1) is straightforward. We will show (2) as follows:
  That (i) $\Rightarrow$ (ii) is clear by the definition of normality. To show (ii) $\Rightarrow$ (iii), let $k \in K$ and seek to show $\theta(k)$ is the identity automorphism on H. For any $h \in H$ we have $(h, 1_K)^{-1} *_\theta (1_H, k) *_\theta (h, 1_K) = (h^{-1}, 1_K) *_\theta ((\theta(k))(h), k)$ $= (h^{-1} \cdot (\theta(k))(h), k)$. Now this belongs to $\{1_H\} \times K$ by (ii). Thus $h^{-1} \cdot (\theta(k))(h) = 1_H$ showing $(\theta(k))(h) = h$ as required.
  To show (iii) $\Rightarrow$ (iv), let $h, h_1 \in H$ and $k, k_1 \in K$. Then $(h, k) *_\theta (h_1, k_1)$ $= (h \cdot (\theta(k))(h_1), kk_1) = (hh_1, kk_1)$ because $\theta(k)$ is the identity automorphism on H.
  Finally, we show (iv) $\Rightarrow$ (i). Let $(1_H, k) \in \{1_H\} \times K$ and let (h_1, k_1) be any element of $H \times_\theta K$. First note that by (iv), $(h_1^{-1}, k_1^{-1}) *_\theta (h_1, k_1) = (1_H, 1_K)$ so that $(h_1, k_1)^{-1} = (h_1^{-1}, k_1^{-1})$. Therefore $(h_1, k_1)^{-1} *_\theta (1_H, k) *_\theta (h_1, k_1)$ $= (h_1^{-1}, k_1^{-1}) *_\theta (1_H, k) *_\theta (h_1, k_1) = (h_1^{-1} 1_H h_1, k_1^{-1} kk_1)$ by (iv) again. That is, conjugation of $(1_H, k)$ by any (h_1, k_1) yields $(1_H, k_1^{-1} kk_1)$ and hence $\{1_H\} \times K \vartriangleleft H \times_\theta K$.

We shall also need the following lemma in proving Theorem 2.3.

LEMMA 2.13 *Let p and q be distinct primes, let* $a = \exp(p, q)$, *and suppose that G is any group of order* $p^a q$. *Then*
(i) $n_q(G) = 1$ *or* p^a.
(ii) *If* $n_q(G) \neq 1$, *then* $Q = N_G(Q)$ *for all Sylow q-subgroups Q of G.*
(iii) *If* $n_q(G) \neq 1$ *and* $a > 2$, *then for all s,* $1 < s < a - 1$, *G has no subgroups of order* $p^s q$; *and G is* BNCLT.
Proof:
  (i) Sylow's third theorem says that $n_q(G) \equiv 1 \pmod{q}$ and $n_q(G) \mid p^a$. Suppose $n_q(G) \neq 1$. Hence $n_q(G) = p^t$ for some positive $t \leq a$ and so $p^t \equiv 1 \pmod{q}$. This and $a = \exp(p, q)$ force $a \leq t$. Conclude $a = t$ so that $n_q(G) = p^a$.

(ii) Let Q be any Sylow q-subgroup of G. Then $n_q(G) = [G : N_G(Q)]$ so $n_q(G) \neq 1$ will imply (by (i)) $p^a = n_q(G) = [G : N_G(Q)]$. But $p^a = [G : Q]$ as well; and $Q \leq N_G(Q)$. Conclude $Q = N_G(Q)$.

(iii) Suppose to the contrary that G has a subgroup B of order $p^s q$ where $1 \leq s \leq a - 1$. Let Q be a Sylow q-subgroup of B. Then $n_q(B) \equiv 1 \pmod{q}$ and $n_q(B) \mid p^s$, say $n_q(B) = p^b$, $b \leq s$. Thus we have $p^b \equiv 1 \pmod{q}$ and $b < a$. Conclude (by definition of exponent) that $p^b = 1$. Hence $n_q(B) = 1$ so $Q \triangleleft B$ and hence $B \leq N_G(Q)$. Now apply (ii) to get $B \leq Q$, a contradiction (p^s divides $|B|$ but $p^s \nmid |Q|$).

It follows that G has no subgroups of order $p^s q$ for $1 \leq s \leq a - 1$. Since $a \geq 2$, it follows that G is both SBP and not CLT.

We are now ready to attack part of Theorem 2.3.

LEMMA 2.14   *For all primes p and all positive integers $a \geq 2$, $|\mathrm{GL}(a, p)|$* $= (p^a - 1)\prod_{i=1}^{a-1}(p^a - p^i)$.

For a proof, see [Scott 1964, 5.7.20].

LEMMA 2.15   *Let p and q be distinct primes such that $a = \exp(p, q) \geq 2$. Then*

(i) *q divides $|\mathrm{Aut}(\mathbb{Z}_p^a)|$.*

(ii) *$\mathbb{Z}_p^a$ has an automorphism T of order q.*

(iii) *There is a monomorphism $\theta : \mathbb{Z}_q \to \mathrm{Aut}(\mathbb{Z}_p^a)$.*

Proof: (i) Since $a = \exp(p, q)$, we have $q \mid (p^a - 1)$; Lemma 2.14 shows that $p^a - 1$ divides $|\mathrm{GL}(a, p)|$. Hence we have q divides $|\mathrm{GL}(a, p)|$, and (i) follows.

(ii) By (i) and Cauchy's theorem.

(iii) Let $T \in \mathrm{Aut}(\mathbb{Z}_p^a)$ have order q as in (ii). Then $\langle T \rangle \cong \mathbb{Z}_q$. Pick any specific isomorphism $f : \mathbb{Z}_q \to \langle T \rangle$, and define a function $\theta_f : \mathbb{Z}_q \to \mathrm{Aut}(\mathbb{Z}_p^a)$ by $\theta_f(x) = f(x)$ for all $x \in \mathbb{Z}_q$. Then θ_f is a monomorphism, and this proves (iii).

We note that (iii) of Lemma 2.15 proves (i) of Theorem 2.3.

LEMMA 2.16   *Let p and q be distinct primes such that $a = \exp(p, q) \geq 2$. Let $\lambda : \mathbb{Z}_q \to \mathrm{Aut}(\mathbb{Z}_p^a)$ be any monomorphism, and let $G = \mathbb{Z}_p^a \times_\lambda \mathbb{Z}_q$. Then* (i) *$|G| = p^a q$.* (ii) *$n_q(G) \neq 1$.* (iii) *$G$ is BNCLT.*

Proof: (i) is clear.

(ii) In G, let Q be the subgroup $\{1\} \times \mathbb{Z}_q$. Since $|Q| = q$, Q is a Sylow q-subgroup of G. Since λ is a monomorphism, we have that λ is not trivial, and it follows from Lemma 2.12 that Q is nonnormal. Consequently, $n_q(G) \neq 1$ as required.

(iii) By Parts (i) and (ii) and Lemma 2.13.

We note that (iii) of Lemma 2.16 proves (2) of Theorem 2.3.

Before we go on to prove (3) and (4) of Theorem 2.3, we pause to point out that we could have given a proof of (2) of Theorem 2.3 which would include our re-tracing some of our previous steps which dealt with certain cyclotomic polynomials and certain irreducible minimum polynomials. We chose to do the proof as we did it in order to illustrate the use of the "counting and packing" arguments in Lemma 2.13. Some of these arguments are very versatile.

We now get back to the job of proving (3) and (4) of Theorem 2.3. The next lemma will give an answer to the following question: Suppose that H and K are any two groups, that $\theta_1, \theta_2 : K \to \mathrm{Aut}(H)$ are any two homomorphisms, that f is any member of $\mathrm{Aut}(H)$, and that g is any member of $\mathrm{Aut}(K)$. If one defines $f \times g : H \times_{\theta_1} K \to H \times_{\theta_2} K$ by $(f \times g)(h, k) = (f(h), g(k))$ (all $h \in H$, $k \in K$), under what circumstances is $f \times g$ a homomorphism? (Note that whenever $f \times g$ is a homomorphism, it is an automorphism as well because both f and g are bijective.)

LEMMA 2.17 *Under the assumptions and notation of the preceding paragraph, $f \times g$ is a homomorphism (and hence an automorphism) iff the diagram*

$$
\begin{array}{ccc}
K & \xrightarrow{\ \theta_1\ } & \mathrm{Aut}(H) \\
{\scriptstyle g}\big\downarrow & & \big\downarrow{\scriptstyle c_f} \\
K & \xrightarrow[\ \theta_2\ ]{} & \mathrm{Aut}(H)
\end{array}
$$

commutes (c_f here denotes conjugation by f, i.e., $c_f(\alpha) = f\alpha f^{-1}$ for all $\alpha \in \mathrm{Aut}(H)$).

Proof: For all $h, h_1 \in H$ and all $k, k_1 \in K$ we have

$$(f \times g)((h,k) *_{\theta_1} (h_1, k_1)) = (f \times g)(h\theta_1(k)(h_1), kk_1)$$
$$= (f(h\theta_1(k)(h_1)), g(kk_1))$$
$$= (f(h)f(\theta_1(k)(h_1)), g(k)g(k_1))$$

while

$$(f \times g)(h,k) *_{\theta_2} (f \times g)(h_1, k_1) = (f(h), g(k)) *_{\theta_2} (f(h_1), g(k_1))$$
$$= (f(h)\theta_2(g(k))(f(h_1)), g(k)g(k_1)).$$

From these two displays we conclude that $f \times g$ is a homomorphism iff $f(\theta_1(k)(h_1)) = \theta_2(g(k))(f(h_1))$ for all h, h_1, k, k_1. But this occurs iff the automorphisms $f \circ \theta_1(k)$ and $\theta_2(g(k)) \circ f$ are identical, i.e., iff $f \circ \theta_1(k) \circ f^{-1} = \theta_2(g(k))$, i.e., iff $c_f \circ \theta_1 = \theta_2 \circ g$.

LEMMA 2.18 *Let H, K be any two finite groups and let $\theta_1, \theta_2 : K \to \mathrm{Aut}(H)$ be monomorphisms. Suppose $\mathrm{Im}(\theta_2)$ is conjugate to $\mathrm{Im}(\theta_1)$ in*

Aut(H). *Then there exist $f \in$ Aut(H) and $g \in$ Aut(K) such that $(f \times g)$:* $H \times_{\theta_1} K \to H \times_{\theta_2} K$ *is an isomorphism.*

Proof: First of all, Im(θ_2) is conjugate to Im(θ_1) in Aut(H), so there exists $f \in$ Aut(H) such that $f\text{Im}(\theta_1)f^{-1} = \text{Im}(\theta_2)$.

We define an appropriate $g \in$ Aut(K) as follows: for all $k \in K$, $\theta_1(k) \in \text{Im}(\theta_1)$ so $f\theta_1(k)f^{-1} \in \text{Im}(\theta_2)$. That is, $f\theta_1(k)f^{-1} = \theta_2(k^*)$ for some $k^* \in K$. We accordingly define $g(k) = k^*$. First we check that g is well-defined: If it happened that $f\theta_1(k)f^{-1} = \theta_2(k^*)$ and $f\theta_1(k)f^{-1} = \theta_2(k^{**})$, then $\theta_2(k^*) = \theta_2(k^{**})$ from whence it follows that $k^* = k^{**}$ by injectivity of θ_2. Conclude $g : K \to K$ is a function.

Next we show g is a homomorphism. Let k, $k_1 \in K$, and suppose $g(k) = k^*$ and $g(k_1) = k_1{}^*$. This means that $f\theta_1(k)f^{-1} = \theta_2(k^*)$ and that $f\theta_1(k_1)f^{-1} = \theta_2(k_1{}^*)$. Consequently, $f\theta_1(kk_1)f^{-1} = f(\theta_1(k)\theta_1(k_1))f^{-1} = (f\theta_1(k)f^{-1})(f\theta_1(k_1)f^{-1}) = \theta_2(k^*)\theta_2(k_1{}^*) = \theta_2(k^*k_1{}^*)$. Conclude that $g(kk_1) = k^*k_1{}^* = g(k)g(k_1)$; hence g is a homomorphism.

Next we show g is injective. Suppose $g(k) = 1$. This means $f\theta_1(k)f^{-1} = \theta_2(1) = 1$ so that $\theta_1(k) = f^{-1}1f = 1$. We now conclude that $k = 1$ by injectivity of θ_1; hence g is injective.

Finally we note that g is surjective because it is injective and K is finite (this is the pigeonhole principle). We conclude that $g \in$ Aut(K).

By the definition of g we have $f\theta_1(k)f^{-1} = \theta_2(g(k))$ for all $k \in K$. Hence $c_f \circ \theta_1 = \theta_2 \circ g$ so that the diagram in Lemma 2.17 commutes. Applying Lemma 2.17 yields the conclusion.

We pause to assure the reader that we have not lost sight of our present goal; we are still heading toward proofs of (3) and (4) of Theorem 2.3. Let us indicate how we intend to proceed. Suppose that the hypotheses of Theorem 2.3 are satisfied; if we can prove the first statement in (3), then the second will follow immediately from Lemma 2.18. To prove the first statement in (3) we will use the following result:

Let p and q be distinct primes such that $a = \exp(p, q) \geq 2$. Then every Sylow q-subgroup of Aut($\mathbb{Z}_p^a$) *is cyclic.*

We again use a series of lemmas to provide a proof of this result. The next lemma is more general than what we need but the proof is no more difficult; the lemma shows that if we have any ring with identity, then there is a copy of the multiplicative group of the ring in the automorphism group of the additive group of the ring. (By multiplicative group of a ring M we mean the group of units, i.e., $U_M = \{x : x \in M$ and there is some $y \in M$ such that $xy = 1 = yx\}$.) Note we are not requiring multiplication be commutative, but we *are* requiring that it be associative.

LEMMA 2.19 *Let M be a ring with identity. For all $a \in U_M$ let $f_a : M \to M$ be the function defined by $f_a(x) = ax$, all $x \in M$. Then (i) $a = b$ iff $f_a = f_b$.*

(ii) $f_a \circ f_b = f_{ab}$. (iii) $f_1 = \mathrm{id}_M$. (iv) $f_a \circ f_{a^{-1}} = \mathrm{id}_M = f_{a^{-1}} \circ f_a$. (v) $f_a \in \mathrm{Aut}(M)$ (Aut(M) *here is the automorphism group of* $(M, +)$.) (vi) *The function* $\alpha : U_M \to \mathrm{Aut}(M)$ *defined by* $\alpha(a) = f_a$ *is a monomorphism.*

Proof: (i) $f_a = f_b$ implies $f_a(1) = f_b(1)$, i.e., $a = b$. The converse is immediate.

(ii), (iii), and (iv) follow immediately from the definition of f_a.

(v) $f_a(x + y) = a(x + y) = ax + ay = f_a(x) + f_a(y)$ for all $x, y \in M$ so $f_a : M \to M$ is a homomorphism. That f_a is bijective follows from (iv). Hence $f_a \in \mathrm{Aut}(M)$.

(vi) α is a homomorphism by (ii). To see α is injective, suppose $\alpha(a) = \mathrm{id}_M$, i.e., $f_a = \mathrm{id}_M$. Then $f_a(1) = \mathrm{id}_M(1)$, i.e., $a = 1$.

LEMMA 2.20 *Let p be a prime and a be a positive integer. Denote the additive group of* $\mathrm{GF}(p^a)$ *by* $B(p,a)$. *Then*
(i) *There is a monomorphism* $\alpha : \mathrm{GF}(p)^* \to \mathrm{Aut}(B(p,a))$.
(ii) *There is an isomorphism* $\beta : B(p,a) \to \mathbb{Z}_p^a$.
(iii) *There is an isomorphism* $\gamma : \mathrm{Aut}(B(p,a)) \to \mathrm{Aut}(\mathbb{Z}_p^a)$.
(iv) *There is a monomorphism* $\delta : \mathrm{GF}(p^a)^* \to \mathrm{Aut}(\mathbb{Z}_p^a)$.
(v) $\mathrm{Aut}(\mathbb{Z}_p^a)$ *has a cyclic subgroup of order* $p^a - 1$.

Proof: (i) This follows from (vi) of Lemma 2.19 because $\mathrm{GF}(p^a)^*$ is the group of units of $\mathrm{GF}(p^a)$.

(ii) The additive group of any finite field is elementary Abelian, and $|B(p,a)| = p^a$.

(iii) follows from (ii).

(iv) follows from (i) and (iii).

(v) The multiplicative group of any finite field is cyclic; thus $\mathrm{GF}(p^a)^*$ is cyclic of order $p^a - 1$. Let δ denote the monomorphism given by (iv). Then $C = \mathrm{Im}(\delta)$ is cyclic of order dividing $p^a - 1$.

LEMMA 2.21 *Let p be a prime and a be a positive integer, $a \geq 2$. Then* $|\mathrm{Aut}(\mathbb{Z}_p^a)| = |\mathrm{GL}(a, p)| = p^u(p^a - 1)(\prod_{i=1}^{a-1}(p^i - 1))$ *where* $u = a(a - 1)/2$.

Proof: From Lemma 2.14 we have $|\mathrm{GL}(a, p)| = (p^a - 1) \prod_{i=1}^{a-1}(p^a - p^i)$. But $\prod_{i=1}^{a-1}(p^a - p^i) = (p^a - p)(p^a - p^2) \ldots (p^a - p^{a-1})$ which is equal to $(p(p^{a-1} - 1))(p^2(p^{a-2} - 1)) \ldots (p^{a-1}(p - 1))$. This in turn is equal to the product of $[p \cdot p^2 \cdot \ldots \cdot p^{a-1}]$ by $[(p^{a-1} - 1)(p^{a-2} - 1) \ldots (p - 1)]$. And this product is equal to $p^{1+2+\cdots+(a-1)} \cdot [\prod_{i=1}^{a-1}(p^i - 1)]$ which finally is equal to $p^u[\prod_{i=1}^{a-1}(p^i - 1)]$. The result follows.

We pause to introduce a notation: $q^t \| n$ will be used to mean $q^t | n$ and $q^{t+1} \nmid n$.

LEMMA 2.22 *Let p and q be distinct primes such that $2 \leq a = \exp(p, q)$. Then*
(i) $q \nmid p^u$ *and* $q \nmid \prod_{i=1}^{a-1}(p^i - 1)$ *where* $u = a(a - 1)/2$.
(ii) $q | (p^a - 1)$ *and* $p^a - 1$ *divides* $|\mathrm{Aut}(\mathbb{Z}_p^a)|$.

(iii) *For all positive integers k, $q^k | (p^a - 1)$ iff q^k divides $|\text{Aut}(\mathbb{Z}_p^a)|$.*

   Now define s and t by $q^s \| (p^a - 1)$ and $q^t \| (|\text{Aut}(\mathbb{Z}_p^a)|)$. Then

(iv) $s = t$.

(v) q^t *is the order of the Sylow q-subgroups of $\text{Aut}(\mathbb{Z}_p^a)$.*

(vi) *All Sylow q-subgroups of $\text{Aut}(\mathbb{Z}_p^a)$ are cyclic.*

Proof: (i) Since $p \neq q$, $q \nmid p^u$. Now by definition of exponent, p^i is not congruent to 1 mod q for any positive $i \leq a - 1$. Thus $q \nmid p^i - 1$ for such i. Hence $q \nmid \prod_{i=1}^{a-1}(p^i - 1)$.

   (ii) $q | (p^a - 1)$ by definition of exponent; the rest follows immediately from Lemma 2.21.

   (iii) The "only if" part follows immediately from the second part of (ii). The "if" part follows from Lemma 2.21 and (i).

   (iv) We have $q^s | (p^a - 1)$ and q^t divides $|\text{Aut}(\mathbb{Z}_p^a)|$, and it follows from (iii) that $q^t | (p^a - 1)$ and q^s divides $|\text{Aut}(\mathbb{Z}_p^a)|$. Since $q^s \| (p^a - 1)$, we must have $t \leq s$; since $q^t \| (|\text{Aut}(\mathbb{Z}_p^a)|)$, we must have $s \leq t$.

   (v) This is immediate from $q^t \| (|\text{Aut}(\mathbb{Z}_p^a)|)$.

   (vi) By Lemma 2.20 $\text{Aut}(\mathbb{Z}_p^a)$ has a cyclic subgroup C of order $p^a - 1$. From $q^s \| (p^a - 1)$ and $s = t$ we conclude $q^t | (p^a - 1)$. Therefore C has a subgroup S of order q^t. Then S is a Sylow q-subgroup of $\text{Aut}(\mathbb{Z}_p^a)$ by (vi). Since $S \leq C$, S is cyclic. Moreover, any Sylow q-subgroup of $\text{Aut}(\mathbb{Z}_p^a)$ is conjugate (hence isomorphic) to S, hence must be cyclic.

We pause to point out a number-theoretic subtlety connected with Lemma 2.22, by giving some examples. First note that $\exp(2,5) = 4$ and $5^1 \| (2^4 - 1)$, so that the Sylow 5-subgroups of $\text{Aut}(\mathbb{Z}_2^4)$ have order 5; since 5 is a prime, it is immediate that these Sylow subgroups are cyclic. In this case, we may dispense completely with Lemmas 2.19, 2.20, and some of 2.22. Why, then, have we gone to so much work? Well, $\exp(3,11) = 5$. Unfortunately, $\exp(3, 11^2)$ is also 5; in fact, we have $3^5 - 1 = 2 \cdot 11^2$, so that $11^2 \| (3^5 - 1)$. In this case, then, the Sylow 11-subgroups of $\text{Aut}(\mathbb{Z}_3^5)$ have order 11^2, and it is now not quite so obvious (without some extra work) that said Sylow subgroups are cyclic.

We wish to give one more example of this sort of behavior; we believe the following is rather spectacular. The reader may check that 239 is a prime, and that $239 - 1 = 2 \cdot 7 \cdot 17$, $239 + 1 = 2^4 \cdot 3 \cdot 5$, $239^2 + 1 = 2 \cdot 13^4$. It follows that 239^2 is not congruent to 1 mod 13, but $239^4 \equiv 1$ mod 13, so that $\exp(239, 13) = 4$; also, we have $13^4 \| (239^4 - 1)$ (!) Thus, the Sylow 13-subgroups of $\text{Aut}(\mathbb{Z}_{239}^4)$ have order 13^4.

Our second example (above) occurs in several number theory texts; we believe that our third example is perhaps not quite as well-known.

We are now very close to proving (3) and (4) of Theorem 2.3; the next two lemmas should indicate why we have gone to so much work.

LEMMA 2.23 *Let G be any finite group and let q be any prime divisor of $|G|$. If every Sylow q-subgroup of G is cyclic, then any two q-subgroups of the same order are conjugate in G.*

Proof: Let C, D be subgroups of order q^s. Then there exist Sylow q-subgroups Q_1, Q_2 of G with $C \leq Q_1$ and $D \leq Q_2$. Moreover, Q_1 and Q_2 are conjugate, say $Q_2 = gQ_1 g^{-1}$. Then gCg^{-1} and D are both subgroups of the cyclic group Q_2 of order q^s. Since finite cyclic groups have a *unique* subgroup for any given divisor of their orders, we conclude $gCg^{-1} = D$.

We pause to point out two examples. First we note that $\mathbb{Z}_2^2$ shows that the assumption requiring the Sylow subgroups to be cyclic in Lemma 2.23 cannot be weakened ($\mathbb{Z}_2^2$ has three subgroups of order 2, no two of which are conjugate since they are all normal in $\mathbb{Z}_2^2$).

On the other hand, A_4 shows the cyclic Sylow requirement is not necessary. This is seen as follows: A_4 has a unique noncyclic Sylow 2-subgroup $V \cong \mathbb{Z}_2^2$ and precisely three subgroups of order 2. Let H be any one of these 2-element subgroups. Then $V = N_{A_4}(H)$ so H has $[A_4 : V] = 3$ conjugates; that is, all subgroups of order 2 are conjugate in A_4.

The next lemma will (finally!) kill off (3) of Theorem 2.3.

LEMMA 2.24 *Let p and q be distinct primes such that $a = \exp(p, q) \geq 2$. If $\theta_1, \theta_2 : \mathbb{Z}_q \to \mathrm{Aut}(\mathbb{Z}_p^a)$ are any two monomorphisms, then*
(i) *$\mathrm{Im}(\theta_1)$ and $\mathrm{Im}(\theta_2)$ are conjugate in $\mathrm{Aut}(\mathbb{Z}_p^a)$.*
(ii) *There exist $f \in \mathrm{Aut}(\mathbb{Z}_p^a)$ and $g \in \mathrm{Aut}(\mathbb{Z}_q)$ such that $f \times g : \mathbb{Z}_p^a \times_{\theta_1} \mathbb{Z}_q \to \mathbb{Z}_p^a \times_{\theta_2} \mathbb{Z}_q$ is an isomorphism. (Recall the definition of $f \times g$; namely, $(f \times g)(h, k) = (f(h), g(k))$.)*
Proof: (i) By Lemma 2.22 all Sylow q-subgroups of $\mathrm{Aut}(\mathbb{Z}_p^a)$ are cyclic. Therefore Lemma 2.23 can be applied to yield conjugacy of $\mathrm{Im}(\theta_1)$ and $\mathrm{Im}(\theta_2)$ since $|\mathrm{Im}(\theta_1)| = q = |\mathrm{Im}(\theta_2)|$.

(ii) Part (1) of Theorem 2.3 (which we have already proved long ago) shows that there exists a monomorphism $\theta : \mathbb{Z}_q \to \mathrm{Aut}(\mathbb{Z}_p^a)$. This together with (i) shows that all of the hypotheses of Lemma 2.18 are satisfied; our conclusion then follows from Lemma 2.18.

We now attack (4) of Theorem 2.3. Suppose A and B are two finite BNCLT groups of the same order. Note that this common order must be ≥ 2 since groups of order 1 are CLT. Now Theorem 2.2 shows that $|A| = p_1^{a_1} q_1$ where p_1, q_1 are distinct primes, $2 \leq a_1 = \exp(p_1, q_1)$, and $A \cong \mathbb{Z}_{p_1}^{a_1} \times_{\theta_1} \mathbb{Z}_{q_1}$ for some monomorphism $\theta_1 : \mathbb{Z}_{q_1} \to \mathbb{Z}_{p_1}^{a_1}$; and that $|B| = p_2^{a_2} q_2$ where p_2, q_2 are distinct primes, $2 \leq a_2 = \exp(p_2, q_2)$, and $B \cong \mathbb{Z}_{p_2}^{a_2} \times_{\theta_2} \mathbb{Z}_{q_2}$ for some monomorphism $\theta_2 : \mathbb{Z}_{q_2} \to \mathbb{Z}_{p_2}^{a_2}$. We now have $p_1^{a_1} q_1 = p_2^{a_2} q_2$. This, the fact that $a_1, a_2 \geq 2$, and uniqueness of factorization of positive integers into prime powers yields $p_1 = p_2$, $a_1 = a_2$, and $q_1 = q_2$. Now set $p = p_1 = p_2$, $q = q_1 = q_2$, and $a = a_1 = a_2$. By (3) of Theorem 2.3 we may now conclude that $\mathbb{Z}_p^a \times_{\theta_1} \mathbb{Z}_q \cong \mathbb{Z}_p^a \times_{\theta_2} \mathbb{Z}_q$. Hence $A \cong B$. This completes the proof of (4) of Theorem 2.3.

Now that we have finally proved all of Theorem 2.3, we shall turn to the topic of McCarthy groups. These groups make up a class of semi-direct products which generalize those which give us BNCLT groups, and were used by McCarthy to prove Theorem 2.1 (see [McCarthy 1970]). The author is taking the liberty of calling these groups "McCarthy groups." We shall endeavor to prove the following theorem (whose statement contains an implicit description of McCarthy groups):

THEOREM 2.25 *Let n be a positive integer and let p be a prime such that $p \nmid n$ and $a = \exp(p,n) \geq 2$. Then there is a monomorphism $\theta : \mathbb{Z}_n \to \operatorname{Aut}(\mathbb{Z}_p^a)$ such that the semi-direct product $G = \mathbb{Z}_p^a \times_\theta \mathbb{Z}_n$ has the following property: for all positive integers $s \leq a - 1$, G has no subgroups of order $p^s n$.*

We shall prove Theorem 2.25 by a series of lemmas but before we get on with the proof we wish to give an example which does the following:

(1) It shows that there exist positive integers n and primes p with n *composite* such that the hypotheses of Theorem 2.25 are satisfied, thus showing that McCarthy groups are semi-direct products which are a *proper* generalization of the semi-direct products which give us BNCLT groups.

(2) It illustrates some of the things that we have done with certain cyclotomic polynomials.

(3) It will give some intuition for why Theorem 2.25 should be true, and it will illustrate the paths that we intend to take in proving Theorem 2.25 (including a subtlety which will be dealt with and dispatched in our next two lemmas).

(4) It will show that there are non-CLT groups which are not only non-BNCLT, but, in fact are "almost" CLT.

Here is the example. Set $p = 5$ and $n = 8$. (Note that 8 is composite.) Then $\exp(5,8) = 2$. Now $|\operatorname{GF}(5^2)^*| = 24$ and $\operatorname{GF}(5^2)^*$ is cyclic. Since $8 \mid 24$, $\operatorname{GF}(5^2)^*$ has a subgroup B of order 8. Since $\operatorname{GF}(5^2)^*$ is cyclic, so is B. Let g generate B. If we set $h = x^8 - [1]_5 \in \operatorname{GF}(5)[x]$, then g is a root of h. Since there are no elements of order 8 in $\operatorname{GF}(5)^*$, we conclude that $\operatorname{GF}(5^2)$ is a minimal splitting field for h over $\operatorname{GF}(5)$. (This illustrates the last sentence of Lemma 2.7.) In $\operatorname{GF}(5)[x]$ we have

$$
\begin{aligned}
x^8 - [1]_5 &= (x^4 - [1]_5)(x^4 + [1]_5) \\
&= (x^2 - [1]_5)(x^2 + [1]_5)(x^4 + [1]_5) \\
&= (x - [1]_5)(x + [1]_5)(x^2 + [1]_5)(x^4 + [1]_5) \\
&= \Phi_1^5 \cdot \Phi_2^5 \cdot \Phi_4^5 \cdot \Phi_8^5
\end{aligned}
$$

(which illustrates (v) of Lemma 2.7). Now we may write

$$
\Phi_1^5 = x - [1]_5, \qquad \Phi_2^5 = x + [1]_5,
$$

$$
\Phi_4^5 = x^2 + [1]_5 = x^2 - [4]_5 = (x - [2]_5)(x + [2]_5)
$$

(2 and -2 are primitive roots of 5, as a number theorist might put it),

$$\Phi_8^5 = x^4 + [1]_5 = x^4 - [4]_5 = (x^2 - [2]_5)(x^2 + [2]_5),$$

and the last two quadratic polynomials are both irreducible in GF(5)[x] (2 and -2 are both quadratic nonresidues mod 5). Note that the above gives us prime factorizations for our four cyclotomic polynomials and illustrates (i)–(v) of Lemma 2.7, because $\varphi(1) = 1 = \varphi(2)$, $\varphi(4) = 2$, $\varphi(8) = 4$, $\exp(5, 1) = \exp(5, 2) = \exp(5, 4) = 1$, and $\exp(5, 8) = 2$.

Now we build a semi-direct product $\mathbb{Z}_5^2 \times_\theta \mathbb{Z}_8$. (Of course, we must first find an appropriate θ.) Let $h_1 = x^2 - [2]_5$ and $h_2 = x^2 + [2]_5$. We showed that $\Phi_8^5 = h_1 h_2$ and that both factors are irreducible in GF(5)[x]. Now we consider the matrix

$$C(h_1) = \begin{pmatrix} [0]_5 & [2]_5 \\ [1]_5 & [0]_5 \end{pmatrix}$$

with all entries in GF(5). (The astute reader will note that $C(h_1)$ is the companion matrix of the polynomial h_1.) A little computation yields

$$C(h_1)^2 = \begin{pmatrix} [2]_5 & [0]_5 \\ [0]_5 & [2]_5 \end{pmatrix}$$

so that h_1 annihilates $C(h_1)$. More than this is true; in fact, h_1 is the minimum polynomial for $C(h_1)$ in GF(5)[x]. This is true for companion matrices over fields in general, and we shall use this fact in our eventual proof of Theorem 2.25, but we should like to give an argument (a "packing argument") which is a special case of one which we shall use later.

Let m denote the minimum polynomial for $C(h_1)$ in GF(5)[x]. Then m has degree 2 and is monic, and (because h_1 annihilates $C(h_1)$), we must have $m | h_1$; but h_1 is irreducible in GF(5)[x], and h_1 is monic, and m is nonconstant, and hence we must have $m = h_1$. Now, in V_5^2, let $b_1 = ([1]_5, [0]_5)$ and $b_2 = ([0]_5, [1]_5)$, and let $D = \{b_1, b_2\}$; then D is a basis for V_5^2, and we consider the unique linear map $T : V_5^2 \to V_5^2$ such that

$$T(b_1) = b_2 \quad \text{and} \quad T(b_2) = [2]_5 * b_1.$$

(The reader will now note that the matrix of T relative to the basis D is $C(h_1)$.) Now, we assert the following: (i) $m_T = h_1$. (ii) m_T is irreducible in GF(5)[x] and $\deg(m_T) = $ the dimension of V_5^2. (iii) $T \in \text{Aut}(\mathbb{Z}_5^2)$. (iv) T has order 8. (v) T is i.s.f. (vi) There is a monomorphism $\theta : \mathbb{Z}_8 \to \text{Aut}(\mathbb{Z}_5^2)$ such that $\text{Im}(\theta) = \langle T \rangle$ and $\mathbb{Z}_5^2 \times_\theta \mathbb{Z}_8$ has no subgroups of order 40. (Note that $|\mathbb{Z}_5^2 \times_\theta \mathbb{Z}_8| = 200$, and $40 | 200$.)

From the way we have rigged things, we must have $m_T = m = h_1$, so that (i) and (ii) come falling out of what we have done earlier in this example. Now $\det(C(h_1)) = [-2]_5 \neq [0]_5$, so that $C(h_1)$ is nonsingular and hence (iii) follows. Now (i) shows that $m_T = h_1$, and our previous work (in this example) shows that $h_1 | \Phi_8^5$, i.e., $m_T | \Phi_8^5$. Since all of the roots of Φ_8^5 (in GF(5^2)) are primitive 8th roots of unity, the same must be true of m_T, and

(iv) follows. Now (ii) shows that (2) of Lemma 2.5 is satisfied, and (v) follows from (1) of that lemma. (In this example, there is a much simpler way; every nontrivial subspace of V_5^2 is one-dimensional, and m_T has no roots in GF(5), so that T has no eigenvalues in GF(5). Unfortunately, not all the vector spaces that we need to deal with are two-dimensional.)

Now we finally define θ. Because of (iii) and (iv), $\langle T \rangle$ is a cyclic subgroup of $\text{Aut}(\mathbb{Z}_5^2)$ and $|\langle T \rangle| = 8$; thus there is a monomorphism $\theta : \mathbb{Z}_8 \to \text{Aut}(\mathbb{Z}_5^2)$ such that $\text{Im}(\theta) = \langle T \rangle$.

Now we face the subtlety which we mentioned in (3) (above). Our next two lemmas will show that if $\mathbb{Z}_5^2 \times_\theta \mathbb{Z}_8$ has a subgroup S of order 40, then it also has a subgroup S_1 of order 40 (possibly different from S) such that there is a subgroup $H_1 \le \mathbb{Z}_5^2$ with $|H_1| = 5$ and $S_1 = H_1 \times \mathbb{Z}_8$ (Cartesian product). But then Lemma 2.4 would say that H_1 must be invariant under $\text{Im}(\theta)$; i.e., invariant under T. But H_1 is a nontrivial subspace of V_5^2 and (v) (above) says that T is i.s.f., and we would have a contradiction. It follows that $\mathbb{Z}_5^2 \times_\theta \mathbb{Z}_8$ has no subgroups of order 40.

Now, it may not be a big deal to have a group of order 200 which has no subgroups of order 40, but in fact, this is best possible: We shall show later that every group of order 200 must have at least one subgroup of order d for any d such that $d \mid 200$ and $d \ne 40$. We do some of this now, indicating what will need to be shown later. Let G be any group of order $200 = 2^3 5^2$. Of course, G must have subgroups with order 1 and order 200, and by Sylow's theorem G has a subgroup of order k for $k = 2, 2^2, 2^3, 5, 5^2$. Now Sylow's theorem also says that $n_5(G) \equiv 1 \bmod 5$ and $n_5(G) \mid 2^3$. It follows that $n_5(G) = 1$, and hence G has a normal Sylow 5-subgroup N (order 5^2). If we pick any $H_2 \le G$ of order 2^2, then $NH_2 \le G$ and $|NH_2| = 2^2 5^2 = 100$. Now comes the part we will do later; we shall show in Section 3 that all groups of order 100 are supersolvable (hence CLT), and hence for $t = 10, 20$, or 50, NH_2 (and hence G) has a subgroup of order t. The reader should check that we have either shown or asserted that G has a subgroup for every divisor of 200 other than 40. We see then that $\mathbb{Z}_5^2 \times_\theta \mathbb{Z}_8$ (as constructed above) is "almost" CLT.

LEMMA 2.26 *Let m and n be any two relatively prime positive integers, let H and K be two solvable groups of orders m and n respectively, and let $\theta : K \to \text{Aut}(H)$ be a homomorphism. Suppose S is a subgroup of $H \times_\theta K$ of order $m_1 n$ where $m_1 \mid m$. Then there exist S_1, H_1, and g such that $S_1 \le H \times_\theta K$, $H_1 \le H$, $g \in H \times_\theta K$, and (1) $gSg^{-1} = S_1$ (2) $|H_1| = m_1$ and $|S_1| = m_1 n$ (3) H_1 is invariant under $\text{Im}(\theta)$.*

Proof: Since both H and K are solvable, so is $H \times_\theta K$, hence so is S. Let M_1 be a Hall π-subgroup of S (existence by Theorem 1 of Appendix A), where π is the set of primes dividing m. Thus (since m_1 and n are relatively prime because m and n are) $|M_1| = m_1$. Let N be a Hall π'-subgroup of S; thus $|N| = n$.

Now note that both N and $\{1\} \times K$ are Hall π'-subgroups of $H \times_\theta K$. Consequently (Theorem 1 of Appendix A again) they are conjugate; i.e., there exists a g such that $gNg^{-1} = \{1\} \times K$. Since M_1 is a π-group, so is gM_1g^{-1} and so (yet a third application of Theorem 1 of Appendix A) gM_1g^{-1} is contained in some Hall π-subgroup of $H \times_\theta K$. But since $H \times \{1\} \lhd H \times_\theta K$, $H \times \{1\}$ is the unique Hall π-subgroup of $H \times_\theta K$. We conclude $gM_1g^{-1} \leq H \times \{1\}$.

We now define $H_1 = \{h : h \in H$ and $(h, 1) \in gM_1g^{-1}\}$. It is easily checked that $H_1 \leq H$. Since $gM_1g^{-1} \leq H \times \{1\}$, we have $gM_1g^{-1} = H_1 \times \{1\}$. Thus $|H_1| = |gM_1g^{-1}| = |M_1| = m_1$.

Now since M_1 is a Hall π-subgroup of S and N is a Hall π'-subgroup of S, $S = M_1N$. Therefore $gSg^{-1} = g(M_1N)g^{-1} = (gM_1g^{-1})(gNg^{-1}) = (H_1 \times \{1\})(\{1\} \times K) = H_1 \times K$. We define S_1 to be $S_1 = H_1 \times K$. Then $gSg^{-1} = S_1$ completing the proof of (1).

As for (2), we have already noted that $|H_1| = m_1$; that $|S_1| = m_1n$ follows from S_1 being conjugate to S.

Finally, (3) follows from an application of Lemma 2.4.

The main reason we went to the trouble of doing Lemma 2.26 is that we need it for the proof of our next lemma:

LEMMA 2.27 *Let m, n, H, K, and θ satisfy the hypotheses of Lemma 2.26. Then for any divisor m_1 of m, $H \times_\theta K$ has a subgroup of order m_1n iff H has a subgroup of order m_1 which is invariant under $\mathrm{Im}(\theta)$.*
Proof: Suppose $H \times_\theta K$ has a subgroup S of order m_1n. Then Lemma 2.26 applies to yield $H_1 \leq H$, $|H_1| = m_1$, and H_1 invariant under $\mathrm{Im}(\theta)$.

Conversely, suppose H has an $\mathrm{Im}(\theta)$-invariant subgroup H_1 of order m_1. Then by Lemma 2.4 $S = H_1 \times K$ is a subgroup of $H \times_\theta K$, and note that $|S| = |H_1| \cdot |K| = m_1n$.

Now that we have Lemma 2.27, we shall attack Theorem 2.25. The proof we shall give is a direct generalization of what we did in the example preceding Lemma 2.26.

Proof of Theorem 2.25 Pick a nonconstant, monic, irreducible polynomial $f \in \mathrm{GF}(p)[x]$ such that $f \mid \Phi_n^p$. From the prime factorization of Φ_n^p given in Lemma 2.7, f must be one of the f_i's given in that lemma, and it follows (same lemma, part (iv)) that $\deg(f) = \exp(p, n) = a =$ dimension of V_p^a as a vector space over $\mathrm{GF}(p)$.

Now we concoct a linear map $T : V_p^a \to V_p^a$ such that $m_T = f$. If $f = c_0 + c_1x + c_2x^2 + \ldots + c_{a-1}x^{a-1} + x^a$, consider the a-by-a matrix

$$C(f) = \begin{pmatrix} 0 & 0 & 0 & \cdots & 0 & -c_0 \\ 1 & 0 & 0 & \cdots & 0 & -c_1 \\ 0 & 1 & 0 & \cdots & 0 & -c_2 \\ 0 & 0 & 1 & \cdots & 0 & -c_3 \\ 0 & 0 & 0 & \cdots & 0 & -c_4 \\ \vdots & \vdots & \vdots & \cdots & \vdots & \vdots \\ 0 & 0 & 0 & \cdots & 1 & -c_{a-1} \end{pmatrix}$$

(where for convenience we have denoted $[0]_p$ by 0 and $[1]_p$ by 1).

Let $B = \{b_1, b_2, \ldots, b_a\}$ be a basis for V_p^a and consider the unique linear map $T: V_p^a \to V_p^a$ such that the matrix of T relative to B is $C(f)$. To be more specific, $T(b_i) = b_{i+1}$ for $i \leq a - 1$ and $T(b_a) = -c_0 * b_1 - c_1 * b_2 - \ldots - c_{a-1} * b_a$. We claim T has the following properties: (i) $m_T = f$. (ii) $m_T | (x^n - 1)$. (iii) T^n is the identity map on V_p^a. (iv) $T \in \mathrm{Aut}(V_p^a)$. (v) T has order n. (vi) m_T is irreducible in $\mathrm{GF}(p)[x]$ and $\deg(m_T) = a = $ dimension of V_p^a as a vector space. (vii) T is i.s.f.

The proofs of (i)–(vii) are as follows:

(i) Let m be the minimum polynomial (in $\mathrm{GF}(p)[x]$) of $C(f)$; it is a classical fact that $m = f$ (for a proof, see [Mac Lane and Birkhoff 1968, p. 315, Proposition 13]). Since the matrix of T (relative to B) is $C(f)$, we have $m_T = m$; (i) follows.

(ii) Recall that $f | \Phi_n^p$; by Lemma 2.7 $\Phi_n^p | (x^n - 1)$. We conclude that $f | (x^n - 1)$, and (ii) follows immediately from (i).

(iii) From (ii) it follows that $x^n - 1$ annihilates T, and (iii) follows.

(iv) We first note that $n \geq 2$ (because $\exp(p, 1) = 1$, and one of our hypotheses requires that $\exp(p, n) \geq 2$); now, from (iii), we may factor the identity map on V_p^a as $T \circ T^{n-1}$ and also as $T^{n-1} \circ T$; and it follows that T is bijective; since T is linear, (iv) follows.

(v) If we set $T = $ the order of T, then (iii) shows that $t | n$ so $t \leq n$. Also, T^t is the identity map on V_p^a, so that $x^t - 1$ annihilates T, and hence $m_T | (x^t - 1)$. Now recall that $f | \Phi_n^p$ hence we get $m_T | \Phi_n^p$ by (i). Now from Lemmas 2.6 and 2.7 we have that $\mathrm{GF}(p^a)$ is a minimal splitting field for Φ_n^p over $\mathrm{GF}(p)$, and that for all $r \in \mathrm{GF}(p^a)$, r is a root of Φ_n^p iff r is a primitive nth root of unity. Now $m_T | \Phi_n^p$ shows that m_T has all its roots in $\mathrm{GF}(p^a)$ and that all of these roots are primitive nth roots of unity. Now let r denote a specific root of m_T in $\mathrm{GF}(p^a)$; since $m_T | (x^t - 1)$, r is a root of $x^t - 1$, and hence $r^t = 1$. Since r is a primitive nth root of unity, r has order n in $\mathrm{GF}(p^a)^*$. Hence $n \leq t$. This and $t \leq n$ combine to prove (v).

(vi) We carefully chose f so that f was irreducible in $\mathrm{GF}(p)[x]$ and $\deg(f) = a$; (i) gives us that $f = m_T$, and (vi) follows.

(vii) By (vi) and Lemma 2.5.

Now we are ready to find a θ. (iv) and (v) above show that the cyclic group $\langle T \rangle$ is a subgroup of $\mathrm{Aut}(\mathbb{Z}_p^a)$ and has order n. It follows that there is a monomorphism $\theta: \mathbb{Z}_n \to \mathrm{Aut}(\mathbb{Z}_p^a)$ such that $\mathrm{Im}(\theta) = \langle T \rangle$. Let $G = \mathbb{Z}_p^a \times_\theta \mathbb{Z}_n$, so that $|G| = p^a n$. We shall finish off the proof of Theorem

2.25 by showing that for all positive integers $s \leq a - 1$, G has no subgroups of order $p^s n$.

Suppose, to the contrary, that G does have a subgroup of order $p^s n$ where s is a positive integer, $s \leq a - 1$. We are finally ready to use Lemma 2.27; in said lemma, we set $m = p^a$, $H = \mathbb{Z}_p^a$, $K = \mathbb{Z}_n$, and $m_1 = p^s$. Note that the hypotheses of Lemma 2.27 are satisfied, and hence H has a subgroup (call it H_1) of order $m_1 = p^s$ which is invariant under $\text{Im}(\theta)$. Then $1 < |H_1| < p^a$, so that H_1 is a proper subspace of V_p^a which is invariant under $\text{Im}(\theta)$ hence invariant under T. It follows that T is not i.s.f. This contradiction completes the proof of Theorem 2.25.

We end this section with a lemma and some comments and numerical results.

From Theorem 2.2 we see that any BNCLT group must have an order which has a factorization of the form $p^a q$, where p, q are distinct primes and $a = \exp(p, q) \geq 2$. From this information, one might get the impression that the class of BNCLT groups is very small relative to the class of all finite groups. While this is true in one sense, the following lemma shows that, in another sense, perhaps the class of all BNCLT groups is not as small as all that.

LEMMA 2.28 *Let r and s be any two primes such that $r < s$ and let $a = \exp(r, s)$ and $b = \exp(s, r)$. Then*
(i) *$a \geq 2$ and there is a BNCLT group of order $r^a s$.*
(ii) *If s is not congruent to 1 mod r, then $b \geq 2$, and there is a BNCLT group of order $s^b r$.*
Proof: We first make the observation that if p and q are any two distinct primes, then $\exp(p, q) = 1$ iff $p \equiv 1$ mod q; it follows that $\exp(p, q) \geq 2$ iff p is not congruent to 1 mod q. Now we prove (i) and (ii).

(i) Since $2 \leq r < s$, we have $0 < r - 1 < s$, so $s \nmid (r - 1)$. Hence r is not congruent to 1 mod s, and our observation shows that $a = \exp(r, s) \geq 2$; the rest of our conclusion follows from Theorem 2.3.

(ii) This is similar to (i).

To avoid some long phrases later on, we make the following definition(s).

DEFINITIONS The positive integer k is said to be a **BNCLT number**, an **McC number**, or an **NCLT number** respectively iff there exists a group of order k which is BNCLT, a McCarthy group, or a non-CLT group, respectively.

These definitions are nonstandard (frequently the positive integer k is called a $\mathcal{C}$ number where $\mathcal{C}$ is some class of groups iff *all* groups of order k are in $\mathcal{C}$) but we will adhere to them throughout the remainder of the chapter.

We wish to emphasize k is an McC number iff it has a factorization of the form $k = p^a n$ where p is a prime, $p \nmid n$, and $2 \leq a = \exp(p, n)$.

To summarize some of our work in this section, let $\mathcal{BNCLT}$, $\mathcal{M}_c\mathcal{C}$, and $\mathcal{NCLT}$ denote, respectively, the classes of all BNCLT, McCarthy, and non-CLT groups. Then $\mathcal{BNCLT} \subset \mathcal{M}_c\mathcal{C} \subset \mathcal{NCLT}$. (The inclusions follow from Theorems 2.2 and 2.25. We have shown the first inclusion is proper with a group of order 200. To show the second inclusion is proper we offer the reader a choice of examples: SL(2, 3) of order 24 or $A_4 \times \mathbb{Z}_5$, order 60.)

Now we give some numerical results. The table below lists all the McC numbers ≤ 1000 (we hope). In other words, the table gives those positive integers ≤ 1000 which have a factorization of the form $p^a n$, where p is a prime, $p \nmid n$, and $2 \leq a = \exp(p, n)$. All BNCLT numbers in the table are underlined.

$\underline{12 = 2^2 \cdot 3}$	$240 = 2^4 \cdot 15$	$600 = 5^2 \cdot 24$
$36 = 3^2 \cdot 4$	$300 = 5^2 \cdot 12$	$702 = 3^3 \cdot 26$
$\underline{56 = 2^3 \cdot 7}$	$\underline{351 = 3^3 \cdot 13}$	$726 = 11^2 \cdot 6$
$72 = 3^2 \cdot 8$	$\underline{363 = 11^2 \cdot 3}$	$784 = 7^2 \cdot 16$
$\underline{75 = 5^2 \cdot 3}$	$392 = 7^2 \cdot 8$	$810 = 3^4 \cdot 10$
$\underline{80 = 2^4 \cdot 5}$	$\underline{405 = 3^4 \cdot 5}$	$\underline{867 = 17^2 \cdot 3}$
$150 = 5^2 \cdot 6$	$484 = 11^2 \cdot 4$	$968 = 11^2 \cdot 8$
$196 = 7^2 \cdot 4$	$576 = 2^6 \cdot 9$	$\underline{992 = 2^5 \cdot 31}$
$200 = 5^2 \cdot 8$	$588 = 7^2 \cdot 12$	

3. Primitive Nonsupersolvable Numbers, Minimal Nonsupersolvable Groups, Pazderski's Theorem, and More Numerical Results

The material in this section has three major goals:

(1) We shall show that there exists a set of positive integers W such that (i) $W \neq \emptyset$. (ii) There exists a nonsupersolvable group of order n iff there is some $w \in W$ such that $w \mid n$. (iii) There exists a non-CLT group of order n only if there is some $w \in W$ such that $w \mid n$. (iv) For all $w \in W$ there exists a non-CLT group of order w. (v) W is minimal in the sense that if $a, b \in W$ with $a \neq b$, then $a \nmid b$.

(2) We shall characterize W in terms of prime factorizations of all the members of W.

(3) We shall characterize all of the positive integers n such that n is divisible by exactly two primes and there exists a nonsupersolvable group of order n.

NOTATION For all positive integers n, let $D(n) = \{d : d$ is a positive integer and $d \mid n\}$ and let $D^1(n) = D(n) \backslash \{n\}$.

DEFINITION AND NOTATION A positive integer n is said to be a **PNSS (primitive nonsupersolvable) number** iff there exists a nonsupersolvable group of order n and for all $d \in D^1(n)$, every group of order d is supersolvable. Let W denote the set of all PNSS numbers.

LEMMA 3.1 $12 \in W$, so that $W \neq \emptyset$.
Proof First of all, A_4 is not supersolvable (since it is not CLT), and $|A_4| = 12$. Now all groups of order 1, 2, or 3 are cyclic, all groups of order 4 are Abelian, and all noncyclic groups of order 6 are isomorphic to S_3. Thus for all $d \in D^1(12)$, all groups of order d are supersolvable.

LEMMA 3.2 *For all positive integers n, there is a nonsupersolvable group of order n iff there is some $w \in W$ such that $w \mid n$.*
Proof: First suppose $w \in W$ such that $w \mid n$, say $n = wk$. Then there is a nonsupersolvable group H of order w. If we set $G = H \times \mathbb{Z}_k$, then G is not supersolvable and $|G| = wk = n$.

Now suppose there exists a nonsupersolvable group of order n. Then set $J(n) = \{d : d \in D(n)$ and there exists a nonsupersolvable group of order $d\}$ is nonempty and hence has a smallest element, which we label w. Since $w \in J(n)$, $w \mid n$. It remains to show that $w \in W$. First, since $w \in J(n)$, there exists a nonsupersolvable group of order w. Next, for any $d \in D^1(w)$, $d \notin J(n)$ (since $d < w$) and hence every group of order d is supersolvable.

LEMMA 3.3 *If there exists a non-CLT group of order n, then there is some $w \in W$ such that $w \mid n$.*
Proof: Any non-CLT group is a nonsupersolvable group, and our conclusion follows immediately from Lemma 3.2.

We pause to make several comments. First of all, Lemma 3.3 gives a necessary condition on any positive integer n for n to be a non-CLT number. As far as we know, it is an unsolved problem as to whether or not said necessary condition is also sufficient; one of the big stumbling blocks here is the fact that the class of all CLT groups is *not* subgroup-closed.

We have shown that (i), (ii), and (iii) of (1) at the beginning of this section are all true; to do (iv) of (1), we introduce a class of finite groups for which supersolvable and CLT are equivalent.

DEFINITION We call the group G **special** iff all of its proper subgroups are supersolvable.

LEMMA 3.4 *Let G be a special group. Then G is CLT iff G is supersolvable.*

Proof: Any supersolvable group is CLT (see Theorem 1.3), so suppose that G is any special CLT group. Then every proper subgroup of G is supersolvable, hence CLT. Therefore every subgroup of G is CLT, so G is supersolvable by Theorem 1.1.

DEFINITION We say that the group G is an **MNSS (minimal nonsupersolvable)** group iff G is special and not supersolvable.

Thus G is a MNSS group iff it is a nonsupersolvable group such that every proper subgroup is supersolvable. We note that if G is any BNCLT group, then G is MNSS.

LEMMA 3.5 *If G is a group such that $|G|$ is PNSS, then G is special.*

Proof: If H is any proper subgroup of G, then $|H| \in D^1(w)$ so that H is supersolvable.

LEMMA 3.6 *If G is a group such that $|G|$ is PNSS and G is not supersolvable, then G is MNSS and non-CLT.*

Proof: Since G is special (by Lemma 3.5) and nonsupersolvable, it is MNSS. Lemma 3.4 shows it is non-CLT.

LEMMA 3.7 *W is minimal in the sense that if $a, b \in W$ and $a \neq b$, then $a \nmid b$.*

Proof: Suppose to the contrary that $a \mid b$. Then $a \in D^1(b)$. This and b is PNSS imply that every group of order a is supersolvable. This contradicts the fact that a is PNSS.

We note that (iv) of (1) at the beginning of this section follows immediately from Lemma 3.6; also, (v) of (1) follows immediately from Lemma 3.7.

We now start a discussion whose ultimate goal is a characterization in terms of prime factorizations of all of the members of W. Our major tool in obtaining said characterization is a beautiful theorem (due to Pazderski), which characterizes all positive integers n such that all groups of order n are supersolvable. This theorem is stated and proved in [Pazderski 1959]; this paper also contains a characterization of all positive integers n such that all groups of order n are nilpotent.

The last two characterizations that we have just mentioned involve a function which we define as follows:

DEFINITION ψ is the function defined on the set of positive integers by the following rules:

(i) $\psi(1) = 1$.
(ii) For p a prime, $\psi(p^a) = \prod_{i=1}^{a}(p^i - 1)$.
(iii) ψ is multiplicative; i.e., $\psi(mn) = \psi(m)\psi(n)$ whenever $(m, n) = 1$.

To indicate the versatility of ψ, we state the following:

THEOREM 3.8
(i) *All groups of order n are cyclic iff n is square-free and $(n, \psi(n)) = 1$.*
(ii) *All groups of order n are Abelian iff n is cube-free and $(n, \psi(n)) = 1$.*
(iii) *All groups of order n are nilpotent iff $(n, \psi(n)) = 1$.*
For a proof, see [Scott 1964, 9.2.7].

The various parts of Theorem 3.8 give short and sweet characterizations; unfortunately, our next characterization is quite a bit more complicated.

THEOREM 3.9 (Pazderski)   *There exists a nonsupersolvable group of order n iff at least of one of the following holds:*
(1) There are primes p, q and a positive integer s such that $p^s \| n$, $q \mid (n, \psi(p^s))$, and $q \nmid (p - 1)$.
(2) There are primes p, q and positive integers s, u such that $p^s \| n$, $q \leq s$, $q^u \mid (n, \psi(p^s))$, and $q^u \nmid (p - 1)$.
(3) There are primes p, q and positive integers s, m such that $p^s \| n$, $q \leq s$, $mq \mid (n, p - 1)$, and $q \mid \varphi(m)$.

LEMMA 3.10   *If n is any positive integer such that there exists a nonsupersolvable group of order n, then at least one of the following must hold:*
(i) *There are distinct primes p, q such that $p^a q \mid n$ and $a \geq 2$ where $a = \exp(p, q)$.*
(ii) *There are distinct primes p, q and a positive integer b such that $q^b \| (p - 1)$ and $p^q q^{b+1} \mid n$.*
(iii) *There are distinct primes, p, q, such that $p \equiv 1 \bmod q^3$ and $p^q q^3 \mid n$.*
(iv) *There are distinct primes p, q, r such that $p \equiv 1 \bmod qr$, $r \equiv 1 \bmod q$, and $p^q q r \mid n$.*
Proof: We have (from Theorem 3.9) that at least one of (1), (2), or (3) of Theorem 3.9 must hold. We will show that (1) forces (i); (2) forces (i) or (ii); and (3) forces (iii) or (iv).

Suppose (1) of Theorem 3.9 holds, i.e., there are primes p, q and a positive integer s such that $p^s \| n$, $q \mid (n, \psi(p^s))$, and $q \nmid (p - 1)$. First we want to show $p \neq q$. Since $q \mid (n, \psi(p^s))$, we get $q \mid \psi(p^s)$. But $\psi(p^s) = \prod_{i=1}^{s}(p^i - 1)$ and since p does not divide any factor of this product, $p \nmid \psi(p^s)$. Therefore $p \neq q$. Now let $a = \exp(p, q)$. Since $q \nmid (p - 1)$, we have $2 \leq a$. Now all that needs to be shown is $p^a q \mid n$. Since $q \mid (n, \psi(p^s))$, we have $q \mid n$. So it suffices to show that $p^a \mid n$. From $q \mid \psi(p^s)$ we have $q \mid \prod_{i=1}^{s}(p^i - 1)$. Hence $q \mid (p^k - 1)$ for some k, $1 \leq k \leq s$. Now $p^k \equiv 1 \bmod q$ implies $a \leq k$ by definition of exponent. Conclude then that $a \leq s$. This and $p^s \mid n$ imply $p^a \mid n$ as we wanted to show.

We now suppose that (2) of Theorem 3.9 holds, i.e., there are primes p, q and positive integers s, u such that $p^s \| n$, $q \leq s$, $q^u | (n, \psi(p^s))$, and $q^u \nmid (p - 1)$. As above we first note that $p \neq q$ (q divides q^u which divides $(n, \psi(p^s))$ which divides $\psi(p^s)$, but $p \nmid \psi(p^s)$). We now consider the two cases: $q \nmid (p - 1)$ and $q | (p - 1)$. In the first case, (1) of Theorem 3.9 holds so that (as we showed above) (i) must hold.

Assume now that $q | (p - 1)$. Define b by $q^b \| (p - 1)$. Now all that needs to be shown in order to demonstrate (ii) is that $p^q q^{b+1} | n$. Since $q \leq s$ and $p^s | n$, we immediately have $p^q | n$. Thus it suffices to show $q^{b+1} | n$. Because $q^u \nmid (p - 1)$ we cannot have $u \leq b$. Thus $b + 1 \leq u$ and so $q^{b+1} | n$ because $q^u | n$.

We now suppose that (3) of Theorem 3.9 holds, i.e., there are primes p, q and positive integers s, m such that $p^s \| n$, $q \leq s$, $mq | (n, p - 1)$, and $q | \varphi(m)$. We first note that $p \neq q$ ($p = q$ would imply $p | mq$ hence $p | (p - 1)$). Now let $m = r_1^{a_1} r_2^{a_2} \ldots r_t^{a_t}$ be the prime factorization of m. Therefore $\varphi(m) = \prod_{i=1}^t \varphi(r_i^{a_i}) = \prod_{i=1}^t r_i^{c_i}(r_i - 1)$ where each $c_i = a_i - 1$. Since $q | \varphi(m)$, we have either $q | r_i^{c_i}$ for some i, $1 \leq i \leq t$ or else $q | (r_i - 1)$ for some i, $1 \leq i \leq t$. We will show that the first case implies (iii), while the second implies (iv).

In the first case note that $q = r_i$. Moreover, $q | r_i^{c_i}$ implies $c_i \neq 0$ so that $2 \leq a_i$. Now from $r_i^{a_i} | m$, $r_i = q$, and $2 \leq a_i$ we can conclude $q^2 | m$. Consequently $q^3 | mq$ so that $q^3 | (p - 1)$, i.e., $p \equiv 1 \bmod q^3$. Also $q^3 | mq$ implies $q^3 | n$. This and $p^q | n$ (because $q \leq s$ and $p^s | n$) imply $p^q q^3 | n$ and so (iii) holds.

We now assume the second case, $q | (r_i - 1)$. Set $r = r_i$. Then $q \nmid r$ so $q \neq r$. Also $p \neq r$ ($p = r$ would imply $p | m$, hence $p | mq$, hence $p | (n, p - 1)$, hence $p | (p - 1)$). Thus p, q, r are distinct. Next note $r | m$ so qr divides mq and hence divides $p - 1$, i.e., $p \equiv 1 \bmod qr$. That $r \equiv 1 \bmod q$ is immediate from $q | (r - 1)$. Finally $p^q | n$ (because $q \leq s$ and $p^s | n$) and $q | n$ (because $mq | (n, p - 1)$) and $r | n$ (because r divides m and m divides mq which divides n) so we conclude that $p^q qr | n$ completing the demonstration of (iv).

Lemma 3.10 will be very important to us in our upcoming characterization of the members of W. As we shall see, there are five different types of positive integers in W; the next two lemmas will give us a description of these types (except for some small subtleties which we shall deal with later).

LEMMA 3.11 *Let n be a positive integer such that* (ii) *of Lemma* 3.10 *holds. Then*:
(i) *If $b = 1$, then $q \| (p - 1)$ and $p^q q^2 | n$.*
(ii) *If $b = 2$, then $p^2 \| (p - 1)$ and $p^q q^3 | n$.*
(iii) *If $b \geq 3$, then* (iii) *of Lemma* 3.10 *holds; i.e., $p \equiv 1 \bmod q^3$ and $p^q q^3 | n$.*
Proof: Parts (i) and (ii) are immediate. Now if $b \geq 3$, then q^3 divides q^b

which divides $p - 1$ and $p^q q^3$ divides $p^q q^{b+1}$ which divides n, so (iii) follows.

LEMMA 3.12 *If there exists a nonsupersolvable group of order n, then at least one of the following must hold*:
(1) *There are distinct primes p, q such that $p^a q \mid n$ and $a \geqslant 2$ where $a = \exp(p, q)$.*
(2) *There are distinct primes p, q such that $q \| (p - 1)$ and $p^q q^2 \mid n$.*
(3) *There are distinct primes p, q such that $q^2 \| (p - 1)$ and $p^q q^3 \mid n$.*
(4) *There are distinct primes p, q such that $p \equiv 1 \bmod q^3$ and $p^q q^3 \mid n$.*
(5) *There are distinct primes p, q, r such that $p \equiv 1 \bmod qr$, $r \equiv 1 \bmod q$, and $p^q q^r \mid n$.*
Proof: Lemma 3.10 shows that at least one of (i), (ii), (iii), or (iv) of Lemma 3.10 must hold. Now, (1) is a restatement of (i) of Lemma 3.10, (4) is a restatement of (iii) of Lemma 3.10, and (5) is a restatement of (iv) of Lemma 3.10. If (ii) of Lemma 3.10 holds, then Lemma 3.11 shows that at least one of (2), (3), or (4) must hold, because (i) of Lemma 3.11 forces (2), (ii) of Lemma 3.11 forces (3), and (iii) of Lemma 3.11 forces (4).

After the above, the following should not come as much of a surprise.

DEFINITIONS The positive integer k is of **type 1** iff there are distinct primes p, q such that $k = p^a q$ and $a \geq 2$ where $a = \exp(p, q)$.
   k is of **type 2** iff there are distinct primes p, q such that $q \| (p - 1)$ and $k = p^q q^2$.
   k is of **type 3** iff there are distinct primes p, q such that $q^2 \| (p - 1)$ and $k = p^q q^3$.
   k is of **type 4** iff there are distinct primes p, q such that $p \equiv 1 \bmod q^3$ and $k = p^q q^3$.
   k is of **type 5** iff there are distinct primes p, q, r such that $p \equiv 1 \bmod qr$, $r \equiv 1 \bmod q$, and $k = p^q q^r$.

NOTATION For $i = 1, \ldots, 5$, let W_i denote $\{k : k$ is a type i positive integer$\}$. Also let $J = W_1 \cup \ldots \cup W_5$.

LEMMA 3.13 *Suppose $q^b \| (p - 1)$, where p, q are distinct primes. Then*
(1) $q \mid (1 + p + p^2 \ldots + p^{q-1})$.
(2) $p^q \equiv 1 \bmod q^{b+1}$.
(3) $\exp(p, q^{b+1}) = q$.
(4) $p^q q^{b+1}$ *is an McC number.*
Proof: (i) Since $q^b \mid (p - 1)$, we have $p \equiv 1 \bmod q^b$, and so $p^t \equiv 1 \bmod q^b$ for all positive integers t. Hence $1 + p + p^2 + \ldots + p^{q-1} \equiv 1 + 1 + \ldots + 1 \bmod q^b$, where the last sum contains q summands. Thus $1 + p + p^2 + \cdots + p^{q-1} \equiv q \bmod q^b$. This and $q \mid q^b$ imply (i).
   (ii) Since $q^b \mid (p - 1)$ and q divides $1 + p + p^2 + \ldots + p^{q-1}$, we get q^{b+1} divides $(p - 1)(1 + p + p^2 + \ldots + p^{q-1}) = p^q - 1$, which proves (ii).

(iii) Since $p \neq q$, it follows that $(p, q^{b+1}) = 1$ and hence it makes sense to set $e = \exp(p, q^{b+1})$. Now, it follows from (ii) that $e \mid q$. Since q is a prime, it follows that $e = 1$ or q. Since $q^b \| (p - 1)$, we must have p not congruent to $1 \bmod q^{b+1}$ and hence $e \neq 1$; it follows that $e = q$.

(iv) This follows immediately from (iii) and the definition of McC number.

LEMMA 3.14
(i) $k \in W_1$ iff k is a BNCLT *number*.
(ii) *If* $k \in W_2 \cup W_3$, *then* k *is an McC number.*
(iii) *If* $k \in W_4$, *then there exists a nonsupersolvable group of order* k.
(iv) *If* $k \in W_5$, *then there exists a nonsupersolvable group of order* k.
(v) *If* $j \in J$, *then there exists a nonsupersolvable group of order* j.
Proof: (i) follows from Theorem 2.3.

(ii) This follows from Lemma 3.13 (set $b = 1$ in that lemma if $k \in W_2$; set $b = 2$ if $k \in W_3$).

We shall do both (iii) and (iv) by using (3) of Theorem 3.9.

(iii) We have $k = p^q q^3$, where p, q are distinct primes such that $p \equiv 1 \bmod q^3$. Set $s = q$ and $m = q^2$, so that $p^s \| k$. Moreover $(k, p - 1) = q^3$ because $q^3 \mid (p - 1)$ and $p \nmid (p - 1)$. Thus $(k, p - 1) = mq \mid (k, p - 1)$. Finally $\varphi(m) = \varphi(q^2) = q(q - 1)$ so that $q \mid \varphi(m)$. Hence k satisfies (3) of Theorem 3.9.

(iv) We have $k = p^q qr$, where p, q, r are distinct primes such that $p \equiv 1 \bmod qr$ and $r \equiv 1 \bmod q$. Set $s = q$ and $m = r$, so that $p^s \| k$. Now we have $(k, p - 1) = qr$ because $qr \mid (p - 1)$ and $p \nmid (p - 1)$. Thus $mq \mid (k, p - 1)$. Finally $\varphi(m) = \varphi(r) = r - 1$ so $q \mid \varphi(m)$. Hence k satisfies (3) of Theorem 3.9.

(v) This follows immediately from (i)–(iv).

In Section 4 we shall use semi-direct products (of types that are different from the ones which we considered in Section 2) to construct a non-CLT group of order k for $k \in W_4 \cup W_5$.

LEMMA 3.15 *There is a nonsupersolvable group of order n iff there exists some $j \in J$ such that $j \mid n$.*
Proof: If there is a nonsupersolvable group of order n, then there is some $j \in J$ such that $j \mid n$ by Lemma 3.12.

Conversely, suppose $j \in J$ and $j \mid n$, say $n = jk$. By Lemma 3.14, there is a nonsupersolvable group H of order j. Then $G = H \times \mathbb{Z}_k$ is the required nonsupersolvable group of order n.

Our next lemma shows that any PNSS number must be of one of the five types.

LEMMA 3.16 $W \subseteq J$.

Proof: Let $w \in W$. Then there is a nonsupersolvable group of order w, so Lemma 3.15 produces a number $j \in J$ with $j \mid w$. It follows from Lemma 3.14 that there exists a nonsupersolvable group of order j. Since $j \mid w$, $j \in D(w)$. But $j \notin D^1(w)$, since there is a nonsupersolvable group of order j. Conclude $j = w$ so that $w \in J$.

Our next lemma shows that W and J are not identical.

LEMMA 3.17 $36 = 3^2 2^2$ *where* 2, 3 *are distinct primes such that* $2 \| (3 - 1)$. *Hence* 36 *is of type* 2, *so* $36 \in J$. *Now,* $12 \mid 36$ *and there is a nonsupersolvable group of order* 12, *namely* A_4, *and it follows that* 36 *is not a PNSS number, i.e.,* $36 \notin W$.

We showed (in the above proof) that $36 \in W_2$, and it follows from (ii) of Lemma 3.14 that 36 is an McC number; the reader may also check that 36 appears in the table at the end of Section 2. Thus, we see that 36 is a nonsupersolvable number, but is not primitive.

Another way of looking at this situation is that there are numbers a, $b \in J$ (namely $a = 12$, $b = 36$) such that $a \mid b$ but $a \neq b$. As a matter of fact, this example is the only one possible; we shall take some time and effort to prove the following theorem:

THEOREM 3.18 *For all* $x, y \in J$ *if* $x \mid y$, *then either* $x = y$ *or* $x = 12$ *and* $y = 36$.

Our major reason for wishing to prove Theorem 3.18 is that the following characterization of W can be obtained from it with just a little work.

THEOREM 3.19 $W = J \backslash \{36\}$.

We shall first indicate how Theorem 3.19 follows from our previous results and Theorem 3.18, and we shall then prove Theorem 3.18 by using a sequence of lemmas.

First of all, Lemmas 3.16 and 3.17 show that $W \subseteq J$ and $36 \notin W$, and it follows that $W \subseteq J \backslash \{36\}$. To get the reverse inclusion, let $j \in J \backslash \{36\}$. Then Lemma 3.14 shows that there exists a nonsupersolvable group of order j, and Lemma 3.2 shows that there is some $w \in W$ with $w \mid j$. Now since $W \subseteq J$, $w \in J$. Thus Theorem 3.18 can be applied to conclude that $w = j$ and hence $j \subset W$.

We now attack Theorem 3.18 with a sequence of lemmas.

LEMMA 3.20 *If* p, q *are primes such that* $p \equiv 1 \bmod q$ *and* $q^2 \equiv 1 \bmod p$, *then* $p = 3$ *and* $q = 2$.

Proof: By the first congruence, $q \leq p - 1$, and hence $q - 1 < p$. By the second congruence, $p \mid (q - 1)(q + 1)$ so that either $p \mid (q - 1)$ or $p \mid (q + 1)$. But $p \mid (q - 1)$ is impossible as it contradicts $q - 1 < p$. Therefore $p \mid (q + 1)$. Thus we have $p \leq q + 1 \leq p$ so that $p = q + 1$. The result follows.

LEMMA 3.21 *If p, q are primes, then the following are equivalent:*
(i) $p \equiv 1 \mod q$ *and* $q^3 \equiv 1 \mod p$.
(ii) $p = q^2 + q + 1$.
Proof: (ii) $\Rightarrow$ (i) By (ii), $p - 1 = q^2 + q$ so that $q \mid (p - 1)$, hence $p \equiv 1 \mod q$. Also, (ii) implies $p(q - 1) = (q^2 + q + 1)(q - 1) = q^3 - 1$ so $q^3 \equiv 1 \mod p$.

(i) $\Rightarrow$ (ii) Since $p \equiv 1 \mod q$, $q \mid (p - 1)$, so that $q \leq p - 1$. It follows that $q - 1 < p$ and hence $p \nmid (q - 1)$. Since $q^3 \equiv 1 \mod p$, $p \mid (q^3 - 1)$, so that $p \mid (q - 1)(q^2 + q + 1)$. This and $p \nmid (q - 1)$ imply $p \mid (q^2 + q + 1)$, say $q^2 + q + 1 = kp$. We endeavor to show $k = 1$ which will complete the proof. Now $q^2 + q + 1 \equiv 1 \mod q$ and hence $kp \equiv 1 \mod q$. Since $p \equiv 1 \mod q$, it follows that $kp \equiv k \mod q$. Consequently we have $k \equiv 1 \mod q$, so that $q \mid (k - 1)$. We now assume $k \neq 1$ and seek a contradiction. This assumption and $q \mid (k - 1)$ imply $q \leq k - 1$ so that $q + 1 \leq k$. This and $q + 1 \leq p$ yield $(q + 1)^2 \leq kp$. Thus $q^2 + 2q + 1 \leq q^2 + q + 1$ and so $2q \leq q$ from whence $q \leq 0$, the desired contradiction.

We pause to note that Lemma 3.21 is not vacuous. Examples of primes that satisfy the relations of this lemma are $p = 7$, $q = 2$, $p = 13$, $q = 3$; and $p = 31$, $q = 5$.

LEMMA 3.22 *If p, q are primes, then we cannot have both $p \equiv 1 \mod q^2$ and $q^3 \equiv 1 \mod p$.*
Proof: Suppose both congruences hold and seek a contradiction. We have $p \equiv 1 \mod q$ and $q^3 \equiv 1 \mod p$, so Lemma 3.21 implies that $p = q^2 + q + 1$. We have then $q^2 + q + 1 \equiv 1 \mod q^2$ from which follows the contradiction $q \mid 1$.

LEMMA 3.23 *If p, q are primes, then $p \equiv 1 \mod q^2$ implies $\exp(q, p) \geq 4$.*
Proof: First we note that $p \neq q$ so $\exp(q, p)$ makes sense ($p = q$ would imply $p \equiv 1 \mod p^2$, hence $p^2 \leq p - 1 < p$). It suffices to show q^i is not congruent to 1 mod p for $i = 1, 2, 3$. We have $q^2 \mid (p - 1)$ and hence

$$q - 1 < q^2 - 1 < q^2 \leq p - 1 < p.$$

This implies q is not congruent to 1 mod p and q^2 is not either. Finally Lemma 3.22 and the congruence $p \equiv 1 \mod q^2$ imply q^3 is not congruent to 1 mod p completing the proof.

The following lemma will save us from a lot of repetition in some of our upcoming work.

LEMMA 3.24 *For primes p and q we have:*
(i) *If* $p \equiv 1 \bmod q$, *then* $p > q$.
(ii) *If* $p > q$, *then* q *is not congruent to* $1 \bmod p$.
(iii) *If* $p > q$, *then* $\exp(q, p) \geq 2$.
Proof: (i) Since $q \mid (p - 1)$, $q \leq p - 1$; it follows that $q < p$.
  (ii) This follows from (i) if we interchange the roles of p and q.
  (iii) This follows from (ii).

We shall be using Lemma 3.24 over and over again (sometimes without explicit mention).

LEMMA 3.25 $n \in W_3 \cup W_4$ *iff* $n = p^q q^3$ *for distinct primes* p, q *such that* $p \equiv 1 \bmod q^2$.
Proof: The only if direction is immediate. Conversely, suppose $n = p^q q^3$, where $p \equiv 1 \bmod q^2$. Thus $q^2 \mid (p - 1)$. Now it is either true that $q^2 \| (p - 1)$, or it isn't. If it is true, then $n \in W_3$. If it isn't, then $q^3 \mid (p - 1)$ so $p \equiv 1 \bmod q^3$ and $n \in W_4$.

We note that Lemma 3.25 gives a single description of all of the members of $W_3 \cup W_4$, so that we may view the positive integers of type 3 or 4 as being of a single type, and thus view J as consisting of positive integers of just four types rather than five types. We shall use the four-types view of J in the rest of this section; we shall indicate later (in Section 4) our reasons for our originally adopting the five-types view of J.

DEFINITION We say that the positive integer n is a **t.p. (two prime) number** iff n is divisible by exactly two primes, i.e., iff $n = p^c q^d$ for distinct primes p, q.

LEMMA 3.26
(1) *If* $j \in W_5$ *and* x *is a t.p. number, then* $j \nmid x$.
(2) *If* $j \in W_5$ *and* $k \in J \setminus W_5$, *then* $j \nmid k$.
(3) *If* $a, b \in W_5$, *then* $a \mid b$ *implies* $a = b$.
Proof: (1) is immediate from the definition of type 5 number and (2) follows from (1) because every number in $J \setminus W_5$ is a t.p. number.
  (3) We are given $a = s^t tu$ and $b = p^q qr$ where s, t, u are distinct primes; p, q, r are distinct primes; $s \equiv 1 \bmod tu$; $p \equiv 1 \bmod qr$; $u \equiv 1 \bmod t$; and $r \equiv 1 \bmod q$. We are also given that $a \mid b$. Since $s \equiv 1 \bmod tu$, $s \equiv 1 \bmod u$, so $s > u$. Since $u \equiv 1 \bmod t$, $u > t$. Similarly, $p > r$ and $r > q$. Since $s^t tu \mid p^q qr$, the four inequalities just established imply that $s = p$, $t = q$, and $u = r$ because of uniqueness of factorization of integers into primes. Conclude $a = b$.

LEMMA 3.27 *If* $x \in W_1$ *and* $y \in W_5$, *then* $x \nmid y$.
Proof: We are given that $x = s^a t$ where s, t are distinct primes such that

$a = \exp(s, t) \geq 2$ and that $y = p^q qr$ where p, q, r are distinct primes such that $p \equiv 1 \bmod qr$ and $r \equiv 1 \bmod q$.

Now suppose $x \mid y$, i.e., $s^a t \mid p^q qr$. Then since a, $q \geq 2$, we get $s = p$ and $t = q$ or r. Now the congruence $p \equiv 1 \bmod qr$ implies both $p \equiv 1 \bmod q$ and $p \equiv 1 \bmod r$. It follows then (from $t = q$ or r) that $s \equiv 1 \bmod t$. This contradiction (to $\exp(s, t) \geq 2$) completes the proof.

LEMMA 3.28 If $x \in W_2$, $y \in W_3 \cup W_4$, $k \in W_1$, and $t \in W_5$, then $x \nmid k$, $y \nmid k$, $x \nmid t$, and $y \nmid t$.

Proof: We are given $x = p^q q^2$ where p, q are distinct primes and $y = r^s s^3$ where r, s are distinct primes. Since q, $s \geq 2$, it follows that $p^2 q^2 \mid x$ and $r^2 s^2 \mid y$. The reader may check that neither k nor t is divisible by the product of the squares of two distinct primes, and our conclusion follows.

To save ourselves a lot of repetition, we put down some notational conventions. From now on (until further notice) we shall use the letters a, b, c, and d to denote positive integers and we shall use the letters s, t, p, and q to denote primes *with the explicit restrictions that $s \neq t$ and $p > q$.*

LEMMA 3.29 *Let x and y be any two* t.p. *numbers say $x = s^a t^b$ and $y = p^c q^d$, and suppose that $x \mid y$. Then*
(i) *Either $s = p$, $t = q$, $a \leq c$, and $b \leq d$; or $s = q$, $t = p$, $a \leq d$, and $b \leq c$.*
(ii) *If s is not congruent to $1 \bmod t$ and $q \| (p - 1)$, then $s = q$, $t = p$, and $s \| (t - 1)$.*
(iii) *If s is not congruent to $1 \bmod t$ and $q^2 \mid (p - 1)$, then $s = q$, $t = p$, and $s^2 \mid (t - 1)$.*
(iv) *If $t \| (s - 1)$, then $s = p$, $t = q$, and $q \| (p - 1)$.*
(v) *If $t^2 \mid (s - 1)$, then $s = p$, $t = q$, and $q^2 \mid (p - 1)$.*
Proof: (i) Since $s^a t^b \mid p^c q^d$, it follows from the uniqueness of factorization of integers into primes that either $s = p$ and $t = q$ or else $s = q$ and $t = p$. In the first case, $p^a q^b \mid p^c q^d$ so $a \leq c$ and $b \leq d$. Similarly the second case implies $a \leq d$ and $b \leq c$.

(ii) Since $q \| (p - 1)$, $p \equiv 1 \bmod q$. This and s not congruent to $1 \bmod t$ mean that it is impossible for both $s = p$ and $t = q$ to hold simultaneously, i.e., the first alternative of (i) cannot hold. Conclude then that $s = q$ and $t = p$. This and $q \| (p - 1)$ further imply $s \| (t - 1)$.

(iii) Similar to (ii).

(iv) Since $t \| (s - 1)$, we have $s \equiv 1 \bmod t$, and hence $s > t$. This and our convention $p > q$ mean that it is impossible for both $s = q$ and $t = p$ to hold simultaneously, i.e., the second alternative of (i) cannot hold. Conclude then that $s = p$ and $t = q$. This and $t \| (s - 1)$ further imply $q \| (p - 1)$.

(v) Similar to (iv).

We assure the reader that we have good reasons for stating Lemma 3.29 as we did; these should become clear as we proceed.

LEMMA 3.30 (1) *Let* $x \in W_3 \cup W_4 \cup W_5$ *and let* y *be any* t.p. *number, say* $y = p^c q^d$, *such that* $q \,\|\, (p - 1)$. *Then* $x \nmid y$.

(2) *If* $x \in W_3 \cup W_4$ *and* $k \in W_2$, *then* $x \nmid k$.

Proof: (1) First note we have already shown $x \in W_5$ implies $x \nmid y$ ((1) of Lemma 3.26). Assume then that $x \in W_3 \cup W_4$. Hence $x = s't^3$ where $t^2 \,|\, (s - 1)$. If $x \,|\, y$, then (v) of Lemma 3.29 would imply $q^2 \,|\, (p - 1)$, a contradiction to our hypothesis $q \,\|\, (p - 1)$. Thus $x \nmid y$.

(2) Since $k \in W_2$, $k = p^c q^2$ with $q \,\|\, (p - 1)$. Thus the hypotheses of (1) are satisfied and we conclude that $x \nmid k$.

LEMMA 3.31 (1) *Let* $x \in W_2 \cup W_5$ *and let* y *be any* t.p. *number, say* $y = p^c q^d$, *such that* $q^2 \,|\, (p - 1)$. *Then* $x \nmid y$.

(2) *If* $x \in W_2$ *and* $m \in W_3 \cup W_4$, *then* $x \nmid m$.

Proof: (1) First note we have already shown $x \in W_5$ implies $x \nmid y$ ((1) of Lemma 3.26). Assume then that $x \in W_2$. Hence $x = s't^2$, where $t \,\|\, (s - 1)$. If $x \,|\, y$, then (iv) of Lemma 3.29 would imply $q \,\|\, (p - 1)$, a contradiction to our hypotheses $q^2 \,|\, (p - 1)$. Thus $x \nmid y$.

(2) Since $m \in W_3 \cup W_4$, $m = p^q q^3$ with $q^2 \,|\, (p - 1)$. Thus the hypotheses of (1) are satisfied and we conclude that $x \nmid m$.

LEMMA 3.32

(i) *Let* $x, y \in W_1$. *Then* $x \,|\, y$ *implies* $x = y$.

(ii) *Let* $k, m \in W_2$. *Then* $k \,|\, m$ *implies* $k = m$.

(iii) *Let* $n, h \in W_3 \cup W_4$. *Then* $n \,|\, h$ *implies* $n = h$.

Proof: (i) We are given $x = s^a t$ and $y = r^b u$ where r and u are distinct primes, $a = \exp(s, t) \geq 2$ and $b = \exp(r, u) \geq 2$. Now if $s^a t \,|\, r^b u$, then $s = r$ and $t = u$. Hence also $a = b$. Thus $x = y$.

(ii) We are given $k = s't^2$ and $m = p^q q^2$ with $t \,\|\, (s - 1)$ and $q \,\|\, (p - 1)$. Now if $k \,|\, m$, then we may apply (iv) of Lemma 3.29 to conclude $s = p$ and $t = q$. Hence $k = m$.

(iii) We are given $n = s't^3$ and $h = p^q q^3$ with $s \equiv 1 \bmod t^2$ and $p \equiv 1 \bmod q^2$. Now if $n \,|\, h$, then we may apply (v) of Lemma 3.29 to conclude $s = p$ and $t = q$. Hence $n = h$.

LEMMA 3.33 *Let* x *be any member of* W_1, *say* $x = s^a t$ *with* $2 \leq a = \exp(s, t)$, *let* y *be any* t.p. *number, say* $y = p^c q^d$, *and suppose that* $x \,|\, y$. *Then*

(i) *If* p *is not congruent to* $1 \bmod q$, *then* $\exp(p, q) \geq 2$, $\exp(q, p) \geq 2$, *and either* $c \geq \exp(p, q)$ *or* $d \geq \exp(q, p)$.

(ii) *If* $p \equiv 1 \bmod q$, *then* $\exp(q, p) \geq 2$, $s = q$, $t = p$, *and* $d \geq \exp(q, p)$.

Proof: (i) Since p is not congruent to $1 \bmod q$, $\exp(p, q) \geq 2$; since $p > q$, $\exp(q, p) \geq 2$. Since x and y are both t.p. numbers and $x \,|\, y$, we may apply (i) of Lemma 3.29 to conclude that either $s = p$, $t = q$, and $a \leq c$; or $s = q$, $t = p$, and $a \leq d$. If the first alternative holds, then $c \geq a = \exp(s, t) = \exp(p, q)$. If the second alternative holds, then $\exp(q, p) = \exp(s, t) = a \leq d$.

(ii) Since $p > q$, $\exp(q, p) \geq 2$. Since $\exp(s, t) \geq 2$ (by hypothesis) it follows that s is not congruent to 1 mod t. Since $p \equiv 1$ mod q, we have that $q \mid (p - 1)$, hence either $q \parallel (p - 1)$ or $q^2 \mid (p - 1)$. We now may apply either (ii) or (iii) of Lemma 3.29 (whichever is appropriate) to conclude that $s = q$ and $t = p$. Since x and y are both t.p. numbers and $x \mid y$, it follows from (i) of Lemma 3.29 that we must have the *second* alternative in (i) of Lemma 3.29, and it follows that $a \leq d$. Finally $\exp(q, p) = \exp(s, t) = a \leq d$.

LEMMA 3.34 (i) *Let* $x \in W_1$ *and* $y \in W_2$. *Then* $x \mid y$ *iff* $x = 12$ *and* $y = 36$.

(ii) *Let* $m \in W_1$ *and* $n \in W_3 \cup W_4$. *Then* $m \nmid n$.

Proof: (i) We may write $x = s^a t$ and $y = p^q q^2$ where $2 \leq a = \exp(s, t)$ and $p \equiv 1$ mod q. Suppose $x \mid y$. Then we may apply (ii) of Lemma 3.33 (with $d = 2$) to conclude that $\exp(q, p) \geq 2$, $s = q$, $t = p$, and $2 \leq \exp(q, p)$. Thus $\exp(q, p) = 2$, and so $q^2 \equiv 1$ mod p. This together with $p \equiv 1$ mod q imply that $p = 3$ and $q = 2$ by Lemma 3.20. Consequently, $s = 2$, $t = 3$, and $a = \exp(2, 3) = 2$. Conclude that $x = 12$ and $y = 36$.

(ii) Suppose to the contrary that $m \mid n$. We may write $m = s^a t$ and $n = p^q q^3$, where $2 \leq a = \exp(s, t)$ and $p \equiv 1$ mod q^2 (so $p \equiv 1$ mod q). We may now apply (ii) of Lemma 3.33 (with $d = 3$) to conclude that $3 \geq \exp(q, p)$. Since $p \equiv 1$ mod q^2, however, Lemma 3.23 says that $\exp(q, p) \geq 4$. This contradiction completes the proof.

LEMMA 3.35 (i) *If* $x \in W_i$ *and* $y \in W_j$, *where* $1 \leq i, j \leq 5$ *and* $i \neq j$, *then* $x \nmid y$ *except when* $i = 1, j = 2$, $x = 12$, $y = 36$.

(ii) *If* $x, y \in W_i$, *where* $1 \leq i \leq 5$, *and* $x \mid y$, *then* $x = y$.

Proof: (i) follows from Lemmas 3.26, 3.27, 3.28, 3.30, 3.31, and 3.34. (ii) follows from Lemmas 3.26 and 3.32. (The reader should check that these lemmas actually *do* cover all of the possible cases which must be considered.)

We now note that Theorem 3.18 follows immediately from Lemma 3.35. Since we have already shown how Theorem 3.19 follows from Theorem 3.18, we have now established both Theorem 3.18 and 3.19.

The reader is invited to check that the following is just Theorem 3.19 stated somewhat differently.

THEOREM 3.36 *The positive integer n is a PNSS number iff $n \neq 36$ and n is one of the five types listed in the definitions following Lemma 3.12.*

We consider that Theorem 3.36 gives a characterization of PNSS numbers (at least in terms of prime factorizations).

The following table lists all of the PNSS numbers ≤ 1000.

$$12 = 2^2 \cdot 3 \qquad\qquad 196 = 7^2 \cdot 2^2 \qquad\qquad 405 = 3^4 \cdot 5$$

$$56 = 2^3 \cdot 7 \qquad\qquad 200 = 5^2 \cdot 2^3 \qquad\qquad 484 = 11^2 \cdot 2^2$$

$$75 = 5^2 \cdot 3 \qquad\qquad 294 = 7^2 \cdot 2 \cdot 3 \qquad\qquad 867 = 17^2 \cdot 3$$

$$80 = 2^4 \cdot 5 \qquad\qquad 351 = 3^3 \cdot 13 \qquad\qquad 992 = 2^5 \cdot 31$$

$$363 = 11^2 \cdot 3$$

We invite the reader to check the following statements:

(1) n is underlined in the table iff n is BNCLT number iff n is of type 1.

(2) 196 and 484 are of type 2.

(3) 200 is of type 3.

(4) 294 is of type 5.

(5) There are no type-4 numbers ≤ 1000; in fact, the smallest type-4 number is $2312 = 17^2 2^3$.

(6) Some of our early work in this section shows that: (i) For $n \leq 1000$, there exists a nonsupersolvable group of order n iff n is divisible by at least one positive integer in the table. (ii) For $n \leq 1000$, a necessary condition for the existence of a non-CLT group of order n is that n be divisible by at least one positive integer in the table. (iii) For every k in the table, there exists a non-CLT group of order k.

(7) 100 is not divisible by any positive integer in the table; it follows that all groups of order 100 are supersolvable. (We shall obtain this result below in a slightly different way.)

The next theorem fulfills a promise which we made in Section 1:

THEOREM 3.37 *If G is any finite group of square-free order, then G is supersolvable (and hence CLT).*

Proof: Suppose G is not supersolvable. Then Lemma 3.15 shows that there is some $j \in J$ such that j divides $|G|$. But a long look at the definition of type 1, type 2, ..., type 5 numbers should convince the reader that no member of J is square-free. Hence there is some prime p such that $p^2 \mid j$ and hence p^2 divides $|G|$ as well. This contradiction shows G must be supersolvable.

In connection with Theorem 3.37 we pause to show that there is a way to make sense out of the phrases: "The probability that a positive integer selected at random is square-free is approximately 3/5" or "Approximately 3/5 of all positive integers are square-free."

Let S denote the set of all square-free positive integers, and for positive integers n, let $I_n = \{x : x$ is a positive integer and $x \leq n\}$. Let $S(n) = |S \cap I_n|$, let R denote the set of all real numbers, and let s be the sequence given by $s(n) = S(n)/n$.

We note that $s(n) = |S \cap I_n|/|I_n|$ for all positive integers n. The point to all of this is that $s(n)$ may be interpreted as the probability that a positive integer selected at random from the set of the first n positive integers is square-free. Now, it is shown in Theorem 11.5, p. 293 of [Niven and Zuckerman 1980] that our sequence s has the limit (relative to the usual topology on R) $6/\pi^2$.

The reader may check that $6/\pi^2$ is approximately 0.608; since 0.608 is approximately $3/5$, we consider that we have shown at least one way of assigning reasonable meaning to our two phrases.

The reader may also check that exactly 608 of the first 1000 positive integers are square-free; i.e., $s(1000) = 0.608$. Combining this with Theorem 3.37 we see that only 392 of the fisrt 1000 positive integers are possible orders of nonsupersolvable groups. It turns out that exactly 128 of these 392 positive integers *are*, in fact, orders of finite nonsupersolvable groups; the reader is invited to use the table preceeding Theorem 3.37 to check that there are exactly 128 positive integers $n \leq 1000$ such that n is divisible by some PNSS number k.

As promised at the beginning of this section, we intend to prove a theorem which contains a characterization in terms of prime factorizations of all t.p. numbers n such that there exists a nonsupersolvable group of order n.

LEMMA 3.38 *Let y be any t.p. number, say $y = p^c q^d$, such that $q \nmid (p - 1)$. Then for all $x \in J \setminus W_1$, $x \nmid y$.*

Proof: We suppose, to the contrary, that $x \mid y$. Now (1) of Lemma 3.26 shows that no member of W_5 can divide y, so $x \notin W_5$. Thus we have either $x \in W_2$ or $x \in W_3 \cup W_4$. If $x \in W_2$, we may write $x = s't^2$, with $t \,\|\, (s - 1)$. We may now apply (iv) of Lemma 3.29 to conclude that $q \,\|\, (p - 1)$, so that $q \mid (p - 1)$, contrary to hypothesis. If $x \in W_3 \cup W_4$, then we may write $x = s't^3$, with $s \equiv 1 \bmod t^2$, so that $t^2 \mid (s - 1)$. Now (v) of Lemma 3.29 applies to show that $q^2 \mid (p - 1)$, so $q \mid (p - 1)$, again contrary to hypothesis.

LEMMA 3.39 *For primes p, q with $p > q$, each of the following gives sufficient conditions for the existence of a nonsupersolvable group of order $p^c q^d$.*
(i) $d \geq \exp(q, p)$.
(ii) $q \nmid (p - 1)$ *and* $c \geq \exp(p, q)$.
(iii) $q \,\|\, (p - 1)$, $c \geq q$, *and* $d \geq 2$.
(iv) $q^2 \mid (p - 1)$, $c \geq q$, *and* $d \geq 3$.
Proof: Let $E = \exp(p, q)$ and $e = \exp(q, p)$.

(i) In this case (iii) of Lemma 3.24 shows that $e \geq 2$; it follows that $q^e p$ is of type 1, so that $q^e p \in J$. Since $d \geq e$, it follows that $q^e p \mid p^c q^d$. Since $q^e p \in J$, it now follows from Lemma 3.15 that there exists a nonsupersolvable group of order $p^c q^d$.

(ii) Here we have p not congruent to 1 mod q, so that $E \geq 2$; it follows that $p^E q$ is of type 1, so that $p^E q \in J$. Since $c \geq E$, it follows that $p^E q \mid p^c q^d$. Since $p^E q \in J$, Lemma 3.15 now shows that there exists a nonsupersolvable group of order $p^c q^d$.

(iii) In this case $p^q q^2$ is of type 2, so that $p^q q^2 \in J$. Since $c \geq q$ and $d \geq 2$, it follows that $p^q q^2 \mid p^c q^d$; since $p^q q^2 \in J$, Lemma 3.15 strikes again.

(iv) The proof here is an easy modification of the proof in (iii).

LEMMA 3.40   *Let p and q be primes with $p > q$, and suppose that there exists a nonsupersolvable group of order $p^c q^d$. Then*
(i) $q \nmid (p - 1)$ *implies either $c \geq \exp(p, q)$ or $d \geq \exp(q, p)$.*
(ii) $q \| (p - 1)$ *implies either $c \geq q$ and $d \geq 2$ or else $d \geq \exp(q, p)$.*
(iii) $q^2 \mid (p - 1)$ *implies either $c \geq q$ and $d \geq 3$ or else $d \geq \exp(q, p)$.*
Proof: Let y denote $p^c q^d$. By Lemma 3.15 there is some $x \in J$ such that $x \mid y$.

(i) In this case we have $x \in W_1$ because if $x \in J \setminus W_1$, then by Lemma 3.38 $x \nmid y$. Since p is not congruent to 1 mod q (because $q \nmid (p - 1)$) we may apply (i) of Lemma 3.33 to conclude the result.

(ii) Here we must have $x \in W_1 \cup W_2$ for if not, then $x \in W_3 \cup W_4 \cup W_5$. But then we could apply (1) of Lemma 3.30 to conclude $x \nmid y$.

If $x \in W_2$, we may write $x = s't^2$ with $t \| (s - 1)$. We may now apply (iv) and (i) of Lemma 3.29 (with $b = 2$) to conclude $s = p$, $t = q$, $t \leq c$, and $2 \leq d$. Therefore $q \leq c$ and $2 \leq d$.

If $x \in W_1$, then since $p \equiv 1$ mod q (because $q \| (p - 1)$) we may apply (ii) of Lemma 3.33 to conclude $d \geq \exp(q, p)$.

(iii) Here we must have $x \notin W_2 \cup W_5$, for if $x \in W_2 \cup W_5$, then we could apply (1) of Lemma 3.31 to conclude that $x \nmid y$. Thus $x \in W_1 \cup W_3 \cup W_4$.

If $x \in W_3 \cup W_4$, then we may write $x = s't^3$ with $s \equiv 1$ mod t^2 so that $t^2 \mid (s - 1)$. We may now apply (i) to Lemma 3.29 (with $b = 3$) to conclude that $s = p$, $a = q$, $a \leq c$, and $3 \leq d$. Hence $q \leq c$ and $3 \leq d$.

If $x \in W_1$, then since $q \mid (p - 1)$, we have $p \equiv 1$ mod q. We may now apply (ii) of Lemma 3.33 to conclude that $d \geq \exp(q, p)$.

We may now combine the sufficient conditions in Lemma 3.39 with the necessary conditions in Lemma 3.40 to obtain:

THEOREM 3.41   *For p, q primes with $p > q$:*
(i) *If $q \nmid (p - 1)$, then there exists a nonsupersolvable group of order $p^c q^d$ iff $c \geq \exp(p, q)$ or $d \geq \exp(q, p)$.*
(ii) *If $q \| (p - 1)$, then there exists a nonsupersolvable group of order $p^c q^d$ iff either $c \geq q$ and $d \geq 2$ or else $d \geq \exp(q, p)$.*
(iii) *If $q^2 \mid (p - 1)$, then there exists a nonsupersolvable group of order $p^c q^d$ iff either $c \geq q$ and $d \geq 3$ or else $d \geq \exp(q, p)$.*

We consider that Theorem 3.41 gives a characterization (at least in terms of prime factorizations) of all t.p. numbers n such that there exists a nonsupersolvable group of order n.

We end this section by inviting the reader to use Lemmas 3.23 and 3.24 in conjunction with Theorem 3.41 to prove:

COROLLARY 3.42   *Let p, q be primes.*
(1) If $p \equiv 1 \bmod q$, *then all groups of order $p^x q$ are supersolvable.*
(2) If $p \equiv 1 \bmod q^2$, *then all groups of order $p^x q^2$ are supersolvable.*
(3) If $p \equiv 1 \bmod q^3$, *then all groups of order $p^x q^3$ are supersolvable iff* $x \le q - 1$.

4.   Finite non-CLT Groups with Orders of Type 4 or 5

Suppose that k is any positive integer of type 1, 2, or 3. Our previous work in Sections 2 and 3 shows that for such k there exists a McCarthy group of order k, and our construction of McCarthy groups shows how to build at least one non-CLT group of order k. Our previous work in Section 3 also shows that if m is any positive integer of type 4 or 5, then there exists a non-CLT group of order m, but we have so far not given any examples of such groups. The major purpose of the present section is to prove a theorem which involves a construction of certain types of semi-direct products; special cases of said construction will give us examples of groups satisfying the title of this section (and other examples as well).

Before we state our theorem, we would like to give an example to motivate the construction which will appear in said theorem. We give the example by stating three lemmas (followed by quick sketches of proofs). We first note that $294 = 7^2 \cdot 2 \cdot 3$, and we invite the reader to use the table following Theorem 3.36 to show that 294 is the smallest positive integer of type 5.

LEMMA 4.1   *If T is an automorphism of $\mathbb{Z}_7^2$ of order 1, 2, 3, or 6, then:*
(i) *There is a basis $B \subseteq V_7^2$ such that the matrix of T relative to B is diagonal.*
(ii) *There is a one-dimensional subspace M of V_7^2 which is invariant under T.*
(iii) *There is a subgroup D of $\mathbb{Z}_7^2$ such that $|D| = 7$ and D is invariant under T.*

LEMMA 4.2   *For any homomorphism $\theta : \mathbb{Z}_6 \to \mathrm{Aut}(\mathbb{Z}_7^2)$, $\mathbb{Z}_7^2 \times_\theta \mathbb{Z}_6$ has a subgroup of order $7 \cdot 2 \cdot 3$.*

LEMMA 4.3   *If G is any group of order $7^2 \cdot 2 \cdot 3$, then:*
(i) *G is solvable and every proper subgroup is supersolvable.*
(ii) *G has a normal subgroup N of order 7^2.*
(iii) *G has subgroups K, L, M of orders $2 \cdot 3$, $7^2 \cdot 2$, and $7^2 \cdot 3$ respectively.*
(iv) *G has subgroups A and B of orders $7 \cdot 2$ and $7 \cdot 3$ respectively.*
(v) *$N \cong \mathbb{Z}_{49}$ or $\mathbb{Z}_7^2$ and $K \cong \mathbb{Z}_6$ or S_3.*
(vi) *There is a homomorphism $\theta : K \to \mathrm{Aut}(N)$ such that $G \cong N \times_\theta K$.*
(vii) *G has a subgroup of order d for every divisor d of $7^2 \cdot 2 \cdot 3$, with the possible exception of $d = 7 \cdot 2 \cdot 3$.*
(viii) *G is supersolvable iff G has a subgroup of order $7 \cdot 2 \cdot 3$.*
(ix) *If $N \cong \mathbb{Z}_{49}$, or $K \cong \mathbb{Z}_6$, or θ is not a monomorphism, then G is supersolvable.*
(x) *If G is not CLT, then there exists a monomorphism $\lambda : S_3 \to \mathrm{Aut}(\mathbb{Z}_7^2)$ such that $G \cong \mathbb{Z}_7^2 \times_\lambda S_3$.*
(xi) *If G is not CLT, then there exist monomorphisms $\beta : \mathbb{Z}_2 \to \mathrm{Aut}(\mathbb{Z}_3)$ and $\alpha : (\mathbb{Z}_3 \times_\beta \mathbb{Z}_2) \to \mathrm{Aut}(\mathbb{Z}_7^2)$ such that $G \cong \mathbb{Z}_7^2 \times_\alpha (\mathbb{Z}_3 \times_\beta \mathbb{Z}_2)$.*

Argument for Lemma 4.1   (i) If $f = x^6 - [1]_7$ in $\mathrm{GF}(7)[x]$, then f annihilates T, so that $m_T \mid f$. Since $6 \mid (7 - 1)$, there exists $w \in \mathrm{GF}(7)^*$ such that w is a primitive sixth root of $[1]_7$; since $7 \nmid 6$, it follows that f splits as a product of pairwise different monic linear polynomials in $\mathrm{GF}(7)[x]$. The same must be true of m_T, and our conclusion follows. (See Theorem 6, p. 204 of [Hoffman and Kunze 1971].)

(ii) If $B = \{b_1, b_2\}$, let M be the subspace spanned by b_1.
(iii) Let D be the additive group of M.

Argument for Lemma 4.2: $\mathrm{Im}(\theta)$ is a cyclic subgroup of $\mathrm{Aut}(\mathbb{Z}_7^2)$ whose order divides 6; say $\mathrm{Im}(\theta) = \langle T \rangle$ where T has order 1, 2, 3, 6. We now apply (iii) of Lemma 4.1 and (ii) and (iii) of Lemma 2.4 to conclude that $D \times \mathbb{Z}_6$ is a subgroup of $\mathbb{Z}_7^2 \times_\theta \mathbb{Z}_6$ of order $7 \cdot 2 \cdot 3$.

Argument for Lemma 4.3:   First note that by Sylow's theorem $n_7(G) \equiv 1$ mod 7 and $n_7(G) \mid 6$, which together force $n_7(G) = 1$. I.e., G has a unique (hence normal) Sylow 7-subgroup N. This proves (ii).

Since G/N has order 6, it is solvable. The subgroup N is also solvable (being a p-group) so G is solvable. Next note that $|G|$ is PNSS so by Lemma 3.5 all proper subgroups of G are supersolvable. This shows (i).

(iii) follows from Theorem 1.2.

(iv) By (i) L and M are both supersolvable, hence CLT. It follows that L has a subgroup A of order $7 \cdot 2$ and M has a subgroup B of order $7 \cdot 3$.

(v) is immediate.

(vi) Since $|N| = 7^2$ and $|K| = 6$, we get $|N \cap K| = 1$ and hence $N \cap K = \{1\}$. Furthermore $|NK| = 7^2 \cdot 6/1 = |G|$ so that $G = NK$. These facts ($N \triangleleft G$, $G = NK$, and $N \cap K = \{1\}$) imply the result.

(vii) Parts (ii), (iii), (iv) produce subgroups of orders 7^2, $2 \cdot 3$, $7^2 \cdot 2$, $7^2 \cdot 3$, $7 \cdot 2$, and $7 \cdot 3$. Cauchy's theorem produces subgroups of orders 2, 3, and 7.

(viii) By (vii), G is CLT iff such a subgroup W exists. Since G is special, now use Lemma 3.4 to conclude that G is supersolvble iff G is CLT.

(ix) First suppose $N \cong \mathbb{Z}_{49}$. Then N cyclic and G/N supersolvable (since $|G/N| = 6$ is square-free) imply G is supersolvable by Theorem 1.2 of Chapter 1.

We may now assume N is not isomorphic to $\mathbb{Z}_{49}$. Thus $N \cong \mathbb{Z}_7^2$. Now suppose $K \cong \mathbb{Z}_6$. Use (vi) to produce a homomorphism $\gamma \colon \mathbb{Z}_6 \to \mathrm{Aut}(\mathbb{Z}_7^2)$ such that $G \cong \mathbb{Z}_7^2 \times_\gamma \mathbb{Z}_6$, then use Lemma 4.2 to produce a 42-element subgroup of G, then use (viii) to conclude that G is supersolvable.

Finally, we suppose that $N \cong \mathbb{Z}_7^2$, $K \cong S_3$, and θ is not a monomorphism. From (vi), we see that there is a nonmonomorphic homomorphism $\delta \colon S_3 \to \mathrm{Aut}(\mathbb{Z}_7^2)$ such that $G \cong \mathbb{Z}_7^2 \times_\delta S_3$. Since A_3 is the one and only proper normal subgroup of S_3, it follows that the kernel of δ must be either A_3 or S_3. We conclude that $|\mathrm{Im}(\delta)|$ is either 2 or 1, and it follows that $\mathrm{Im}(\delta)$ is a cyclic subgroup of $\mathrm{Aut}(\mathbb{Z}_7^2)$ whose order divides 6. The argument we used for Lemma 4.2 can be used again now to show that $G \cong \mathbb{Z}_7^2 \times_\delta S_3$ has a subgroup of order $7 \cdot 2 \cdot 3$, and it follows from (viii) that G is supersolvable.

(x) If G is not CLT, then it is not supersolvable. Now apply (v), (vi), and (ix).

(xi) This follows from (x) and the fact that S_3 can be represented as a semi-direct product $S_3 \cong \mathbb{Z}_3 \times_\beta \mathbb{Z}_2$.

The construction of the "big" semi-direct product in (xi) of Lemma 4.3 can be generalized. If G is any non-CLT group of order $294 = 7^2 \cdot 2 \cdot 3$, and se set $p = 7$, $q = 2$, and $n = 3$, then $p \equiv 1 \bmod n$, $q \mid \varphi(n)$, $|G| = p^q qn$, and G has no subgroups of order pqn. Hopefully we have now provided some motivation for the material which follows.

THEOREM 4.4 *Let p and q be any two primes and let n be any positive integer such that $p \equiv 1 \bmod n$ and $q \mid \varphi(n)$. Then there exist monomorphisms $\beta \colon \mathbb{Z}_q \to \mathrm{Aut}(\mathbb{Z}_n)$ and $\alpha \colon (\mathbb{Z}_n \times_\beta \mathbb{Z}_q) \to \mathrm{Aut}(\mathbb{Z}_p^q)$ such that $G = \mathbb{Z}_p^q \times_\alpha (\mathbb{Z}_n \times_\beta \mathbb{Z}_q)$ has order $p^q qn$ and has no subgroup of order pqn.*

We first state four corollaries of Theorem 4.4, then prove the corollaries, and lastly prove the theorem.

COROLLARY 4.5 *Let p and q be primes such that $p \equiv 1 \bmod q^2$. Then the construction in Theorem 4.4 can be used to exhibit a finite group G of order $p^q q^3$ having no subgroups of order pq^3.*

COROLLARY 4.6 *Let p, q and r be any three primes such that $p \equiv 1$ mod r and $r \equiv 1$ mod q. Then the construction in Theorem 4.4 can be used to exhibit a finite group G of order $p^q q r$ having no subgroups of order pqr.*

COROLLARY 4.7 *If c is any positive integer of type 4, then the construction in Theorem 4.4 can be used to exhibit a finite non-CLT group of order c.*

COROLLARY 4.8 *If d is any positive integer of type 5, then the construction in Theorem 4.4 can be used to exhibit a finite non-CLT group of order d.*

We stated Corollaries 4.5 and 4.6 as we did because it is possible to pull more out of Theorem 4.4 than just what is indicated in Corollaries 4.7 and 4.8. This will be discussed further after the proof of Theorem 4.4.

Argument for Corollary 4.5: Set $n = q^2$ in Theorem 4.4.

Argument for Corollary 4.6: Set $n = r$ in Theorem 4.4.

Argument for Corollary 4.7: If c is any positive integer of type 4, then there exist primes p and q such that $p \equiv 1$ mod q^3 and $c = p^q q^3$. It follows that $p \equiv 1$ mod q^2, and we may apply Corollary 4.5.

Argument for Corollary 4.8: If d is any positive integer of type 5, then there exist primes p, q, and r such that $p \equiv 1$ mod qr, $r \equiv 1$ mod q, and $d = p^q q r$. It follows that $p \equiv 1$ mod r and $r \equiv 1$ mod q, and we may apply Corollary 4.6.

We will now polish off Theorem 4.4 with an omnibus lemma.

LEMMA 4.9 *If all of the hypotheses of Theorem 4.4 hold, then:*
(i) There is an element $w \in \mathrm{GF}(p)^*$ of order n.
If U_n denotes the multiplicative group of the ring of integers modulo n, then there is some positive integer k such that:
(ii) $[k]_n \in U_n$ and $[k]_n$ has order q; i.e., $(k, n) = 1$ and $\exp(k, n) = q$.
(iii) $k^q \equiv 1$ mod n and k is not congruent to 1 mod n.
(iv) If s and t are any two integers such that $0 \le s \le t \le q - 1$, then $w^{k^t} = w^{k^s}$ iff $t = s$; also, $w^{k^q} = w$.
Furthermore:
(v) There exists $S, T \in \mathrm{Aut}(\mathbb{Z}_p^q)$ such that S has order q, T has order n, and $S^{-1} \circ T \circ S = T^k$.
If $C = \langle T, S \rangle$, then:
(vi) There is a monomorphism $\beta : \mathbb{Z}_q \to \mathrm{Aut}(\mathbb{Z}_n)$ such that $C \cong \mathbb{Z}_n \times_\beta \mathbb{Z}_q$.
(vii) There is a monomorphism $\alpha : (\mathbb{Z}_n \times_\beta \mathbb{Z}_q) \to \mathrm{Aut}(\mathbb{Z}_p^q)$ such that $\mathrm{Im}(\alpha) = C$.
(viii) If L is any one-dimensional subspace of V_p^q, then L is not invariant under C.

(ix) *If M is any subgroup of $\mathbb{Z}_p^q$ such that $|M| = p$, then M is not invariant under C.*

Furthermore, all of the following must hold:

(x) $2 \leq q \leq \varphi(n) \leq n - 1 < n \leq p - 1 < p$, *so that* $2 \leq q < n < p$.

(xi) $(p^q, nq) = 1$.

(xii) *If* $H = \mathbb{Z}_p^q$ *and* $K = \mathbb{Z}_n \times_\beta \mathbb{Z}_q$, *then* $(|H|, |K|) = 1$.

(xiii) *If* $G = H \times_\alpha K$, *then* $|G| = p^q qn$ *and G has no subgroups of order pqn.*

(xiv) *The conclusion of Theorem 4.4 must hold.*

Argument for Lemma 4.9: (i) n divides $p - 1 = |GF(p)*|$ and $GF(p)*$ is cyclic. Hence $GF(p)*$ has an element of order n.

(ii) $|U_n| = \varphi(n)$ which q divides. Cauchy's theorem produces $[k]_n \in U_n$ of order q. Now $[k]_n \in U_n$ implies $(k, n) = 1$. Also $\exp(k, n) =$ the order of $[k]_n$, which is q.

(iii) This follows from $\exp(k, n) = q$.

(iv) Using the facts that $(k, n) = 1$, $t - s \geq 0$, $\exp(k, n) = q$, and w has order n, we get $w^{k^t} = w^{k^s}$ iff $k^t \equiv k^s \mod n$ iff $k^{t-s} \equiv 1 \mod n$ iff $\exp(k, n) | (t - s)$ iff $q | (t - s)$ iff $t \equiv s \mod q$ iff $t = s$. Finally, since $k^q \equiv 1 \mod n$, it follows that $w^{k^q} = w^1 = w$ because w has order n.

In the rest of this argument we shall use id to denote the identity map on V_p^q and we shall use I to denote the q-by-q multiplicative identity matrix (with all of its entries in $GF(p)$). Now, let $B_1 = \{b_1, b_2, \ldots, b_q\}$ be a basis for V_p^q, and let $S : V_p^q \to V_p^q$ and $T : V_p^q \to V_p^q$ be the linear maps defined by

$$S(b_i) = b_{i+1} \quad \text{for } i < q; \qquad S(b_q) = b_1,$$

$$T(b_i) = w^{k^{i-1}} * b_i \quad \text{all } i.$$

(v) The matrices of T and S relative to B_1 are $A = \text{diag}(w, w^k, w^{k^2}, \ldots, w^{k^{q-1}})$ and

$$L = \begin{pmatrix} 0 & 0 & 0 & \ldots & 0 & 1 \\ 1 & 0 & 0 & \ldots & 0 & 0 \\ 0 & 1 & 0 & \ldots & 0 & 0 \\ 0 & 0 & 1 & \ldots & 0 & 0 \\ 0 & 0 & 0 & \ldots & 0 & 0 \\ \cdot & \cdot & \cdot & \ldots & \cdot & \cdot \\ \cdot & \cdot & \cdot & \ldots & \cdot & \cdot \\ \cdot & \cdot & \cdot & \ldots & \cdot & \cdot \\ 0 & 0 & 0 & \ldots & 1 & 0 \end{pmatrix}$$

respectively, where for convenience we are using 0 to denote $[0]_p$ and 1 to denote $[1]_p$. Now $w^n = 1$, and it follows that $A^n = \text{diag}(1, 1, 1, \ldots, 1) = I$, which shows that $T^n = \text{id}$. Thus A is nonsingular, so that $T \in \text{Aut}(V_p^q)$. Letting u denote the order of T, we have $u | n$ because $T^n = \text{id}$. On the other hand, id $= T^u$, so that $\text{diag}(1, 1, \ldots, 1) = I = A^u = \text{diag}(w^u, \ldots)$, and it follows that $w^u = 1$; since w has order n, we get $n | u$. Thus $u = n$ so that T has order n. Now, we note that L is the companion matrix for the polynomial

$$f = -1 - 0x - 0x^2 - \ldots - 0x^{q-1} + x^q \in GF(p)[x].$$

It follows that $m_S = f = x^q - 1$; since f annihilates S, it follows that $S^q = $ id. Therefore S is invertible (so that $S \in \text{Aut}(V_p^q)$) and has order q. It remains to show that $S^{-1} \circ T \circ S = T^k$. We first note that L is a permutation matrix (as well as a companion matrix); we rigged S so that $S|_{B_1}$ is a permutation of B_1. The reader is invited to check that $S|_{B_1}$ "looks like" the q-cycle $(1, 2, 3, \ldots, q)$. We now use L as a "change-of-basis" matrix in the following way: We first "reorder B_1 by $(1, 2, 3, \ldots, q)$," viz.: let $c_i = b_{i+1}$ for $i < q$ and let $c_q = b_1$. Now let $B_2 = \{c_1, c_2, \ldots, c_{q-1}, c_q\}$. Then $B_2 = B_1$, so that B_2 is a basis for V_p^q. Also, we note that $\text{id}(c_i) = c_i = b_{i+1}$ for $i < q$ and that $\text{id}(c_q) = c_q = b_1$, and it follows that the matrix of id relative to (B_2, B_1) is precisely L. We now conclude that L is nonsingular, and that L^{-1} is the matrix of id relative to (B_1, B_2); since A is the matrix of T relative to (B_1, B_1), we also conclude that the matrix of T relative to (B_2, B_2) is $L^{-1}AL$. Now a glance back to our definition of T shows that $T(c_i) = T(b_{i+1}) = w^{k^i} * b_{i+1} = w^{k^i} * c_i$ for $i < q$, and $T(c_q) = T(b_1) = w * b_1 = w * c_q$; it follows that the matrix of T relative to (B_2, B_2) is

$$A_1 = \text{diag}(w^k, w^{k^2}, \ldots, w^{k^{q-1}}, w). \tag{1}$$

We may now conclude that

$$L^{-1}AL = A_1. \tag{2}$$

If we recall the definition of A, then we note that our change of basis "effects the permutation $(1, 2, 3, \ldots, q)$ on the diagonal entries of A," and we also note from Equation (1) and the definition of A that

$$A^k = A_1. \tag{3}$$

From Equations (2) and (3), we get what we were after all along, viz., $L^{-1}AL = A^k$. Translating back from matrices to automorphisms we now have $S^{-1} \circ T \circ S = T^k$ as was to be shown.

(vi) If we set $a = T$, and $d = S^{-1}$, then

$$C = \langle T, S^{-1} \rangle = \langle a, d \rangle, \tag{4}$$

and it follows from (v) and (iii) that

$$a^n = \text{id}, \qquad d^m = \text{id}, \tag{5}$$
$$dad^{-1} = a^k, \quad \text{and} \quad k^m \equiv 1 \bmod n.$$

Equations (4) and (5) give us exactly the generators and defining relations that we need to assert that there exists a homomorphism $\beta : \mathbb{Z}_q \to \text{Aut}(\mathbb{Z}_n)$ such that $C \cong \mathbb{Z}_n \times_\beta \mathbb{Z}_q$. (See the discussion contained in the last eight lines on p. 461 and the first eight lines on p. 462 of [Mac Lane and Birkhoff 1968].) It only remains to show that β is injective; since β is a homomorphism with $\mathbb{Z}_q$ as its domain, it suffices to show that β is not trivial. Suppose, to the contrary, that β is trivial. Then $C \cong \mathbb{Z}_n \times \mathbb{Z}_q$, so that C is Abelian. It follows that $T \circ S = S \circ T$, so that $T^k = S^{-1} \circ T \circ S = T$. Since T has order n, it follows that $k \equiv 1 \bmod n$, and this contradicts (iii).

(vii) is immediate from (vi).

(viii) Let $A_2 = \{i : i \text{ is a positive integer and } i \leq q\}$, and for all $i \in A_2$, let $c_i = w^{k^{i-1}}$. For all $x \in V_p^q$ such that $x \neq 0$, let $\langle x \rangle$ be the one-dimensional subspace of V_p^q spanned by x.

We now suppose that (viii) is false, and seek a contradiction. Thus, we assume that there is a one-dimensional subspace $J = \langle x \rangle$ of V_p^q which is invariant under C. It follows that $\langle x \rangle$ is invariant under both T and S, and we conclude that x is a common eigenvector for both T and S in V_p^q. Recall that B_1 is a basis for V_p^q, and the matrix of T relative to B_1 is $A = \text{diag}(c_1, c_2, \ldots, c_q)$; it follows that if E is any overfield of $\text{GF}(p)$, then the set of all eigenvalues of T in E is $\{c_1, c_2, \ldots, c_q\}$. Now, $w \in \text{GF}(p)$ and it follows that $\{c_1, c_2, \ldots, c_q\} \subseteq \text{GF}(p)$; furthermore, it follows from (iv) that $c_i \neq c_j$ for $i, j \in A_2$ such that $i \neq j$. We conclude that T has all of its eigenvalues in $\text{GF}(p)$ and has no repeated eigenvalues. We may now conclude that if t is any eigenvector of T in V_p^q, then there is some $i \in A_2$ such that $t \in \langle b_i \rangle$ and $t \neq 0$. Since x is an eigenvector of T in V_p^q, we conclude that $x \in \langle b_i \rangle$ for some i, and hence $J = \langle x \rangle = \langle b_i \rangle$. (Recall that $x, b_i \neq 0$.) Since J is invariant under S, there is some $c \in \text{GF}(p)$ such that $S(b_i) = c * b_i$. Now, a glance back to our definition of S shows that $S(b_i) = b_j$ for some $j \in A_2$, $j \neq i$, and it follows that $c * b_i = b_j$, so that $\{b_i, b_j\}$ is linearly dependent over $\text{GF}(p)$. Since $\{b_i, b_j\} \subseteq B_1$ and B_1 is a basis for V_p^q, we have our desired contradiction.

(ix) M is a one-dimensional subspace of V_p^q; apply (viii).

(x) The hypotheses of Theorem 4.4 give us that q is a prime and $q \mid \varphi(n)$, and it follows that

$$2 \leq q \leq \varphi(n). \tag{6}$$

Since $\varphi(n) \geq 2$ and $\varphi(1) = 1 = \varphi(2)$, it follows that $n \geq 3$; we may now conclude that

$$\varphi(n) \leq n - 1. \tag{7}$$

Now, the hypotheses of Theorem 4.4 give us that p is a prime and $p \equiv 1$ mod n, so that $n \mid (p - 1)$; it follows that

$$n - 1 < n \leq p - 1 < p. \tag{8}$$

Now combine inequalities (6), (7), and (8).

(xi) It follows from (x) that $n < p$ and $q < p$, so that $p \nmid n$ and $p \nmid q$; since p is a prime, we conclude that $(p, n) = 1 = (p, q)$, and it follows that $(p^q, nq) = 1$.

(xii) follows from (xi).

(xiii) That $|G| = p^q qn$ is immediate. Now suppose that G has a subgroup of order pqn. If we set $H = \mathbb{Z}_p^q$, $K = \mathbb{Z}_n \times_\beta \mathbb{Z}_q$, $\theta = \alpha$, $m = p^q$, $m_1 = p$, and replace n by nq in Lemma 2.27, we may apply Lemma 2.27 to conclude that $\mathbb{Z}_p^q$ has a subgroup of order p which is invariant under $\text{Im}(\alpha) = C$. This contradicts (ix). It follows that G has no subgroups of order pqn.

(xiv) follows immediately from (xiii).

We shall finish this section with the following result:

THEOREM 4.10 *If c is any positive integer of type 3, then there exist two finite groups G_1 and G_2 such that:*
(i) $|G_1| = c = |G_2|$.
(ii) G_1 *and* G_2 *are both non-CLT.*
(iii) G_1 *is not isomorphic to* G_2.

Argument for Theorem 4.10: (i) and (ii) By definition of type 3, there are distinct primes p and q such that $q^2 \| (p - 1)$ and $c = p^q q^3$. If we set $b = 2$ in Lemma 3.13, then we may use (iii) and (iv) of Lemma 3.13 to conclude that $\exp(p, q^3) = q$ and c is a McCarthy number. If we set $n = q^3$ and $a = q$ in Theorem 2.25, then we may use Theorem 2.25 to conclude that there is a monomorphism $\theta : \mathbb{Z}_{q^3} \to \text{Aut}(\mathbb{Z}_p^q)$ such that the semi-direct product $G_1 = \mathbb{Z}_p^q \times_\theta \mathbb{Z}_{q^3}$ is not CLT, and we note that $|G_1| = p^q q^3 = c$. On the other hand, we have that $q^2 \| (p - 1)$, so that $p \equiv 1 \mod q^2$. If we set $n = q^2$ in Theorem 4.4, then $q | \varphi(n)$ and we may use Theorem 4.4 and Corollary 4.5 to conclude that there are monomorphisms $\beta : \mathbb{Z}_q \to \text{Aut}(\mathbb{Z}_{q^2})$ and $\alpha : (\mathbb{Z}_{q^2} \times_\beta \mathbb{Z}_q) \to \text{Aut}(\mathbb{Z}_p^q)$ such that the semi-direct product $G_2 = \mathbb{Z}_p^q \times_\alpha (\mathbb{Z}_{q^2} \times_\beta \mathbb{Z}_q)$ is not CLT, and we note that $|G_2| = p^q q^3 = c$.

(iii) The Sylow q-subgroups of G_1 are isomorphic to $\mathbb{Z}_{q^3}$, hence cyclic. The Sylow q-subgroups of G_2 are isomorphic to $\mathbb{Z}_{q^2} \times_\beta \mathbb{Z}_q$ hence non-Abelian (because β is a monomorphism, hence nontrivial). Thus G_1 and G_2 have nonisomorphic Sylow q-subgroups, so we conclude G_1 is not isomorphic to G_2.

Miscellaneous Classes

John F. Humphreys and David Johnson

1. Solvable Linear Groups

Let G be a finite solvable group and N be a minimal normal subgroup of G. Then N is an elementary Abelian p-group for some prime p and so may be regarded as a vector space over the field with p elements. In this context, the elements of $G/C_G(N)$ may be thought of as invertible linear transformations of N and $G/C_G(N)$ as a linear group. Special techniques are available for investigating linear groups in general and solvable linear groups in particular. These techniques can often be exploited to give useful information about finite solvable groups. In this section, an outline of the theory of solvable linear groups is given.

DEFINITION Let F be any field and V be a vector space of dimension d over F. A subgroup of $\mathrm{GL}(V)$, the group of invertible linear transformations of V into itself, is called a **linear group** of degree d over F.

DEFINITIONS Let G be a subgroup of $\mathrm{GL}(V)$. A subspace W of V is **G-invariant** if $Wg \subseteq W$ for all $g \in G$. The linear group G is **irreducible** if the only G-invariant subspaces of V are $\{0\}$ and V itself.

EXAMPLE Let V be a 2-dimensional vector space over the field of residue classes modulo 5 and let $\{x, y\}$ be a basis for V. Let A, B be linear transformations of V into itself whose matrices with respect to this basis are

$$A = \begin{pmatrix} 2 & 0 \\ 0 & 3 \end{pmatrix} \text{ and } B = \begin{pmatrix} 0 & 1 \\ -1 & 0 \end{pmatrix}$$

Then $A^4 = I$, the 2-by-2 identity matrix, $A^2 = B^2$, and $B^{-1}AB = A^{-1}$ so that $G = \langle A, B \rangle$ is isomorphic to the quaternion group of order 8.

The subspaces V_1, V_2 of V generated by x and y respectively are easily seen to be the only one-dimensional subspaces invariant under A. Since neither of these is invariant under B, G is an irreducible linear group.

DEFINITION A subgroup G of $GL(V)$ is **completely reducible** if there exist minimal G-invariant subspaces $V_1, \ldots, V_n$ of V such that

$$V = V_1 \oplus \ldots \oplus V_n.$$

A well-known result of A. H. Clifford relates representations of a group to those of a normal subgroup. One version of this result has already been given (Theorem 1.1 of Chapter 2). In our context, Clifford's theorem gives the following result.

THEOREM 1.1 *Let V be a vector space of dimension n, G an irreducible subgroup of $GL(V)$, and N a normal subgroup of G. Let W be a minimal N-invariant subspace of V. Then, for all $g \in G$, Wg is a minimal N-invariant subspace of V and there exist $g_1, \ldots, g_k \in G$, with $g_1 = 1$, such that*

$$V = Wg_1 \oplus \ldots \oplus Wg_k.$$

In particular, N is completely reducible and the minimal N-invariant subspaces of V each have dimension n/k.
Proof: For any $g \in G$, $x \in N$

$$Wgx = Wgxg^{-1}g = Wg$$

since $gxg^{-1} \in N$ and W is N-invariant. Thus Wg is N-invariant and the existence of an N-invariant proper subspace of Wg would imply the existence of such a subspace of W. Hence Wg is a minimal N-invariant subspace.

Now let $\{ g_1 = 1, g_2, \ldots, g_k \}$ be a set of elements of G, maximal with respect to the property that V_1, the subspace generated by $Wg_1, \ldots, Wg_k$ is a direct sum of $Wg_1, \ldots, Wg_k$. For any $g \in G$, Wg being a minimal N-invariant subspace of V implies that $V_1 \cap Wg$ is either Wg or $\{0\}$. However, choice of $\{ g_1, \ldots, g_k \}$ implies that $V_1 \cap Wg$ cannot be $\{0\}$, so that for all $g \in G$, $Wg \subseteq V_1$. It therefore follows that V_1 is a G-invariant subspace of V and so irreducibility of G forces $V_1 = V$.

DEFINITIONS Let G be an irreducible subgroup of $GL(V)$. A **system of imprimitivity** for G is a set $\{ W_1, W_2, \ldots, W_r \}$ of subspaces of V such that
(i) $r > 1$,

(ii) $V = W_1 \oplus \ldots \oplus W_r$,

(iii) for each $g \in G$, the map $W_i \to W_i g$ $(1 \leq i \leq r)$ is a permutation of the set $\{W_1, \ldots, W_r\}$.

A **primitive** linear group is one which has no system of imprimitivity.

The analysis of imprimitive linear groups is provided by the following result.

THEOREM 1.2   *Let G be an imprimitive subgroup of* GL(V), *where V has dimension n, and $\{W_1, \ldots, W_r\}$ be a system of imprimitivity for G. For $g \in G$, let $g\pi$ denote the permutation $W_i \to W_i g$ and let $G_i = \{g \in G : W_i g = W_i\}$ $(1 \leq i \leq r)$. Then π is a homomorphism from G into a transitive permutation group of degree r and $\ker \pi = \bigcap_{i=1}^{r} G_i$. Also $\dim W_i = n/r$ $(1 \leq i \leq r)$ and $\ker \pi$, as a linear group, is isomorphic to a subgroup of $N_1 \times \ldots \times N_r$, where N_i is $\ker \pi$ regarded as a linear group on W_i.*

Proof: It is easily checked that π is a homomorphism and

$$\ker \pi = \{g \in G : W_i g = W_i \ (1 \leq i \leq r)\} = \bigcap_{i=1}^{r} G_i.$$

If the image of G under π were not transitive, there would be a subset of $\{W_1, \ldots, W_r\}$ which was invariant under action by G. Thus the irreducibility of G implies that the image of G under π is transitive, so that for any i there exists $g_i \in G$ such that $W_i = W_1 g_i$ $(1 \leq i \leq r)$. This means that $\dim W_i = \dim W_1$ and so $n = \dim V = r \dim W_1$.

Since $G_i \pi$ is the stabilizer of W_i in the transitive permutation group $G\pi$, $[G\pi : G_i\pi] = r$ and since $G_i \geq \ker \pi$, $[G : G_i] = r$. We now show that W_i is a minimal G_i-invariant space. For if not, let W be a minimal G_i-invariant subspace of W_i and let $g_1, \ldots, g_r$ be a set of right coset representatives for G_i in G. Thus for any $g \in G$, $g_j g$ may be written in the form $x g_k$ for some k $(1 \leq k \leq r)$ and some $x \in G_i$. Hence $W g_j g = W x g_k = W g_k$ and so

$$U = \sum_{i=1}^{r} W g_i$$

is a G-invariant subspace of V. The irreducibility of V therefore forces $U = V$ and so

$$n = \dim U \leq r \dim W \leq r \dim W_i = n.$$

It follows that W_i is a minimal G_i-invariant subspace of V. Also $V = \bigcap_{j=1}^{r} W_i g_j$ and so, on grounds of dimension,

$$V = W_i g_1 \oplus \ldots \oplus W_i g_r.$$

Hence V is completely reducible as a $\ker \pi$ space. Finally, for $x \in \ker \pi$, let $x\tau_i$ denote the linear transformation induced by x on W_i. The map $x \to (x\tau_1, \ldots, x\tau_r)$ provides an isomorphism from $\ker \pi$ into a subgroup of $N_1 \times \ldots \times N_r$, where N_i is $\ker \pi$ regarded as a linear group on W_i $(1 \leq i \leq r)$.

The analysis of primitive solvable linear groups is provided by a result of Suprunenko. Unfortunately, this result only holds over algebraically closed fields, so we must first consider a "change of ring" situation. We require a preliminary result.

PROPOSITION 1.3 *Let F be a field of characteristic p and G be an irreducible linear group of degree n over F. Then G has no nontrivial normal p-subgroups.*

Proof: Let P be a normal p-subgroup of G and suppose $P > \{1\}$. By Clifford's theorem (Theorem 1.1), P is completely reducible. Let W be a minimal P-invariant subspace. Since $P > \{1\}$, $Z(P) > \{1\}$ so we can chose $z \in Z(P)$ with $z \neq 1$. Since $zg = gz$ for all $g \in P$, Schur's lemma ([Scott 1964, 12.1.5]) may be applied to the linear transformation $z - 1$ on W, where 1 denotes the identity linear transformation. Thus $z - 1$ is either invertible or zero. However, z has order a power of p, so $z^{p^k} = 1$ for some k. Thus, by the Cayley–Hamilton theorem, the characteristic polynomial of z divides $x^{p^k} - 1$. Since F has characteristic p, $x^{p^k} - 1 = (x - 1)^{p^k}$, so that all the eigenvalues of z are 1. Hence $z - 1$ is not invertible and so is zero and $z = 1$ on W. Clifford's theorem now implies that $z = 1$ on V, which is impossible, so $P = \{1\}$.

We can now establish our result on change of field.

THEOREM 1.4 *Let G be a finite completely reducible subgroup of $\mathrm{GL}(V)$ where V is a vector space over F. Let E be an extension field of F and $V^E = V \otimes_F E$, so that V^E is V regarded as a vector space over E. Then G is isomorphic to a completely reducible subgroup of $\mathrm{GL}(V^E)$.*

Proof: The result is trivially true by Maschke's theorem ([Scott 1964, 12.1.2]) if F has characteristic zero. We therefore suppose F has characteristic $p > 0$. It is clearly sufficient to prove the result when G is irreducible so that, by Proposition 1.3, G has no nontrivial normal p-subgroups. Let

$$V^E = V_0 \supseteq V_1 \supseteq \ldots \supseteq V_s = \{0\}$$

be a G-composition series for V^E. Let U_i be any subspace of V_{i-1} such that $V_{i-1} = U_i \oplus V_i$, so that U_i is isomorphic to V_{i-1}/V_i as a vector space ($1 \leq i \leq s$). Define a map $\rho : G \to \mathrm{GL}(V^E)$ as follows: firstly define ρ on U_i by

$$ux\rho = u'$$

for all $u \in U_i$, $x \in G$ where

$$(u + V_i)x = u' + V_i$$

and now extend ρ using the fact that $V^E = U_1 \oplus U_2 \oplus \ldots \oplus U_s$. It is easily checked that ρ is a homomorphism with kernel

$$K = \{x \in G : x \text{ acts identically on } V_{i-1}/V_i \ (1 \leq i \leq s)\}.$$

Thus, choosing a basis of V adopted to the decomposition $V^E = U_1 \oplus U_2 \oplus \ldots \oplus U_s$, we see that each element of K is represented in this basis by a triangular matrix with 1's down the main diagonal. Thus K is a normal p-subgroup of G and so is $\{1\}$, which means that ρ is an isomophism from G into a completely reducible subgroup of $GL(V^E)$.

The proof of Suprenenko's theorem is too lengthy to be included here but may be found in [Dixon 1971].

THEOREM 1.5 *Let F be an algebraically closed field and G be a primitive solvable linear group of degree n over F. Then*
(1) $Z(G)$ *is the unique maximal normal Abelian subgroup of G.*
(2) $\mathrm{Fit}(G)/Z(G)$ *is the unique maximal normal Abelian subgroup of $G/Z(G)$.*
(3) $[\mathrm{Fit}(G):Z(G)] = d^2$ *for some divisor d of n.*
(4) *The Sylow p-subgroups of $\mathrm{Fit}(G)/Z(G)$ are Abelian.*
(5) *Let $p_1^{e_1} \ldots p_s^{e_s}$ be the prime decomposition of d. Then $G/\mathrm{Fit}(G)$ is isomorphic to a subgroup of the direct product of the symplectic groups $\mathrm{Sp}(2e_i, p_i)$.*

Note The definition of the symplectic group $\mathrm{Sp}(2e, p)$ may be found in [Dixon 1971], but for our purposes, we only need to know that this group has order

$$p^{e^2}(p^{2e} - 1)(p^{2(e-1)} - 1) \ldots (p^2 - 1).$$

Theorems 1.2, 1.4, and 1.5 provide us with a strategy for proving results about solvable linear groups. This is perhaps best illustrated by giving an application of this method.

THEOREM 1.6 *Let p be an odd prime and F be a field of characteristic p. Let G be a finite completely reducible solvable linear group of degree n over F. Suppose that a Sylow p-subgroup of G has order $p^{\lambda(n)}$. Then $\lambda(n) \leq n - 1$ with equality only if $n = 1$ or $n = 2$ and $p = 3$.*
Proof: We induce on n. Since $GL(F)$ is Abelian, an irreducible subgroup of it is Abelian, and so is of order coprime to p by Proposition 1.3. Thus the result holds in the case $n = 1$.

Notice that Theorem 1.4 allows us to suppose that F is algebraically closed without loss of generality. We divide the rest of the proof into three steps.

Step 1: *Suppose G is reducible.* In this case, G is a subgroup of $G_1 \times G_2$, where G_i is completely reducible of degree n_i ($i = 1, 2$) and $n = n_1 + n_2$. Thus,

$$\lambda(n) \leq \lambda(n_1) + \lambda(n_2) \leq n_1 - 1 + n_2 - 1 < n - 1,$$

equality being impossible here.

Step 2: Suppose G is irreducible and imprimitive. This is the case to which Theorem 1.2 applies and we see that G has a normal subgroup N which is a subgroup of a direct product of r copies of $\mathrm{GL}(d, F)$, the group of linear transformations of an r-dimensional vector space over F, where $n = dr$. Furthermore G/N is isomorphic to a subgroup of the symmetric group $S(r)$. By induction, a Sylow p-subgroup of N has order p^{μ} where $\mu \le r(d - 1) = n - r$.

The Sylow p-subgroup of $S(r)$ (see [Hall 1959, page 81]) has order p^{ν} where

$$ \nu = [r/p] + [r/p^2] + \ldots + [r/p^k] $$

where $[x]$ denotes the greatest integer less than or equal to x, and k is the largest integer such that $p^k \le r$. Hence

$$ \nu \le (r/p^k)(1 + p + \ldots + p^{k-1}) = \frac{r(p^k - 1)}{p^k(p - 1)} \le \frac{r(p^k - 1)}{2p^k} $$

$$ < \frac{rp^k}{2p^k} = \frac{r}{2} \le r - 1. $$

Thus $\nu \le r - 1$ and, in fact, equality is impossible. It therefore follows that $\lambda(n)$ satisfies

$$ \lambda(n) \le \mu + \nu < n - r + r - 1 = n - 1. $$

Step 3: Suppose G is primitive, so that Theorem 1.5 applies. By Proposition 1.3, $Z(G)$ has order coprime to p. Let p^e be the exact power of p dividing d (we do not exclude the case $e = 0$). By Theorem 1.5(3), $|\mathrm{Fit}(G)|$ is divisible by p^{2e}. However, a Sylow p-subgroup of $\mathrm{Fit}(G)$ would be a characteristic subgroup of $\mathrm{Fit}(G)$ since $\mathrm{Fit}(G)$ is nilpotent and so would be normal in G. Proposition 1.3 therefore implies that $e = 0$ and so that p does not divide d.

Not let $q_1^{e_1} \ldots q_s^{e_s}$ be the prime decomposition of d. By Theorem 1.5(5), we can estimate the power of p dividing $|G/\mathrm{Fit}(G)|$ (and therefore $|G|$) by estimating the power of p dividing each of $|\mathrm{Sp}(2e_i, q_i)|$. Thus let q be any prime different from p and p^{ν} be the power of p dividing

$$ |\mathrm{Sp}(2e, q)| = q^{e^2}(q - 1)(q + 1)(q^2 - 1)(q^2 + 1) \ldots (q^e - 1)(q^e + 1). $$

Since $q^i - 1 > q^{i-1} + 1$ unless $q^i - 1 = 3$, we get p^{ν} divides $3(q^e + 1)!$.

Hence if $p \ge 5$ and k is the largest integer such that $p^k \le q^e + 1$,

$$ \nu \le \left[\frac{q^e + 1}{p} \right] + \ldots + \left[\frac{q^e + 1}{p^k} \right] \le \frac{q^e + 1}{p^k} \frac{p^k - 1}{p - 1} $$

$$ \le \frac{(q^e + 1)(p^k - 1)}{4p^k} < \frac{q^e + 1}{4} < q^e - 1. $$

While for $p = 3$,

$$v \leq 1 + \left[\frac{q^e + 1}{3}\right] + \ldots + \left[\frac{q^e + 1}{3^k}\right] \leq 1 + \frac{q^e + 1}{2} < q^e - 1$$

provided $q^e > 5$. In the exceptional cases, $q^e = 5$, the power of 3 dividing $|\text{Sp}(2,5)|$ is 3, for $q^e = 4$, the power of 3 dividing $|\text{Sp}(4,2)|$ is 9 and for $q^e = 2$, the power of 3 dividing $|\text{Sp}(2,2)|$ is 3. Notice that q^e cannot equal 3 since q is not 3.) Thus in every case

$$v \leq q^e - 1.$$

with equality only if $q^e = 2$ and $p = 3$.

Now let p^λ be the highest power of p dividing $|G|$. By Theorem 1.5(5) and the above

$$\lambda \leq \sum_{i=1}^{s} q^{e_i} - 1 \leq d - 1 \leq n - 1$$

as required. Equality is only possible here if $n = d = q^e$ for some prime q and so only if $q^e = 2$ and $p = 3$.

This method of proof has been used by several authors to obtain results about solvable linear groups. See, for example, [Dixon 1968], [Dornhoff 1970], and [Fisher 1974].

Although Theorem 1.6 is not valid when $p = 2$, results on the order of a Sylow 2-subgroup of a finite solvable linear group have been obtained in [Dixon 1967].

2. Groups all of whose Homomorphic Images are CLT-groups

DEFINITION A group G is a **QCLT-group** if every homomorphic image of G is a CLT-group.

The example of the symmetric group on 4 symbols shows that a QCLT-group need not be supersolvable. The main result of this section is that a group of odd order which is a QCLT-group is supersolvable. We need some preliminary results.

LEMMA 2.1 *Let G be a finite group of the form $G = NK$ where N is a normal Abelian subgroup of G and $N \cap K = \{1\}$. Let L be a subgroup of G with $LN \lhd G$. Then there exists a subgroup L_1 of G with $|L_1| = |L|$ and $L_1 = N_1 K_1$ where $N_1 = N \cap L$ and $K_1 \lhd K$.*

Proof: Define $K_1 = K \cap LN$ so that $K_1 \lhd K$. Then

$$NK_1 = N(K \cap LN) = NK \cap LN = LN$$

and $N \cap K_1 = \{1\}$. Thus

$$|K_1| = |LN|/|N| = |L|/|L \cap N|.$$

Now define $N_1 = L \cap N$. Since $N \lhd G$, $N_1 \lhd L$ and since N is Abelian, $N_1 \lhd N$ and so $N_1 \lhd LN$. It therefore follows that

$$\langle N_1, K_1 \rangle = N_1 K_1$$

and also that $N_1 \cap K_1 = \{1\}$. Thus if $L_1 = N_1 K_1$, we have

$$|L_1| = |N_1 K_1| = \frac{|N_1| \cdot |K_1|}{|N_1 \cap K_1|} = \frac{|L \cap N| \cdot |L|}{|L \cap N|} = |L|$$

as required.

LEMMA 2.2 *Let V be a vector space over a field of characteristic p and G be an irreducible subgroup of $\mathrm{GL}(V)$. Let N be a normal subgroup of G and W be a minimal N-invariant subspace of V. Let*

$$\ker W = \{ g \in N : wg = w \text{ for all } w \in W \}.$$

Then no nontrivial normal subgroup of G is contained in $\ker W$. Further, if G has Abelian Sylow p-subgroups and G/N is a p-group, then $\ker W$ has order coprime to p.

Proof: By Theorem 1.1, there exist elements $1 = g_1, \ldots, g_k$ of G such that

$$V = W \oplus Wg_2 \oplus \ldots \oplus Wg_k.$$

Let $K \lhd G$ with $K \leq \ker W$. Then for all $w \in W$, $x \in K$, and $1 \leq i \leq k$,

$$wg_i(g_i^{-1} x g_i) = w x g_i = wg_i$$

so that $g_i^{-1} x g_i \in \ker Wg_i$. Hence $K \leq \ker Wg_i$ $(1 \leq i \leq k)$, and so $K \leq \ker V = \{1\}$.

Now suppose G has Abelian Sylow p-subgroups and G/N is a p-group and let $x \in \ker W$ have order a power of p. Choose a Sylow p-subgroup P, say, containing x and note that $G = NP$ since G/N is a p-group. For any $g \in G$, we may write $g = g_1 g_2$ with $g_1 \in N$ and $g_2 \in P$ and so

$$Wg = Wg_1 g_2 = Wg_2$$

since W is N-invariant. Since $x \in \ker W$, $g_2^{-1} x g_2 \in \ker Wg_2$ because

$$wg_2(g_2^{-1} x g_2) = w x g_2 = wg_2$$

for all $w \in W$. However, since P is Abelian and $g_2, x \in P$, $g_2^{-1} x g_2 = x$ so that $x \in \ker Wg$ for all $g \in G$. It therefore follows that $x \in \ker V = \{1\}$ as required.

We now come to the main result of this section.

THEOREM 2.3 *Let G be a QCLT-group of odd order. Then G is supersolvable.*

Proof: We may suppose G has a unique minimal normal subgroup N. (If N_1, N_2 were distinct minimal normal subgroups, then G/N_1, G/N_2 would be supersolvable by induction, hence G would be also as it can be embedded in $G/N_1 \times G/N_2$.) Let $|N| = p^\alpha$, say, with $\alpha > 1$.

We may further assume, because of Corollary 3.2 of Chapter 1, that $\Phi(G) = \{1\}$. Hence there is a maximal subgroup K, say, of G not containing N. Thus $G = NK$ and $N \cap K$ is normal in K and in N (since N is Abelian) and so $N \cap K \lhd G$. Minimality of N now forces $N \cap K = \{1\}$ so that N is complemented in G by K. Now apply Lemma 2.3 and Theorem 2.6 of Appendix C to get $N = \mathrm{Fit}(G)$ and $N = C_G(N)$.

Thus $G = NK$ with $|N| = p^\alpha$ and K a supersolvable group. Also $K \cong G/N = G/C_G(N)$ so K is isomorphic to an irreducible subgroup of $\mathrm{GL}(\alpha, p)$. Theorem 1.6 therefore implies that if p^β is the order of a Sylow p-subgroup of K, then

$$\beta < \alpha - 1$$

unless $\alpha = 2$ and $p = 3$. However, $|\mathrm{GL}(2,3)| = 48$ so a supersolvable subgroup of $\mathrm{GL}(2,3)$ of order divisible by 3 has a normal Sylow 3-subgroup. Proposition 1.3 now implies that an irreducible supersolvable subgroup of $\mathrm{GL}(2,3)$ is a 2-group and so $\beta < \alpha - 1$ in all cases.

Since K is irreducible, Proposition 1.3 implies that K has no normal p-subgroups. Hence $\mathrm{Fit}(K)$ is of order coprime to p. However, $K' \leq \mathrm{Fit}(K)$ since K is supersolvable (Chapter 1, Theorem 1.6) and so K has a normal p-complement T, say, and K/T is Abelian. Thus every subgroup of K of p-power index must contain T and be normal in K.

Since G is a QCLT-group, G has a subgroup H of index $p^{\beta+1}$. Since the order of a Sylow p-subgroup of H is therefore $p^{\alpha-1}$, $H \cap N$ must be a proper subgroup of N. If $H \cap N = \{1\}$, the order of a Sylow p-subgroup of HN would be $p^{2\alpha-1}$ which is impossible since the order of a Sylow p-subgroup of G is $p^{\alpha+\beta}$ and $\beta < \alpha - 1$. Thus

$$\{1\} < H \cap N < N \tag{1}$$

Since $[G:H]$ is a power of p, $[G/N : HN/N]$ is also. However, $G/N \cong K$ so $HN/N \lhd G/N$ and $HN \lhd G$. Lemma 2.1 now implies that there exists a subgroup H_1 of G with $|H| = |H_1|$ and $H_1 = K_1 N_1$ where $K_1 \lhd K$ and $N_1 = H \cap N$.

By Clifford's theorem, K_1 is completely reducible and if M is a minimal K_1-invariant subspace of N, M has order p^γ where γ divides α. Since K' is a group of order coprime to p, the Sylow p-subgroups of K are Abelian and so by Lemma 2.2, $\ker M = C_{K_1}(M)$ has order coprime to p. Also $K_1/C_{K_1}(M)$ is isomorphic to an irreducible supersolvable subgroup of

$\mathrm{GL}(\gamma, p)$ so by Theorem 1.6 and an argument already used, either $\gamma = 1$ or the order of a Sylow p-subgroup of K_1 is p^λ where $\lambda < \gamma - 1$.

First suppose that $\gamma > 1$, so that $\lambda < \gamma - 1$. Then since N_1 is a K_1-invariant subspace of N,

$$|N_1| = p^{n\gamma}$$

where $0 < n < \alpha/\gamma$ by (1). Thus the power of p dividing $|H_1|$ is less than p^e where

$$e = n\gamma + \gamma - 1 = \gamma(n + 1) - 1.$$

However, the power of p dividing $|H_1|$ is $p^{\alpha - 1}$ so

$$\alpha - 1 < \gamma(n + 1) - 1$$

and $\alpha/\gamma < n + 1$. Thus α/γ is an integer satisfying $n < \alpha/\gamma < n + 1$ wihch is impossible, so this case cannot arise.

We may therefore suppose that $\gamma = 1$, so that $K_1/C_{K_1}(M)$ has order dividing $p - 1$ and K_1 is of order coprime to p. Let q be the largest prime dividing $|K|$, so that a Sylow q-subgroup Q of K is normal in K since K is supersolvable. If $q > p$, Q would be contained in $C_{K_1}(M)$, since $[K : K_1]$ is a power of p and $[K_1 : C_{K_1}(M)]$ divides $p - 1$. Lemma 2.2 would then imply that $Q = \{1\}$, so this case cannot arise. Proposition 1.3 implies that q cannot equal p, so we must have $q < p$ so that $|K|$ has order coprime to p, since q is the largest prime dividing $|K|$. Then $K = K_1$ and M, being a K_1-invariant subspace of N, is equal to N. Thus $|N| = p$ and G is supersolvable as required.

EXAMPLE   The holomorph of an elementary Abelian 3-group of order 9 is a QCLT-group of even order. This group has a 3-chief factor of order 9, so it is not the case that every p-chief factor, with p odd, of a QCLT-group is cyclic.

Theorem 2.3 is proved in [Humphreys 1974]. For related results see also [Brewster and Ottaway 1976].

Finally in this section, we prove a result to show that the class of QCLT-groups is closely connected with the class of supersolvable groups.

THEOREM 2.4   *Let n be a positive integer such that every group of order n is a QCLT-group. Then every group of order n is a supersolvable group.*

Proof: Let d be a divisor of n with $n = kd$. Suppose that there is a group H of order d which is not QCLT. Then if C is cyclic of order k, $G_1 = H \times C$ would be a group of order n with a quotient group H which is not QCLT. The definition of n therefore implies that for every divisor d of n, every group of order d is QCLT. It therefore follows that G, and every subgroup is CLT and so G is supersolvable by Theorem 4.1 of Chapter 1.

3. Joins of Normal Supersolvable Subgroups

In this section we consider those finite groups which are the join of two normal supersolvable subgroups. Recall from Chapter 1 (see Theorem 1.13 and the example preceding it) that such a group need not be supersolvable, and moreover will be supersolvable if and only if its commutator subgroup is nilpotent.

A weaker condition than supersolvability will, however, always hold for such groups:

THEOREM 3.1 *Let G be the join of two normal supersolvable subgroups. Then G satisfies the Sylow tower property.*

Since supersolvable groups satisfy the Sylow tower property (Theorem 1.8 of Chapter 1), Theorem 3.1 is a simple corollary of the more general result:

THEOREM 3.2 *Let G be the join of normal subgroups N_1 and N_2. If N_1 and N_2 have the Sylow tower property, then so does G.*
Proof: (i) Write $G = N_1 \# N_2$ to mean that G is the join of the normal subgroups N_1, N_2 and for any class $\mathcal{C}$, let $\mathcal{C}^\# = \{ N_1 \# N_2 : N_1, N_2 \in \mathcal{C} \}$. Thus, denoting the class of groups which satisfy the Sylow tower property by $\mathcal{STP}$, the theorem just asserts that $\mathcal{STP}^\# \subseteq \mathcal{STP}$.

(ii) If $\mathcal{C}$ is a class closed under homomorphic images, then so is $\mathcal{C}^\#$. For if N_1, $N_2 \in \mathcal{C}$, $G = N_1 \# N_2$, and N is a normal subgroup of G, then $G/N = N_1 N/N \# N_2 N/N$ and $N_i N/N \cong N_i/N_i \cap N \in \mathcal{C}$ $(i = 1, 2)$.

(iii) The class $\mathcal{STP}$ is closed under homomorphic images. For if $G = G_0 > G_1 > \ldots > G_r = \{1\}$ is a Sylow tower for G with $[G_i : G_{i+1}] = p_i^{n_i}$, then $G/N > G_1 N/N > \ldots > N/N = \{1\}$ is a Sylow tower for G/N since $[G_i N/N : G_{i+1} N/N] = [G_i N : G_{i+1} N] = [G_i : G_i \cap G_{i+1} N]$ and so divides $p_i^{n_i}$.

(iv) By (ii) and (iii), $\mathcal{STP}^\#$ is closed under homomorphic images so using induction on $G \in \mathcal{STP}^\#$, it only remains to show that G has a normal Sylow p-subgroup where p is the largest prime dividing $|G|$. Let P_i be a Sylow p-subgroup of N_i $(i = 1, 2)$, so that P_i is a characteristic subgroup of N_i and hence a normal subgroup of G. Thus $P = P_1 P_2$ is a normal p-subgroup of G. In fact, P is the required normal Sylow p-subgroup of G. To see this first observe that $P_1 \cap P_2$ is a Sylow p-subgroup of $N_1 \cap N_2$. This is because $P_1 \cap N_2$ and $P_2 \cap N_1$ are both Sylow p-subgroups of $N_1 \cap N_2$, so if P_0 is a Sylow p-subgroup of $N_1 \cap N_2$, we have that P_0 is contained in both P_1 and P_2. On the other hand $P_1 \cap P_2$ is a (normal) p-subgroup of $N_1 \cap N_2$, so $P_0 = P_1 \cap P_2$. Now

$$[G : P] = \frac{|G|}{|P|} = \frac{|N_1 N_2|}{|P_1 P_2|} = \frac{|N_1|}{|P_1|} \frac{|N_2|}{|P_2|} \frac{|P_1 \cap P_2|}{|N_1 \cap N_2|}$$

so that $[G : P]$ is not divisible by p and P is a Sylow p-subgroup of G.

The second result of this section establishes a sufficient condition on normal supersolvable subgroups whose join is G so that G' will be nilpotent.

THEOREM 3.3 *Let G be the join of normal supersolvable subgroups N_1 and N_2 such that $[G : N_1]$ and $[G : N_2]$ are coprime. Then G' is nilpotent.*

This and Theorem 1.13 of Chapter 1 immediately combine to yield

THEOREM 3.4 *Let G be the join of normal supersolvable subgroups N_1 and N_2 such that $[G : N_1]$ and $[G : N_2]$ are coprime. Then G is supersolvable.*

It remains to prove Theorem 3.3. Here again, it is no more difficult to relax the supersolvability hypothesis on N_1 and N_2 and prove instead

THEOREM 3.5 *Let G be the join of normal subgroups N_1 and N_2 such that N_1', N_2' are nilpotent and $[G : N_1]$ and $[G : N_2]$ are coprime. Then G' is nilpotent.*

Proof: First we will show that $G' = [N_1, N_2]N_1'N_2'$ (this holds, actually, in any group $G = N_1 \# N_2$ with no further requirements on N_1 and N_2). It suffices to show that a typical generator of G', which has the form $g = (x_1 x_2, y_1 y_2)$, $x_i, y_i \in N_i$, is contained in $[N_1, N_2]N_1'N_2'$. We have

$$g = x_2^{-1}(x_1, y_2)x_2(x_2, y_2)y_2^{-1}x_2^{-1}(x_1, y_1)x_2 y_2 y_2^{-1}(x_2, y_1))y_2.$$

Now N_1', N_2', being characteristic in N_1 and N_2 respectively, are normal in G. Also, $[N_1, N_2] \lhd G$ by [Scott 1964, 3.4.6]. It follows then that the factors $x_2^{-1}(x_1, y_2)x_2$, (x_2, y_2), $y_2^{-1}x_2^{-1}(x_1, y_1)x_2 y_2$, and $y_2^{-1}(x_2, y_1)y_2$ of g above belong to $[N_1, N_2]$, N_2', N_1', and $[N_2, N_1]$ $(= [N_1, N_2])$ respectively. Thus $g \in [N_1, N_2]N_1'N_2'$ as claimed.

Next we apply the hypothesis of coprime indices to show that every $g \in G$ is a product of *commuting* elements

$$g = s_1 s_2 = s_2 s_1, \qquad s_1 \in N_1, s_2 \in N_2.$$

Indeed, this hypothesis yields that $1 = in_1 + jn_2$ for some integers i, j, where $n_1 = [G : N_1]$ and $n_2 = [G : N_2]$. Letting $s_1 = g^{in_1}$ and $s_2 = g^{jn_2}$ establishes the claim.

Now let $t = x_1^{-1}x_2^{-1}x_1 x_2$, $x_1 \in N_1$, $x_2 \in N_2$, be a typical generator of $[N_1, N_2]$. By the above, $x_2^{-1}x_1 = s_1 s_2 = s_2 s_1$, where $s_1 \in N_1$, $s_2 \in N_2$. Thus $t = x_1^{-1}s_1 s_2 x_2 = s_1 x_1^{-1}(x_1^{-1}, s_1)(s_2^{-1}, x_2^{-1})x_2 s_2$. Letting $w = s_1 x_1^{-1}(x_1^{-1}, s_1) \cdot (s_2^{-1}, x_2^{-1})x_1 s_1^{-1}$, we have $w \in N_1'N_2'$ and $t = w s_1 x_1^{-1}x_2 s_2 = w s_1(x_2^{-1}x_1)^{-1}s_2 = w s_1(s_2 s_1)^{-1}s_2 = w$. Thus $t \in N_1'N_2'$ for an arbitrary typical generator of $[N_1, N_2]$ and so $[N_1, N_2] \leq N_1'N_2'$. This in conjunction with the result of the first paragraph gives $G' = N_1'N_2'$. Thus, since N_1', N_2' are normal nilpotent subgroups of G, $G' = N_1'N_2'$ is also nilpotent.

Remarks Theorem 3.4 is a special case of a theorem in [Kappe 1969]. Kappe shows that if $\mathcal{F}$ is a saturated formation defined locally by $\{\mathcal{F}(p)\}$, then the join of two normal $\mathcal{F}$-subgroups of coprime index is an $\mathcal{F}$-group provided that the join of two normal $\mathcal{F}(p)$-subgroups of coprime index is an $\mathcal{F}(p)$-group for all primes p. This result is also proved in [Slepova 1977]. The result for supersolvable groups was given in [Friesen 1971], but Friesen's proof is incorrect!

For a final result in this section we consider the situation $G = N_1 \# N_2$, where N_1, N_2 are assumed to be not only supersolvable, but moreover to belong to class $\mathcal{Y}$. (Class $\mathcal{Y}$ was introduced in Section 4 of Chapter 1 and will be looked at again later in Section 5 of this chapter. Recall that all groups in class $\mathcal{Y}$ are supersolvable.)

THEOREM 3.6 *Let G be the join of normal subgroups N_1 and N_2. If N_1, $N_2 \in \mathcal{Y}$, then G is supersolvable.*
Proof: As a first case we assume that one of the factors, N_1 say, is nilpotent. If $N_1 \cap N_2 = \{1\}$, then $G = N_1 \oplus N_2$ is surely supersolvable, so we may assume nontriviality of $N_1 \cap N_2$. Therefore $Z(N_1) \cap (N_1 \cap N_2)$ $= Z(N_1) \cap N_2$ is nontrivial by nilpotency of N_1. Now $Z(N_1)$ is characteristic in N_1, hence normal in G. Thus $Z(N_1) \cap N_2 \lhd N_2$. Let R be a minimal normal subgroup of N_2 contained in $Z(N_1) \cap N_2$. Since N_2 is supersolvable, R is cyclic. Now $R \leq Z(N_1)$ implies $R \lhd N_1$. This and $R \lhd N_2$ mean that $R \lhd G$. Moreover $G/R = N_1/R \# N_2/R$ so induction applies to yield G/R is supersolvable. This and R cyclic imply G is supersolvable.

We now consider the general case, N_1, $N_2 \in \mathcal{Y}$. By Theorem 4.3 of Chapter 1, N_i $(i = 1, 2)$ has a normal Hall subgroup H_i such that H_i and N_i/H_i are nilpotent and $N_i = H_i N_{N_i}(T)$ for all $T \leq H_i$. As a consequence of this last equation we have $T \lhd N_i$ for all $T \lhd H_i$.

Since normal Hall subgroups are characteristic, we get H_1, $H_2 \lhd G$. If $H_i = \{1\}$ for either $i = 1, 2$, then N_i is nilpotent so G is supersolvable by the first case above. Assume then that H_1, $H_2 \neq \{1\}$. Now note that G/H_1 $= N_1/H_1 \# N_2 H_1/H_1$ so G/H_1 is supersolvable by induction and likewise is G/H_2. Hence $G/H_1 \cap H_2$, which can be embedded in $G/H_1 \times G/H_2$, is supersolvable. Thus if $H_1 \cap H_2 = \{1\}$, we get G is supersolvable.

Assume then that $H_1 \cap H_2 \neq \{1\}$. Hence $Z(H_1) \cap (H_1 \cap H_2)$ $= Z(H_1) \cap H_2$ is nontrivial by nilpotency of H_1. Then $Z(H_2) \cap (Z(H_1) \cap H_2) = Z(H_1) \cap Z(H_2)$ is nontrivial by nilpotency of H_2. Let $z \in Z(H_1) \cap Z(H_2)$, $z \neq 1$. Then (see the last sentence of the second paragraph) $\langle z \rangle$ $\lhd H_i$ implies $\langle z \rangle \lhd N_i$ for $i = 1, 2$. Hence $\langle z \rangle \lhd G$. But since $G/\langle z \rangle$

$= N_1/\langle z \rangle \# N_2/\langle z \rangle$, we have $G/\langle z \rangle$ is supersolvable by induction. Conclude G is supersolvable.

4. Groups whose Lattice of Subgroups is Lower Semi-modular

DEFINITION The lattice of subgroups of a group G is said to be **lower semi-modular** if for every pair of subgroups A, B of G such that A is a maximal subgroup of $\langle A, B \rangle$, then $A \cap B$ is a maximal subgroup of B. We call G an **LM-group** when its lattice of subgroups is lower semi-modular.

PROPOSITION 4.1 *Let G be an* LM-*group. Then every subgroup and every homomorphic image of G is an* LM-*group.*
Proof: The proof of these facts follows easily from the definitions.

PROPOSITION 4.2 *The following conditions on a finite group G are equivalent*:
(a) *G is an* LM-*group*
(b) *For every pair M_1, M_2 of distinct maximal subgroups of a subgroup of G, $M_1 \cap M_2$ is a maximal subgroup of M_1 and of M_2.*
Proof: (a)$\Rightarrow$(b) Let G be an LM-group and M_1, M_2 be distinct maximal subgroups of a subgroup H of G. Then $\langle M_1, M_2 \rangle = H$ and so $M_1 \cap M_2$ is a maximal subgroup of both M_1 and M_2.

(b)$\Rightarrow$(a) Let A, B be subgroups of G with A a maximal subgroup of $\langle A, B \rangle$. Choose a maximal subgroup B_1 of $\langle A, B \rangle$ containing B. Then $\langle A, B \rangle = \langle A, B_1 \rangle$ and so by condition (b), $A \cap B_1$ is maximal in both A and B_1. Now choose a maximal subgroup B_2 of B_1 containing B, so that $\langle A \cap B_1, B_2 \rangle = B_1$. Condition (b) now implies that $A \cap B_2$ is maximal in both B_2 and $A \cap B_1$. Continue in this way to get a maximal chain

$$B_1 > B_2 > \ldots > B_r = B.$$

Then $A \cap B_r = A \cap B$ is a maximal subgroup of $B_r = B$, proving condition (a).

THEOREM 4.3 *An* LM-*group is supersolvable.*
Proof: Let G be an LM-group. We prove by induction on $|G|$ that G has the equichain condition and so is supersolvable by Theorem 2.4 of Chapter 1. By Proposition 4.1, we may suppose that every proper subgroup of G is supersolvable and so has the equichain condition. Let B be any subgroup

of G and

$$G = G_0 > G_1 > \ldots > G_r = B$$
$$G = H_0 > H_1 > \ldots > H_s = B$$

be two chains of maximal subgroups from G to B. We want to show that $r = s$ and induction allows us to suppose that $H_1 \neq G_1$. Since G is an LM-group, (b) of Proposition 4.2 implies that $H_1 \cap G_1$ is a maximal subgroup of both G_1 and H_1. Thus, if t is the length of a maximal chain from $H_1 \cap G_1$ to B, then there is a maximal chain of length $t + 1$ from G_1 to B. Since G_1 is supersolvable, $t + 1 = r - 1$. Similarly, since H_1 is supersolvable, $t + 1 = s - 1$ and so $r = s$ as required.

EXAMPLE Not every finite supersolvable group is LM. For let $G = \langle x, y : y^7 = x^6 = 1, x^{-1}yx = y^3 \rangle$, which can be realized as the holomorph of $\mathbb{Z}_7$. Then G is supersolvable since $\langle y \rangle$ is a normal cyclic subgroup of G with cyclic quotient group. However, G is not LM. For let $U = \langle x \rangle$ so that U is maxmial in G since $[G : U] = 7$. Let $V = \langle yx \rangle$ so that V also has order six and index seven and is maximal in G. It follows that $\langle U, V \rangle = G$, but $U \cap V = \{1\}$ so $U \cap V$ is not maximal in either U or V.

The class of LM-groups has been characterized in [Suzuki 1956, Chapter 1] based on the work of N. Itô ([Itô 1951]). We now give this characterization.

THEOREM 4.4 *A finite group G is an LM-group if and only if G is supersolvable and for all chief factors H/K of G, $\mathrm{Aut}_G(H/K)$ is cyclic of order dividing a prime (i.e., either cyclic of prime order or trivial).*
Proof: Suppose G is an LM-group. Then by Theorem 4.3, G is supersolvable. We show by induction on $|G|$ that the group of automorphisms induced on each chief factor has the required property. Let N be a minimal normal subgroup of G, so that $|N| = p$ for some prime p. Since G/N is an LM-group by Proposition 4.1, we only need show that $G/C_G(N)$ is cyclic of order dividing some prime q. If G had a minimal normal subgroup of prime order r with $r \neq p$, induction and the Jordan–Hölder theorem would give the result. Thus we may suppose that p is the largest prime dividing $|G|$ and also that $\mathrm{Fit}(G)$ is a p-group.

Let P be a Sylow p-subgroup and H be a Sylow p-complement of G. Then $Z(P) \cap N \neq \{1\}$ so $C_G(N) \geq P$. Since

$$C_{HN}(N) = C_G(N) \cap HN,$$

we have

$$HN/C_{HN}(N) = HN/C_G(N) \cap HN \cong HNC_G(N)/C_G(N) = G/C_G(N)$$

so $\mathrm{Aut}_G(N) \cong \mathrm{Aut}_{HN}(N)$.

Now let $C = C_{HN}(N)$ so that $C = N(C \cap H)$. Since G is an LM-group, Proposition 4.1 implies that $HN/C \cap H$ is also an LM-group. Since $C/C \cap H$ is a chief factor of order p, induction allows us to suppose that $HN = G$ and $C \cap H = \{1\}$. Thus G is of the form

$$G = AB = \langle x, y : x^p = 1 = y^e, x^y = x^i \text{ where } e \text{ divides } p - 1$$
$$\text{and } i \text{ has order } e \bmod p \rangle$$

where $A = \langle x \rangle$ and $B = \langle y \rangle$, the case $|B| = 1$ being the trivial case of the result. We therefore wish to show that B has prime order. Consider $B^x = \langle y^x \rangle$. We now show $B^x \cap B = \{1\}$. For if $g \in B^x \cap B$, we have that $y^n = (y^x)^m = (y^m)^x$ for some integers n, m. However, $y^x = yx^{i-1}$ and so $(yx^{i-1})^m = y^m x^k$ for some integer k. It therefore follows that $n = m$ and x centralizes y^n. Since x has order p and $x^y = x^i$ with $i \neq 1$, this is impossible unless $y^n = 1$ so $B^x \cap B = \{1\}$. Since B, B^x have index p in $AB = G$, they are both maximal and so, since G is an LM-group, $B^x \cap B$ is a maximal subgroup of B. This implies that B must be cyclic of prime order.

Conversely, let G be a finite supersolvable group such that the group of automorphisms induced on each chief factor of G is cyclic of order dividing a prime. Let H be any subgroup of G. The series obtained by intersecting a chief series of G term-by-term with H is, after deleting repetitions, a chief series for H. Furthermore the group of automorphisms induced on each chief factor of this series for H is a subgroup of the corresponding automorphism group induced by G. Thus H has the same property as G and so induction and Proposition 4.2(b) mean that we only need to show that for any pair M_1, M_2 of distinct maximal subgroups of G, $M_1 \cap M_2$ is maximal in M_1 and in M_2.

Suppose that there is a minimal normal subgroup N of G contained in M_1. Then G/N is supersolvable and the automorphism group induced by G/N on each chief factor of G/N is cyclic of order dividing a prime, so by induction G/N is an LM-group. Since M_2 is a maximal subgroup of G, NM_2 is either M_2 or G. If $NM_2 = M_2$, then M_1/N and M_2/N are maximal subgroups of the LM-group G/N so $M_1 \cap M_2/N$ is a maximal subgroup of both M_1/N and M_2/N proving that $M_1 \cap M_2$ is a maximal subgroup of both M_1 and M_2, as required. On the other hand, if $NM_2 = G$, $N \cap M_2$ is normal in M_2 and in N since N is Abelian, so $N \cap M_2$ is a normal subgroup of G. Minimality of N then forces $N \cap M_2 = \{1\}$ so that $[G:M_2]$ is a prime p. Then $M_1 = G \cap M_1 = NM_2 \cap M_1 = N(M_2 \cap M_1)$ so $[M_1 : M_1 \cap M_2] = p$ and $M_1 \cap M_2$ is a maximal subgroup of M_1. Also since G is supersolvable, $[G:M_1]$ is a prime q and so

$$[G . M_1][M_1 . M_1 \cap M_2] = pq = [G : M_2][M_2 : M_1 \cap M_2]$$

which implies that $M_1 \cap M_2$ is also a maximal subgroup of M_2. We may therefore suppose that neither M_1 nor M_2 contain a nontrivial normal subgroup of G.

Now let p be the largest prime dividing $|G|$ and P be a Sylow p-subgroup of G. Since G is supersolvable, P is a normal subgroup of G and so P is not contained in M_1 and P is not contained in M_2. This implies that $[G : M_1] = p = [G : M_2]$. Also $G = PM_1 = PM_2$. Now $P \cap M_1$ is normal in M_1 and, since it has index p in P, $P \cap M_1$ is normal in P. Therefore $P \cap M_1$ is normal in G and so by the above, $P \cap M_1 = \{1\}$ and P must therefore have order p. Hence M_1, M_2 are Sylow p-complements in G. Writing C for $C_G(P)$, we have that $[G : C]$ is, by assumption, either 1 or q for some prime q. Since M_1, M_2 are distinct, $[G : C]$ is not 1 and so $[G : C] = q$. Also

$$C = P(M_2 \cap C)$$

so that $M_2 \cap C$ being a normal Sylow p-complement in C, is in fact characteristic in C and therefore normal in G. However, M_2 contains no nontrivial normal subgroups of G, so $M_2 \cap C = \{1\}$. Since $G = CM_2$, $[G : C] = q$ and $[G : M_2] = p$. We deduce that $|G| = pq$ and so

$$[M_2 : M_1 \cap M_2] = q = [M_1 : M_1 \cap M_2].$$

Thus $M_1 \cap M_2$ $(= \{1\})$ is a maximal subgroup of both M_1 and M_2, as required.

COROLLARY 4.5 *If G_1, G_2 are LM-groups, then so is $G_1 \times G_2$.*

COROLLARY 4.6 *A group G is an LM-group if and only if the quotient group of G by its hypercenter is an LM-group.*
Proof: By Proposition 4.1 and the fact that each chief factor of G below its hypercenter is central.

5. *The Classes $\mathcal{X}$ and $\mathcal{Y}$*

It is obvious that a maximal subgroup M of a group G cannot be written as a proper intersection of subgroups of G. Proper subgroups with this property are called **primitive**. In other terminology they are the meet-irreducible elements of the subgroup lattice. Primitive subgroups arise naturally in the study of CLT-groups, minimal permutation representations, and elsewhere. For example, every subgroup of G can be written as an intersection of primitive subgroups, and the set of all primitive subgroups of G is characterized by its minimality with respect to this property.

Conditions on maximal subgroups are related to the normal structure of a group. Thus the maximal subgroups of a solvable group all have prime-power index, and those of a supersolvable group have prime index. While the converse to the first of these results is false (every maximal subgroup of the simple group PSL(2, 7) has index 7 or 8), that of the second (Theorem 3.1 of Chapter 1) was proved by B. Huppert. As might be expected, the imposition of conditions on the primitive subgroups leads to stronger conclusions, and an example is provided by the main result of this section, Theorem 5.1, which is from [Johnson 1971].

We let $\mathcal{X}$ denote the class of groups G such that all primitive subgroups of G have prime-power index.

THEOREM 5.1 *If $G \in \mathcal{X}$, then G is supersolvable.*
Proof:
1. The first of our three steps is to establish a basis for induction. This consists of the following three assertions:
   (a) *If H is a subgroup of G and K is a primitive subgroup of H, then there is a primitive subgroup X of G with $K = X \cap H$.*
   (b) *If $G \in \mathcal{X}$ and $N \lhd G$, then $G/N \in \mathcal{X}$.*
   (c) *If $G \in \mathcal{X}$ and $N \lhd G$, then $N \in \mathcal{X}$.*

To prove (a), write K as an intersection of primitive subgroups X_i of G, $1 \le i \le n$, so that

$$K = K \cap H = \bigcap (X_i \cap H).$$

Since K is primitive in H, $K = X_i \cap H$ for some i. As to (b), it is clear that if H/N is primitive in G/N, then H is a primitive subgroup of G of the same index. To prove (c), let H be primitive in N and write $H = X \cap N$ with X primitive in G in accordance with (a). Then

$$[N : H] = [N : X \cap N] = [NX : X]$$

is a divisor of the prime prower $[G : X]$.

2. We now claim that every $G \in \mathcal{X}$ is solvable. Assume that this is false and let G be a minimal counterexample. Then G is a non-Abelian simple group because of (b) and (c). Let S be a Sylow p-subgroup of G, where p is the smallest prime dividing $|G|$, with $|S| = p^n$ say. Let M be a maximal subgroup of S and N the normalizer of M in G (the case $n = 1$ does not need separate treatment). Then $M \le S \le N$ and p divides $|N/M|$ to the first power only so we can invoke Burnside's transfer theorem ([Scott 1964, 6.2.11]) to assert the existence of a normal complement C/M for S/M in N/M. Now $[N : C] = p$ and so C is maximal, and hence primitive, in N. By (a) above, there is a primitive subgroup X of G with $C = X \cap N$. By hypothesis, there is a prime q such that $[G : X] = q^v$, and we distinguish two cases.

First assume $p = q$, so that $v = 1$ since p^{n-1} divides $|X|$. Hence G/K can be embedded in the symmetric group S_p, where $K = \text{Core}_G(X)$. It follows that $[X : K]$ has order dividing both $|X|$ and $(p - 1)!$. Since p is the smallest prime dividing $|G|$, these numbers are coprime, and we conclude that $X = K \lhd G$, a contradiction.

In the second case ($p \neq q$) we can find a Sylow p-subgroup T of G lying in X and containing M. Since T normalizes M, $T \leq X \cap N = C$. Thus $[G : C]$ is coprime to p, contrary to the fact that $[N : C] = p$.

3. Finally we prove that every $G \in \mathfrak{X}$ is supersolvable. Since all quotient groups of G are in $\mathfrak{X}$, all proper quotient groups are supersolvable by induction. We may therefore assume G has a unique minimal normal subgroup N. We may further assume that $\Phi(G)$ is trivial, since nontriviality would imply $G/\Phi(G)$ is supersolvable, hence so too is G by Corollary 3.2 of Chapter 1. Now uniqueness of N and $\Phi(G) = \{1\}$ yield $\mathrm{Fit}(G) = N$ by Theorem 2.3 of Appendix C.

Let $|N| = p^s$.

Case 1: $p \nmid |G/N|$. Let H be a maximal subgroup of N so that $[N : H] = p$. Let X be a primitive subgroup of G so that $H = X \cap N$ (existence by (a) above) and observe that both X and N normalize H. Furthermore, $[G : X]$ is a prime-power divisible by $[NX : X] = p$ and is thus a power of p. So too then is $[G : NX]$, hence $G = NX$ by case hypothesis. This implies $H \lhd G$ so $H = \{1\}$ by minimality of N. We now have N cyclic and G/N supersolvable. Hence G is supersolvable.

Case 2: $p \,|\, |G/N|$. Since G/N is supersolvable, it satisfies the Sylow tower property (Theorem 1.8 of Chapter 1) so has a normal Sylow q-subgroup K/N for its largest prime divisor q. If $p = q$, then K is a p-group. Hence $K \leq \mathrm{Fit}(G) = N$ so K/N is trivial, a contradiction.

We will show $p \neq q$ also leads to a contradiction, which will serve to eliminate Case 2 altogether and complete the proof. If $p \neq q$, then $p \,|\, [G : N]$ implies $p \,|\, [G : K]$ so that $K < G$. Now $K \lhd G$ implies $K \in \mathfrak{X}$ so K is supersolvable by induction. Since $p \,|\, [G : N]$, $p < q$ by definition of q. Then since K satisfies the Sylow tower property, K has a normal Sylow q-subgroup Q. But then $Q \lhd G$ contradicting the fact that N is the unique minimal normal subgroup and $|N| = p^s$, $p \neq q$.

EXAMPLE Since $S_3 \in \mathfrak{X}$, $\mathfrak{X}$ properly contains the class of nilpotent groups. Moreover, the supersolvable group $S_3 \times \mathbb{Z}_3$ contains a primitive subgroup of order 3, and so the converse of the theorem is false. This example also shows that $\mathfrak{X}$ is not closed under direct products, and thus is not a formation.

It is clear that $\mathfrak{X}$ consists of those groups whose subgroups are intersections of subgroups of prime-power indices. Consider the stronger condition: every subgroup can be written as an intersection of subgroups of pairwise coprime prime-power indices. For notational convenience we denote the class of groups satisfying this stronger condition by $\mathfrak{Y}^*$. The connection with class $\mathfrak{Y}$ introduced in Chapter 1 is established by the following theorem of McLain ([McLain 1957]), which will be proved here by a slightly different method.

THEOREM 5.2 $\mathscr{Y}^* = \mathscr{Y}$.

Proof: That $\mathscr{Y} \subseteq \mathscr{Y}^*$ is an immediate consequence of the elementary but ubiquitous fact that subgroups of coprime indices m and n intersect in a subgroup of index mn. To prove $\mathscr{Y}^* \subseteq \mathscr{Y}$, assume the result false, let G be a minimal counterexample, and H a contrary subgroup of G of minimal index. Since $G \in \mathscr{Y}^*$, it follows from the minimality of $[G:H]$ that $[G:H]$ is a prime-power, and that G has no subgroup containing H as a subgroup of prime index. Since $\mathscr{Y}^* \subseteq \mathscr{X}$, G is supersolvable by Theorem 5.1 and so a minimal normal subgroup N of G has prime order. If $N \leq H$, the quotient-closure of $\mathscr{Y}^*$ and the minimality of $|G|$ provide a contradiction. But otherwise, $[NH : H]$ is a prime, and this is also impossible.

Let us denote by $\mathscr{S}\mathscr{X}$ the class of groups whose subgroups all belong to $\mathscr{X}$, and define $\mathscr{S}\mathscr{Y}$ similarly. (This, we note in passing, is nonstandard usage of the closure operations as defined, e.g., in [Hall 1963]; cf. the class $\mathscr{Q}\mathscr{C}\mathscr{L}\mathscr{T}$). The class $\mathscr{S}\mathscr{Y}$ was studied by McLain under the name B1. The next result establishes a strong connection between $\mathscr{X}$ and $\mathscr{Y}$.

THEOREM 5.3 $\mathscr{S}\mathscr{X} = \mathscr{S}\mathscr{Y}$.

Proof: Since $\mathscr{Y} \subseteq \mathscr{X}$, it is clearly sufficient to prove that $\mathscr{S}\mathscr{X} \subseteq \mathscr{Y}$, for then $\mathscr{S}\mathscr{X} = \mathscr{S}\mathscr{S}\mathscr{X} \subseteq \mathscr{S}\mathscr{Y} \subseteq \mathscr{S}\mathscr{X}$. We assume the result to be false and let G be a minimal counterexample. Let H be a contrary subgroup of minimal index in G, so that H cannot be the intersection of a pair of subgroups of coprime indices. If p denotes the largest prime dividing $|G|$, then G has a normal subgroup N of order p because supersolvable groups satisfy the Sylow tower property (Theorem 1.8 of Chapter 1). If $N \leq H$, the quotient-closure of $\mathscr{S}\mathscr{X}$ and the minimality of $|G|$ lead to a contradiction. Thus, NH contains H as a subgroup of index p and we can write $[G:H] = p^\alpha m$ where $\alpha \geq 1$ and $m \neq 1$ is coprime to p. By the minimality of $[G:H]$, $NH = M \cap P$, where $[G:M] = m$ and $[G:P] = p^{\alpha-1}$. If $\alpha > 1$, P is a proper subgroup of G and so has a subgroup Q of index p containing H by induction. This gives the contradiction $H = M \cap Q$, whence $\alpha = 1$.

Now write $m = qm'$, where q is a prime $q \neq p$. Since $G/N \in \mathscr{S}\mathscr{X}$, $G/N \in \mathscr{Y}$ by induction so that (Theorem 5.2) G has a subgroup X of index q containing NH. Again by induction, X has a subgroup Y of index p containing H. If $Y \neq H$, the minimality of $[G:H]$ ensures the existence of a subgroup Z of index p in G containing Y, and we have the contradiction $H = M \cap Z$. Thus $Y = H$ so that $[G:H] = pq$. Using the property $\mathscr{X}$ (for the first and only time), the subgroup H cannot be primitive, and so $H = NH \cap K$ with $N < K$ (by (a) of Theorem 5.1). Now $[G:NH] = q$, so that $[G:K] \neq p$ by hypothesis. Hence, $[G:K] = q$ and we have

$$p = [NH : H] = [NH : NH \cap K] = |NHK|/|K| \leq [G : K] = q,$$

contradicting the fact that p is the largest prime dividing $|G|$.

Other links between $\mathcal{X}$ and $\mathcal{Y}$ include the result (proved in [Humphreys/Johnson 1973]) that if $G \in \mathcal{X}$ and has fourth-power-free order, then $G \in \mathcal{Y}$. However, the question of the equality of $\mathcal{X}$ and $\mathcal{Y}$ is settled by the following example.

EXAMPLE ([Humphreys 1974]) Let P be the split extension of an elementary Abelian group $\langle u_1, u_2, x_1, x_2 \rangle$ of order 3^4 by an elementary Abelian group $\langle x_3, x_4 \rangle$ of order 3^2, where the action is given by

$$x_3 \rightarrow \begin{pmatrix} 1 & 0 & 0 & 0 \\ 0 & 1 & 0 & 0 \\ 1 & 0 & 1 & 0 \\ 0 & 1 & 0 & 1 \end{pmatrix} \qquad x_4 \rightarrow \begin{pmatrix} 1 & 0 & 0 & 0 \\ 0 & 1 & 0 & 0 \\ 0 & 1 & 1 & 0 \\ 1 & 1 & 0 & 1 \end{pmatrix}$$

Then it is easily shown that P is a special group of exponent 3, and that if $1 \neq z \in Z(P) = \langle u_1, u_2 \rangle$, then $P/\langle z \rangle$ is extra-special. Now let G be the split extension of P by the involutory automorphism that inverts the generators x_i, $1 \leq i \leq 4$, and centralizes u_1, u_2. It is not hard to show that the subgroup $\langle x_1 x_3, x_2 x_4 \rangle$ of G is normalized by no element of order 2, so that $G \notin \mathcal{Y}$. On the other hand, the proof that $G \in \mathcal{X}$ is by no means a simple matter, so we suppress the details and refer the reader to [Humphreys 1974].

We conclude this section by stating a local version of Theorem 5.1 obtained by M. S. Chen ([Chen 1977]).

THEOREM 5.4 *Let p be a prime and G a group in which every primitive subgroup has p-power index or p'-index. If G has a solvable (supersolvable) p-complement, then G is solvable (supersolvable).*

6. Digression: Minimal Permutation Representations

We digress briefly to discuss the part played by primitive subgroups in the study of minimal permutation representations. For a finite group G, define the degree $\mu(G)$ of G to be the minimal cardinality of a set admitting a faithful G-action. Thus $\mu(G)$ is the least number in the set $\{n: S_n$ has a subgroup isomorphic to $G\}$. Alternatively, a simple application of the orbit stabilizer theorem shows that $\mu(G)$ is the minimum value of $\sum_{j=1}^{n}[G:G_i]$, where $\{G_i: 1 \leq i \leq n\}$ ranges over all collections of subgroups of G whose cores intersect trivially. It is not hard to see that this minimum is achieved at a collection having primitive subgroups only.

As might be expected ([Johnson 1971]), the degree of an Abelian group is just the sum of its elementary divisors, and in general we have $\mu(G \times H) \leq \mu(G) + \mu(H)$, with equality when $|G|$ and $|H|$ are coprime. That equality prevails when G and H are both p-groups is a deeper result ([Wright 1976]), requiring among other things a nontrivial result from the theory of matroids. In its most general form, Wright's theorem may be stated as follows:

THEOREM 6.1 *Let $\mathfrak{D}$ be the class of groups G having a nilpotent subgroup N such that $\mu(G) = \mu(N)$. Then G is closed under direct products and*

$$\mu(H \times K) = \mu(H) + \mu(K)$$

for all $H, K \in \mathfrak{D}$.

It can be shown that $\mathfrak{D}$ is not comparable with the class $\mathfrak{S}$ of solvable groups: it is easy to show that $\mathfrak{D}$ contains all symmetric, alternating, and dihedral groups, and on the other hand an example is given in [Wright 1976] of a group G of order 150 with $\mu(G) = \mu(G \times \mathbb{Z}_5) = 15$. The paper [Berkovič and Engels 1979] gives examples of p-groups A, B and a proper quotient group G of $A \times B$ such that $\mu(G) > \mu(A) + \mu(B)$. The reference [Ljubič 1979] contains an alternate proof of Wright's theorem, and also a subgroup of $\mathbb{Z}_2 \wr \mathbb{Z}_5$ not in $\mathfrak{D}$.

7. Semi-nilpotent Groups

It is well known that a group whose proper subgroups are all nilpotent must be solvable ([Scott 1964, 6.5.7]). Our purpose in this section is to describe a considerable strengthening and sharpening of this result obtained by Sah. He defined ([Sah 1957]) the class $\mathfrak{S}\mathfrak{N}$ of **semi-nilpotent** groups to consist of those groups G satisfying either of the following two equivalent conditions:

(a) If H is a nonnormal nilpotent subgroup of G, then $N_G(H)$ is nilpotent.
(b) If P is a nonnormal p-subgroup of G, then $N_G(P)$ is nilpotent.

To show that (a) is a consequence of (b), let H be a nonnormal nilpotent subgroup of G. Then all the Sylow subgroups of H are normal in H, but at least one of them—call it P—fails to be normal in G. Since P is characteristic in H, $N_G(P)$ contains $N_G(H)$, which is thus nilpotent, as required.

An arbitrary group G has two canonical nilpotent subgroups of particular importance in this context. The first of these is the Fitting subgroup $\text{Fit}(G)$, best thought of here in the following way: if $P_1, \ldots, P_n$ is a list of Sylow subgroups of G containing one for each prime dividing $|G|$, then $\text{Fit}(G)$ is the direct product of the subgroups $\text{Core}_G(P_i)$, $1 \leq i \leq n$, the cores of the P_i. The other useful subgroup is the hypercenter $Z^*(G)$, defined as the terminus of the upper central series of G. See Appendix C for basic results concerning these two subgroups.

From now on, let $G \in \mathfrak{S}\mathfrak{N}$. The subgroups of G on which attention is focussed throughout are its maximal nilpotent subgroups. Note that such a subgroup is either normal of self-normalizing, and in the former case it coincides with $\text{Fit}(G)$. In any case, the connection between these subgroups and the normal structure of G is illustrated by the following fundamental result.

LEMMA 7.1 *If A and B are two distinct maximal nilpotent subgroups of the semi-nilpotent group G, then $A \cap B \lhd G$.*

Proof: For a fixed A and B let $C \neq A$ be a maximal nilpotent subgroup of G with $A \cap B \leq A \cap C = D$, chosen in such a way that D is maximal with respect to this property. We claim that $D \lhd G$. Assuming this to be false, $N_G(D)$ must be nilpotent and there is a maximal nilpotent subgroup T of G with $N_G(D) \leq T$. Now distinct maximal objects cannot be comparible, whence $D < A$ and it follows from the nilpotence of A that $D < N_A(D)$ $= A \cap N_G(D) \leq A \cap T$. The maximality of D now implies that $T = A$. Since C is nilpotent, we have $D < N_G(D) = N_G(D) \cap C \leq T \cap C = A \cap C = D$, which is the required contradiction. Hence, $A \cap B \leq D \lhd G$, and so $A \cap B \leq \text{Core}_G(A)$. By symmetry, $A \cap B \leq \text{Core}_G(B)$, whence $A \cap B = \text{Core}_G(A) \cap \text{Core}_G(B) \lhd G$, completing the proof.

COROLLARY 7.2 *If A is a maximal nilpotent subgroup of $G \in \mathfrak{S}\mathfrak{N}$ and $x \notin A$, then $A \cap A^x = \text{Core}_G(A)$.*

Proof: If $A \lhd G$ the result is obvious, while if not, $N_G(A)$ is nilpotent and equal to A be maximality. Hence, A and A^x are distinct and $A \cap A^x \lhd G$ by Lemma 7.1. Thus, $\text{Core}_G(A) = \text{Core}_G(A) \cap \text{Core}_G(A^x) \leq A \cap A^x$ $\leq \text{Core}_G(A)$, proving the corollary.

LEMMA 7.3 (i) *If H is a nilpotent Hall subgroup of $G \in \mathfrak{S}\mathfrak{N}$ and $K \lhd H$ but K is not normal in G, then $N_G(K)$ is a maximal nilpotent subgroup of G.*

(ii) *Let A be a maximal nilpotent subgroup of G and B a Hall subgroup of A. If none of the Sylow subgroups of B is normal in G, then B is a Hall subgroup of G.*

Proof: (i) Let T be a nilpotent subgroup of G containing $N_G(K)$. Since H is a Hall subgroup of T, it must be a direct factor, whence $K \lhd T$. Thus $T = N_G(K)$ as required.

(ii) Let P be a Sylow p-subgroup of B. This and B is a Hall subgroup of A imply that P is a Sylow subgroup of A. It follows that $A = N_G(P)$ and $[N_G(P):P]$ is coprime to p. Now let Q be a Sylow p-subgroup of G containing P and assume that $P < Q$. Then $P < N_Q(P) \leq N_G(P)$ provides a contradiction, so that $P = Q$. Thus every Sylow p-subgroup of B is a Sylow p-subgroup of G which proves the lemma.

COROLLARY 7.4 *If H is a nilpotent Hall subgroup of $G \in \mathcal{SN}$ and $x \notin N_G(H)$, then $\mathrm{Core}_G(H) = H \cap H^x = H \cap \mathrm{Core}_G(N_G(H))$.*
Proof: We show first that $H \cap \mathrm{Core}_G(N_G(H))$ is a normal subgroup of G. Since $N_G(H)$ is nilpotent and H is a Hall subgroup, we can write $\mathrm{Core}_G(N_G(H)) = S_1 \times S_2$, where S_1 is the direct product of the Sylow subgroups lying in H and S_2 that of those in its normal complement. Since each Sylow factor of S_1 is characteristic in $\mathrm{Core}_G(N_G(H))$, it follows that $H \cap \mathrm{Core}_G(N_G(H)) = S_1$ is normal in G. Hence $H \cap \mathrm{Core}_G(N_G(H)) \leq \mathrm{Core}_G(H)$, and the reverse inclusion is obvious. Further, $N_G(H)$ is maximal nilpotent by Lemma 7.3(i) and so $\mathrm{Core}_G(N_G(H)) = N_G(H) \cap N_G(H)^x$, by Corollary 7.2. Thus,

$$H \cap H^x = N_G(H) \cap N_G(H^x) \cap H \cap H^x = N_G(H) \cap N_G(H)^x \cap H \cap H^x$$

$$= \mathrm{Core}_G(N_G(H)) \cap H \cap H^x = \mathrm{Core}_G(H) \cap H^x = \mathrm{Core}_G(H).$$

The main tool in Sah's proof is the famous theorem of Frobenius ([Scott 1964, 12.5.11]), together with some of its consequences, restated here for convenience.

THEOREM OF FROBENIUS *Let G be a group and H a subgroup of G such that $H \cap H^x = \{1\}$ for all $x \in G \setminus H$. Then the elements of G not conjugate to an element of H, together with the identity, comprise a normal subgroup N of G.*

Consequences ([Scott 1964, 12.6.1 and 12.6.2]): It is clear that H and N are complementary subgroups of G, and that H acts fixed-point-freely on N. A simple counting argument shows that $|H|$ and $|N|$ are coprime, whence a nilpotent subgroup of G either lies in N or has order dividing $|H|$. It follows that N contains every normal p-subgroup of G, and thus contains $\mathrm{Fit}(G)$.

LEMMA 7.5 *If A is a maximal nilpotent subgroup of $G \in \mathcal{SN}$, then there is a unique subgroup A^* of G satisfying the following conditions:*
(a) $A^* \triangleleft G$,
(b) $A^* A = G$,
(c) $A^* \cap A = \mathrm{Core}_G(A)$,
(d) $A^* \geq \mathrm{Fit}(G)$.

Proof: If $A \lhd G$, then G is clearly the required subgroup, so assume A is not normal in G. In view of Corollary 7.2 we can apply Frobenius's theorem to $G/\mathrm{Core}_G(A)$ to obtain the Frobenius kernel $A^*/\mathrm{Core}_G(A)$. Now A^* is the unique subgroup of G satisfying (a), (b), and (c), and since $\mathrm{Core}_G(A)\mathrm{Fit}(G)/\mathrm{Core}_G(A) \leq \mathrm{Fit}(G/\mathrm{Core}_G(A))$, it follows from the consequences of Frobenius's theorem that (d) also holds.

THEOREM 7.6 *If G is a semi-nilpotent group and $F_0(G)$ denotes the direct product of its normal Sylow subgroups, then $G/F_0(G)$ is nilpotent.*
Proof: It is clear that $F_0(G)$ is a direct factor of $\mathrm{Fit}(G)$; we show first that $G/\mathrm{Fit}(G)$ is nilpotent. Let $S_1, \ldots, S_n$ be a set of representatives for the nonnormal Sylow subgroups of G. Assume that $n \geq 1$, otherwise G is nilpotent and there is nothing to prove. By Lemma 7.3(i), $A_i = N_G(S_i)$ is a maximal nilpotent subgroup of G for each i. Letting A_i^* be the subgroup given by Lemma 7.5, each G/A_i^* is nilpotent. Putting $K = \bigcap_{i=1}^{n} A_i^*$, it follows that G/K is nilpotent and $K \geq \mathrm{Fit}(G)$. Assuming for a contradiction that $K > \mathrm{Fit}(G)$, it follows that K is not nilpotent and thus has a nonnormal Sylow p-subgroup T. If S is a Sylow p-subgroup of G containing T, then $\mathrm{Core}_G(S) \leq \mathrm{Fit}(G)$ and so $\mathrm{Core}_G(S) \leq T$, whence $\mathrm{Core}_G(T) = \mathrm{Core}_G(S) < T \leq S$. Hence, S is conjugate to some S_i, say $S^x = S_i$. Using Corollary 7.4, we obtain:

$$\mathrm{Core}_G(S) = \mathrm{Core}_G(T) < T^x = (K \cap S)^x = K \cap S_i = K \cap A_i^* \cap A_i \cap S_i$$

$$= K \cap \mathrm{Core}_G(A_i) \cap S_i = K \cap \mathrm{Core}_G(S_i) = K \cap \mathrm{Core}_G(S) = \mathrm{Core}_G(S),$$

which is the required contradiction.

The next step is to show that if A is a maximal nilpotent subgroup of G with A not normal in G, then $G = A\mathrm{Fit}(G)$. Lemma 7.3(ii) asserts that every Sylow subgroup of A is either normal in G or a Sylow subgroup of G. Thus, A contains a Sylow p-subgroup S of G such that $A = N_G(S)$. Now $S\mathrm{Fit}(G)/\mathrm{Fit}(G)$ is a Sylow p-subgroup of $G/\mathrm{Fit}(G)$, so by the first part of the proof $S\mathrm{Fit}(G) \lhd G$. Thus, for any $x \in G$, S and S^x are Sylow p-subgroups of $S\mathrm{Fit}(G)$, and there is a $y \in \mathrm{Fit}(G)$ such that $S^y = S^x$. Hence, $x \in N_G(S)y = Ay \leq A\mathrm{Fit}(G)$, as required.

Now let B denote the direct product of all the Sylow subgroups of A that are not normal in G. Then $G = B\mathrm{Fit}(G)$ by the above, and B is a Hall subgroup of G by Lemma 7.3(ii). Thus $G/\mathrm{Fit}(G)$ only involves primes dividing $|B|$, so the Sylow subgroups for other primes are all normal. Hence $G = BF_0(G)$ and $G/F_0(G)$ is nilpotent.

COROLLARY 7.7 *If A is a maximal nilpotent subgroup of $G \in \mathfrak{S}\mathfrak{N}$ such that A is not normal in G, then $G = A\mathrm{Fit}(G)$ and $A \cap \mathrm{Fit}(G) = \mathrm{Core}_G(A)$.*
Proof: The first assertion is just the second step of the previous proof. For the rest, $\mathrm{Core}_G(A) \leq \mathrm{Fit}(G)$ by definition, so that $\mathrm{Core}_G(A) \leq A \cap \mathrm{Fit}(G) \leq A \cap A^* = \mathrm{Core}_G(A)$, by Lemma 7.5 (so that $A^* = \mathrm{Fit}(G)$ in all cases, by uniqueness).

THEOREM 7.8 *If G is a semi-nilpotent group, then:*
(a) *Fit(G) is a maximal nilpotent subgroup of G.*
(b) *If A is a maximal nilpotent subgroup of G and A is not normal in G, then* $\text{Core}_G(A) = Z^*(G)$.

Proof: (a) is obvious when G is nilpotent, and otherwise we can assume that G has a maximal nilpotent subgroup A such that $\text{Fit}(G) < A < G$. Corollary 7.7 yields an immediate contradiction. Turning to (b), Theorem 6.2 of Appendix C asserts that $Z^*(G)$ is contained in every maximal nilpotent subgroup. Thus $Z^*(G) \leq A$ and since $Z^*(G) \lhd G$, $Z^*(G) \leq \text{Core}_G(A)$. For the reverse inclusion, let P be a Sylow p-subgroup of $\text{Core}_G(A)$. Then $P \lhd G$ since $\text{Core}_G(A)$ is nilpotent. Now let B and C be the normal p-complements in $\text{Fit}(G)$ and A, respectively so that $[B, P] = [C, P] = \{1\}$, since $\text{Fit}(G)$ and A are nilpotent. Furthermore, $B \lhd G$ and so BC is a subgroup of $C_G(P)$. But

$$[\text{Fit}(G) : B][A : C] = [A\text{Fit}(G) : BC][A \cap \text{Fit}(G) : B \cap C],$$

so that $[G : BC]$ is a p-power by Corollary 7.7. Hence $P \leq Z^*(G)$ by Theorem 6.3 of Appendix C and so $Z^*(G)$ contains the Sylow subgroups of $\text{Core}_G(A)$, as required.

COROLLARY 7.9 *For* $G \in \mathcal{S}\mathcal{N}$, $\text{Fit}(G) = F_0(G)Z^*(G)$.

Proof: Write $\text{Fit}(G) = F_0(G) \times F_1(G)$ and let P be a Sylow subgroup of $F_1(G)$, so it is sufficient to prove that $P \leq Z^*(G)$. If S is a Sylow subgroup of G containing P, then $P = \text{Core}_G(S) < S$. Hence $A = N_G(S)$ is a maximal nilpotent subgroup of G by Lemma 7.3(i). Moreover, A is not normal in G and we can apply Theorem 7.8(b) to deduce that $P = \text{Core}_G(S) \leq \text{Core}_G(A) = Z^*(G)$, as required.

THEOREM 7.10 *The class* $\mathcal{S}\mathcal{N}$ *is subgroup-closed and quotient-closed.*

Proof: Let H be a subgroup of a semi-nilpotent group G and K a nonnormal nilpotent subgroup of H. Then K is not normal in G and $N_G(K)$ is nilpotent, showing that $N_H(K) = H \cap N_G(K)$ is nilpotent. For quotient-closure, let $G \in \mathcal{S}\mathcal{N}$ and let M be a minimal normal subgroup of G; it is clearly sufficient to prove that $G/M \in \mathcal{S}\mathcal{N}$. Since G is solvable (Theorem 7.6), M is an elementary Abelian p-group for some prime p. If P/M is a nonnormal primary subgroup of G/M, we must show that its normalizer is nilpotent. Now $N_{G/M}(P/M) = N_G(P)/M$ and when P/M is a p-group, this is nilpotent by hypothesis. Thus we can assume that P/M is a q-group, with $q \neq p$, and write $P = MQ$ where Q is a Sylow q-subgroup of P and Q is not normal in G. By a familiar argument, $N_G(P) = N_G(MQ) = MN_G(Q)$ and since $N_G(Q)$ is nilpotent, so is $N_{G/M}(P/M)$.

As further properties of semi-nilpotent groups, we mention the following: $G/\text{Fit}(G)$ is actually cyclic, and the nonnormal maximal nilpotent subgroups of G are all conjugate. These follow from further consequences of Frobenius's theorem; for example, the second assertion follows from Theorem 7.8(b) together with the fact that the Frobenius complements are all conjugate. Alternatively, we note the benefit of hindsight enables us to apply Carter's theorem (Chapter 5, Theorem 1.7) once we have proved that semi-nilpotent groups are solvable.

Classes of Finite Solvable Groups

Gary L. Walls

In this chapter we are concerned with classes of finite solvable groups. The study of such classes evidently began with W. Gaschütz who, in the paper [Gaschütz 1963], generalized a result of [Carter 1959]. Carter had shown that every finite solvable group contained a self-normalizing nilpotent subgroup and that all such subgroups were necessarily conjugate. In order to try and abstract the conditions necessary to show the existence and conjugacy of a such class of subgroups in every finite solvable group, Gaschütz was led to the concepts of saturated formations and $\mathcal{F}$-projectors. The class of nilpotent groups being a particular saturated formation and the self-normalizing nilpotent subgroups being the projectors for this class.

In the first section of this chapter we will define saturated formations of finite solvable groups and study their properties. We will show how the "$\mathcal{F}$-projectors" generalize the "self-normalizing nilpotent" subgroups studied by Carter. Also we will study another conjugacy class of subgroups of solvable groups associated with a saturated formation $\mathcal{F}$, the $\mathcal{F}$-normalizers. For the formation of nilpotent groups the $\mathcal{F}$-normalizers are simply the system normalizers. We then look at the particular saturated formation of supersolvable groups and see how our general results apply in this case.

In the next section we study Fitting classes of finite solvable groups. The notion of a Fitting class, which was introduced by B. Fischer in his dissertation, is dual to that of a formation. We show that for a Fitting class $\mathcal{F}$, every finite solvable group contains an $\mathcal{F}$-injector and all such $\mathcal{F}$-

injectors are conjugate. We look at a few examples of particular Fitting classes.

In the final section we briefly consider homomorphs and normal homomorphs. These are generalizations of the concept of a saturated formation. We consider how these classes have properties necessary for the existence and conjugacy of projectors.

Throughout the chapter we will denote classes of groups by script letters, $\mathcal{F}$, $\mathcal{G}$, etc. By "class of groups" we will always mean a collection of groups that is closed under isomorphisms.

1. Formations

In this section we study classes of finite solvable groups called formations. We first state the definition.

DEFINITION Let $\mathcal{F}$ be a class of finite solvable groups. We call $\mathcal{F}$ a **formation** provided:
(i) If $G \in \mathcal{F}$ and $N \triangleleft G$, then $G/N \in \mathcal{F}$.
(ii) If $N_1, N_2 \triangleleft G$, such that $G/N_1, G/N_2 \in \mathcal{F}$, then $G/N_1 \cap N_2 \in \mathcal{F}$.

Remark Condition (ii) can be replaced by the weaker
(ii)′ If $N_1, N_2 \triangleleft G$, such that $G/N_1, G/N_2 \in \mathcal{F}$ and $N_1 \cap N_2 = \{1\}$, then $G \in \mathcal{F}$.

Clearly (ii) implies (ii)′. To see that (ii)′ implies (ii), let $N_1, N_2 \triangleleft G$ such that $G/N_1, G/N_2 \in \mathcal{F}$. Then $(G/N_1 \cap N_2)/(N_j/N_1 \cap N_2) \cong G/N_j \in \mathcal{F}$ for $j = 1, 2$. Also $(N_1/N_1 \cap N_2) \cap (N_2/N_1 \cap N_2) = \{1\}$. Therefore $G/N_1 \cap N_2 \in \mathcal{F}$.

In order to generalize Carter's results, Gaschütz needed to consider a special kind of formation which he defined (in [Gaschütz 1963]) as follows:

DEFINITION Let $\mathcal{F}$ be a nonempty formation. Then $\mathcal{F}$ is said to be **saturated** provided the following condition is satisfied: if $G \notin \mathcal{F}$ and M is a minimal normal subgroup of G such that $G/M \in \mathcal{F}$, then M has a complement and all such complements are conjugate in G.

In order to get a better understanding of saturated formations we prove the following characterization of them which is from [Gaschütz and Lubeseder 1963].

THEOREM 1.1 *Let $\mathcal{F}$ be a nonempty formation. Then, the following are equivalent:*
(i) *$\mathcal{F}$ is saturated;*
(ii) *if $G \notin \mathcal{F}$, M is a minimal normal subgroup of G, and $G/M \in \mathcal{F}$, then M has a complement in G; and*
(iii) *if $G/\Phi(G) \in \mathcal{F}$, then $G \in \mathcal{F}$.*
Proof: Clearly (i) implies (ii). Thus, we begin the proof by showing that (ii) implies (iii).

(ii) $\Rightarrow$ (iii) Hence, we are assuming (ii) and that $G/\Phi(G) \in \mathcal{F}$. We want to conclude that $G \in \mathcal{F}$.

Suppose that $G \notin \mathcal{F}$ and let M be a minimal normal subgroup of G contained in $\Phi(G)$. Now we know that $\Phi(G/M) = \Phi(G)/M$ and that $(G/M)/\Phi(G/M) = (G/M)/(\Phi(G)/M) \cong G/\Phi(G) \in \mathcal{F}$ and so it follows (by induction on the order of G) that $G/M \in \mathcal{F}$ ($M \neq \{1\}$ as $G \notin \mathcal{F}$). It follows from condition (ii) that M has a complement. This is contrary to the fact that $M \leq \Phi(G)$. The result follows.

(iii) $\Rightarrow$ (i) We are supposing that (iii) holds and that $G \notin \mathcal{F}$, M is a minimal normal subgroup of G, and $G/M \in \mathcal{F}$. We need to show that M has a complement and that any two complements of M are conjugate.

Since (iii) holds and $G \notin \mathcal{F}$ we can conclude that $G/\Phi(G) \notin \mathcal{F}$.

If $\Phi(G) \neq \{1\}$, the result follows by induction. (As $G/M \in \mathcal{F}$, it follows that M is not contained in $\Phi(G)$ and hence that $M \cap \Phi(G) = \{1\}$. Then $M\Phi(G)/\Phi(G)$ is a minimal normal subgroup of $G/\Phi(G)$ and so it has a complement $L/\Phi(G)$. It follows that $G = ML$ and $M \cap L = \{1\}$. Also, any two complements of M must be maximal subgroups and hence, again by induction, are conjugate in G.)

Now we can assume that $\Phi(G) = \{1\}$. Hence, there exists a maximal subgroup H of G so that $G = MH$. Now $M \cap H \lhd MH = G$, as M is Abelian. Hence H is a complement of M in G. It only remains to show that any two complements of M in G are conjugate.

Thus we suppose that H and K are complements of M in G. By induction we can assume that M is the unique minimal normal subgroup of G. For if N is a minimal normal subgroup of G so that $N \cap M = \{1\}$, it follows that $G/N \notin \mathcal{F}$ (if G/N, $G/M \in \mathcal{F}$, then $G/N \cap M = G \in \mathcal{F}$) and that $N \leq H$ and K (since H is a maximal subgroup, either $NH = H$ or $NH = G$; but $H \cong G/M \in \mathcal{F}$, so $NH = G$ would imply $G/N \cong H/H \cap N \in \mathcal{F}$, which was just seen to be false; so $NH = H$ instead). Thus H/N and K/N are complements of NM/N in G/N. It follows (by induction) that H/N and K/N are conjugate in G/N and hence that H and K are conjugate in G.

It follows from Lemma 2.3 of Appendix C that $M = \text{Fit}(G)$ (recall we are assuming $\Phi(G) = \{1\}$) and, as G is solvable, that $C_G(M) = M$ (Theorem 2.6 of Appendix C). Suppose $|M| = p^k$ for a prime p. Thus we can conclude that $O_p(G/M) = \{1\}$ (or else, the inverse image of $O_p(G/M)$

in G is a normal p-group strictly containing $M = \text{Fit}(G)$) and hence that $O_{p'}(G/M) \neq \{1\}$.

Thus letting R/M be a minimal normal subgroup of G/M contained in $O_{p'}(G/M)$, it follows that $(|R/M|, |M|) = 1$. Now R/M is an elementary Abelian q-subgroup for some prime $q \neq p$. Now $R = M(R \cap H) = M(R \cap K)$ so we can conclude that $R \cap H$ and $R \cap K$ are Sylow q-subgroups of R.

Hence there exists $m \in M$ so that $R \cap H = (R \cap K)^m = R \cap K^m$. As $R \lhd G$, we have that $R \cap H \lhd H$ and $R \cap H = (R \cap K)^m \lhd K^m$. Hence $N_G(R \cap H) \geq H, K^m$. Since M is the unique minimal normal subgroup of G, it follows that $R \cap H$ is not normal in G and that $\langle H, K^m \rangle$ is not all of G. As H and K^m are maximal subgroups of G we can conclude that $H = K^m$ as desired. This completes the proof.

Our next objective is to describe a method (due to Gaschütz) for constructing all saturated formations.

DEFINITION For each prime p we associate some formation $\mathfrak{F}(p)$. ($\mathfrak{F}(p)$ could possibly be empty.) We say that $\mathfrak{F}$ is the **local formation, locally defined** by $\{\mathfrak{F}(p)\}$ provided $G \in \mathfrak{F}$ if and only if for each prime p dividing $|G|$ and each p-chief factor H/K of G, $\text{Aut}_G(H/K) \in \mathfrak{F}(p)$.

The next theorem actually shows that the above method can be used to define a saturated formation.

THEOREM 1.2 *The class $\mathfrak{F}$ of finite solvable groups given in the above definition is a saturated formation.*
Proof: We first show that $\mathfrak{F}$ is a formation.

1. Suppose that $G \in \mathfrak{F}$ and $N \lhd G$. We must show that $G/N \in \mathfrak{F}$. So we suppose that p divides $|G/N|$ and $(H/N)/(K/N)$ is a p-chief factor of G/N. Then we know that H/K is a p-chief factor of G. Since $G \in \mathfrak{F}$, it follows that $\text{Aut}_G(H/K) \in \mathfrak{F}(p)$ and as $\text{Aut}_{G/N}((H/N)/(K/N)) \cong \text{Aut}_G(H/K)$ (Theorem 2 of Appendix B), it follows that $G/N \in \mathfrak{F}$.

2. Suppose $N, M \lhd G$ and $G/N, G/M \in \mathfrak{F}$. We need to show that $G/N \cap M \in \mathfrak{F}$. By the remark immediately following the definition of formation, it is sufficient to assume that $N \cap M = \{1\}$. Now we suppose that H/K is a p-chief factor of G. We will induce on $|G| + |N| + |M|$. The proof proceeds in the following steps.

(i) *Either $G \in \mathfrak{F}$ or N, M are minimal normal subgroups of G.* Suppose that $\{1\} \neq L < N$ and L is normal in G. Then $(G/L)/(N/L) \in \mathfrak{F}$ and $(G/L)/(ML/L) \cong G/ML \cong (G/M)/(ML/M) \in \mathfrak{F}$ by part (1). Then as $(N/L) \cap (ML/L) = \{1\}$; it follows by induction that $G/L \in \mathfrak{F}$. Since $G/M \in \mathfrak{F}$, it also follows by induction that $G \in \mathfrak{F}$ as required to complete the result.

(ii) *Either $Aut_G(H/K) \in \mathfrak{F}(p)$ or $K \cap N = \{1\}$.* If $K \cap N \neq \{1\}$, then $N \leq K$ and (using Theorem 2 of Appendix B again) $Aut_G(H/K) \cong Aut_{G/N}((H/N)/(K/N)) \in \mathfrak{F}(p)$.

(iii) *Either $Aut_G(H/K) \in \mathfrak{F}(p)$ or $H \cap N = \{1\}$.* Suppose $H \cap N \neq \{1\}$. Then $N \leq H$ and it may be assumed that $H = KN$ (else $K \cap N = N$, and so (ii) can be used to give $Aut_G(H/K) \in \mathfrak{F}(p)$). Now by Theorems 1 and 3 of Appendix B, $G/C_G(N) \cong Aut_G(N/\{1\}) \cong Aut_G(H/K)$. Moreover, the fact that $C_{G/M}(MN/M) = C_G(N)/M$ is easy to establish. It follows that MN/M is a p-chief factor of G/M and thus $Aut_G(H/K) \cong G/C_G(N) \cong (G/M)/(C_G(N)/M) = (G/M)/C_{G/M}(MN/M) \cong Aut_{G/M}(MN/M) \in \mathfrak{F}(p)$ as was required to complete the proof of this step.

(iv) $Aut_G(H/K) \in \mathfrak{F}(p)$. By virtue of the previous step we may assume $H \cap N = \{1\}$. This implies $H \cap KN = K$ and further implies HN/KN is a p-chief factor of G. Hence $(HN/N)/(KN/N)$ is a p-chief factor of G/N and so $Aut_{G/N}((HN/N)/(KN/N)) \in \mathfrak{F}(p)$. Apply Theorems 2 and 3 of Appendix B to get $Aut_{G/N}((HN/N)/(KN/N)) \cong Aut_G(HN/KN) \cong Aut_G(H/K)$. Conclude that $Aut_G(H/K) \in \mathfrak{F}(p)$.

Since this holds for every p-chief factor of G and every prime p dividing $|G|$, it follows that $G \in \mathfrak{F}$ as desired.

All that remains is to show that $\mathfrak{F}$ is saturated. Thus we suppose that $G \notin \mathfrak{F}$, and that $G/N \in \mathfrak{F}$ where N is a minimal normal subgroup of G. We must show that N has a complement in G. We use induction on the order of G. The proof proceeds in the following steps.

(i) *Either N has a complement in G or N is the unique minimal normal subgroup of G.* Let M be a minimal normal subgroup of G so that $N \cap M = \{1\}$. Then we know that $G/M \notin \mathfrak{F}$ and NM/N is a minimal normal subgroup of G/M such that $(G/M)/(NM/M) \cong G/NM \cong (G/N)/(NM/N) \in \mathfrak{F}$. Thus by induction NM/M has a complement L/M in G/M. Hence $G = (NM)L$ where $NM \cap L = M$. But then $G = NL$ and $N \cap L \leq NM \cap L = M$ and as $N \cap M = \{1\}$ we must have $N \cap L = \{1\}$.

Since $G \notin \mathfrak{F}$ we let p be a prime dividing $|G|$ so that there exists a p-chief factor H/K of G so that $Aut_G(H/K) \notin \mathfrak{F}(p)$.

(ii) $K = \{1\}$ *and* $H = N$, *or N has a complement in G.* If $N \leq K$, then $Aut_G(H/K) \cong Aut_{G/N}((H/N)/(K/N)) \in \mathfrak{F}(p)$ as $G/N \in \mathfrak{F}$. Then H is a minimal normal subgroup of G so $H = N$.

(iii) *N has a complement in G.* By virtue of the two previous steps we may suppose that N is the unique minimal normal subgroup of G and that $H = N$ and $K = \{1\}$. This means that $G/C_G(N) \notin \mathfrak{F}(p)$ where the order of N is a power of p.

We let $L/N = O_{p', p}(G/N)$. Now
$$G/L = (G/N)/(L/N) = (G/N)/O_{p', p}(G/N) \in \mathfrak{F}(p).$$

(As $G/N \in \mathfrak{F}$, it follows that $(G/N)/C_{G/N}(X/Y) \in \mathfrak{F}(p)$ for all p-chief factors X/Y of G/N. Thus $(G/N)/\bigcap\{C_G(X/Y):X/Y$ is a p-chief factor of $G/N\} \in \mathcal{F}(p)$. The result follows from Theorem 3.2 of Appendix C.) Consequently L is not contained in $C_G(N)$; for otherwise $G/C_G(N) \cong (G/L)/(C_G(N)/L) \in \mathcal{F}(p)$, contrary to assumption. We can conclude that $Z(L) = \{1\}$ (else $N \le Z(L)$, N being the unique minimal normal subgroup).

Let F/N be $O_{p'}(G/N)$. Now $F > N$ (otherwise L is a p-group and $Z(L) \ne \{1\}$.) Let R be a Hall p'-subgroup of F. Note that $R \cong F/N$ so that $R \ne \{1\}$. We note that N is not contained in $Z(F)$ for $N \le Z(F)$ would imply $R \triangleleft F$ and then, as normal Hall subgroups are characteristic, $R\,\mathrm{char}\,F \triangleleft G$ so $R \triangleleft G$. This is contrary to the fact that $N \cap R = \{1\}$. Since N is the unique minimal normal subgroup of G, N not being contained in $Z(F)$ in turn implies $Z(F) = \{1\}$.

It is easily seen that $N_F(R) = [N \cap N_F(R)] \oplus R$. This implies $[N \cap N_F(R), R] = \{1\}$ from which it follows that $N \cap N_F(R) \le Z(F) = \{1\}$.

The Frattini argument shows $G = FN_G(R) = NN_G(R)$ (we're making use here of the fact that any two π-Hall subgroups of a solvable group are conjugate—see Theorem 1 of Appendix A). Then since $N \cap N_G(R) = N \cap N_F(R) = \{1\}$, $N_G(R)$ is the complement of N we have been seeking.

The next remark is really a portion of the above argument. We make it again to establish some notation and to give another condition for a group to be in a local formation.

Remark Let $\mathcal{F}$ be a formation locally defined by $\{\mathcal{F}(p)\}$. We let $\mathcal{F}_p$ denote the class of groups $\{G : G/O_{p',p}(G) \in \mathcal{F}(p)\}$. Reasoning as in the above proof we can conclude that $\mathcal{F}_p$ is precisely the class of groups for which $\mathrm{Aut}_G(H/K) \in \mathcal{F}(p)$ for all the p-chief factors H/K of G. It follows that $\mathcal{F} = \bigcap_p \mathcal{F}_p$ since this is the class of groups so that for all p dividing $|G|$, $\mathrm{Aut}_G(H/K) \in \mathcal{F}(p)$ for all p-chief factors H/K of G.

We will postpone consideration of the converse of Theorem 1.2 until we look at some examples of formations. After that we will prove that every saturated formation can be locally defined. This was a result that U. Lubeseder proved in her dissertation, [Lubeseder 1963].

EXAMPLES (i) Let $\mathcal{C}$ be the class of finite Abelian groups. Then $\mathcal{C}$ is clearly a formation, as factor groups of Abelian groups are Abelian and if $G/N, G/M$ are Abelian, it follows that $G' \le N \cap M$, so that $G/N \cap M$ is Abelian. For any non-Abelian p-group G, we know that $G/\Phi(G) \in \mathcal{C}$ but $G \notin \mathcal{C}$. Thus $\mathcal{C}$ is not saturated.

(ii) Let $\mathcal{N}$ be the class of finite nilpotent groups. This is easily seen to be a formation. Moreover, Theorem 1.1 shows that $\mathcal{N}$ is saturated since a group G is nilpotent if and only if all maximal subgroups of G are normal in G which occurs if and only if $G/\Phi(G)$ is nilpotent.

As an alternative demonstration that $\mathfrak{N}$ is saturated, we can explicitly exhibit $\mathfrak{N}$ as a locally defined formation as follows. Let $\mathfrak{N}'$ be the formation locally defined by $\{\mathfrak{F}(p)\}$, where for each prime p we let $\mathfrak{F}(p)$ be the formation of trivial (i.e., 1-element) groups. Thus $G \in \mathfrak{N}'$ if and only if for each p-chief factor H/K of G, $G/C_G(H/K) = \{1\}$. In other words, $G \in \mathfrak{N}'$ if and only if each chief factor H/K is central in G/K. Thus $G \in \mathfrak{N}'$ is equivalent to G being nilpotent, and so $\mathfrak{N}' = \mathfrak{N}$.

(iii) Let π be a set of primes. We will let $\mathcal{G}_\pi$ denote the class of solvable π-groups. It is easily verified that $\mathcal{G}_\pi$ is a formation. For each prime p we let $\mathfrak{F}(p) = \mathcal{G}_\pi$ if $p \in \pi$, and let $\mathfrak{F}(p) = \varnothing$ if $p \notin \pi$. It follows rather easily that $\mathcal{G}_\pi$ is locally defined by $\{\mathfrak{F}(p)\}$.

(iv) Let $\mathcal{S}$ be the class of all solvable groups. Let q be a prime and for each prime p define $\mathfrak{F}(p)$ to be $\mathcal{S}$ if $p = q$, and to be $\mathcal{G}_{q'}$ if $p \neq q$. Then the formation locally defined by the $\{\mathfrak{F}(p)\}$, which we will denote by $\mathfrak{F}_{q'}$, is the class of all solvable groups with normal Sylow q-subgroups.

For $G \in \mathfrak{F}_{q'}$ if and only if for all $p \neq q$, $G/O_{p',p}(G) \in \mathfrak{F}(p)$ if and only if for all $p \neq q$ a Sylow q-subgroup of G is contained in $O_{p',p}(G)$ if and only if $O_{p'}(G)$ contains a Sylow q-subgroup of G for all primes $p \neq q$ if and only if a Sylow q-subgroup of G is normal in G.

(v) Let $\mathcal{SS}$ be the class of finite supersolvable groups. This is a formation by Theorem 1.1 of Chapter 1. Since a group G is supersolvable if and only if $G/\Phi(G)$ is supersolvable (Corollary 3.2 of Chapter 1), $\mathcal{SS}$ is saturated by Theorem 1.1.

In this example we can explicitly exhibit $\mathcal{SS}$ as a locally defined formation: $\mathcal{SS}$ is the formation locally defined by $\{\mathfrak{F}(p)\}$ where for each prime p, $\mathfrak{F}(p)$ is the class of Abelian groups of exponent dividing $p - 1$. This was proven in Corollary 1.5 of Chapter 1.

(vi) Let $\mathcal{C}$ be any class of solvable groups. Clearly there is a smallest formation containing $\mathcal{C}$, usually denoted by Form($\mathcal{C}$). If $\mathcal{C} = \{G\}$, we would write Form(G).

Remark   It is easy to see that Form($\mathcal{C}$) $= \{G : G$ is a homomorphic image of a subdirect product of elements of $\mathcal{C}\}$. This follows easily from the fact that if X is a subdirect product of groups $X_1/N_1, X_2/N_2, \ldots, X_r/N_r$ where for each $i = 1, \ldots, r$, $X_i \in \mathcal{C}$, then X is a homomorphic image of a subdirect product of $X_1, \ldots, X_r$ under the natural homomorphism of $X_1 \times X_2 \times \ldots \times X_r$ onto $(X_1/N_1) \times (X_2/N_2) \times \ldots \times (X_r/N_r)$.

In the next theorem we will show that for every formation $\mathfrak{F}$ there is a unique minimual saturated formation containing $\mathfrak{F}$. The construction in this result will enable us to show that every saturated formation is locally defined—the result of Lubeseder to which we have already referred.

THEOREM 1.3 *Let $\mathfrak{F}$ be a formation. Let $\hat{\mathfrak{F}} = \bigcap\{\mathcal{K}: \mathcal{K}$ is a locally defined formation containing $\mathfrak{F}\}$. Then*
(i) $\hat{\mathfrak{F}}$ is a locally defined formation containing $\mathfrak{F}$.
(ii) $\hat{\mathfrak{F}} = \mathfrak{F}$ if and only if $\mathfrak{F}$ is saturated.
(iii) For any formation $\mathfrak{F}$, $\hat{\mathfrak{F}}$ is the unique minimal saturated formation containing $\mathfrak{F}$.

Proof: (i) Clearly $\mathfrak{F} \subseteq \hat{\mathfrak{F}}$.

Now $G \in \hat{\mathfrak{F}}$ if and only if $G \in \mathcal{K}$ for every locally defined formation $\mathcal{K}$ containing $\mathfrak{F}$.

Suppose $\mathcal{K}$ is locally defined by $\{\mathcal{K}(p)\}$. Then $G \in \hat{\mathfrak{F}}$ if and only if for each prime p, $G/O_{p',\,p}(G) \in \mathcal{K}(p)$ for all locally defined formations $\mathcal{K} \supseteq \mathfrak{F}$.

Now let $\mathcal{W}(p) = \bigcap\{\mathcal{K}(p): \mathcal{K}$ is a locally defined formation containing $\mathfrak{F}\}$. Clearly $\mathcal{W}(p)$ is a formation. From the above discussion $G \in \hat{\mathfrak{F}}$ if and only if for each prime p, $G/O_{p',\,p}(G) \in \mathcal{W}(p)$. It follows from the remark following Theorem 1.2 that $\hat{\mathfrak{F}}$ is locally defined by the $\{\mathcal{W}(p)\}$.

(ii) If $\mathfrak{F} = \hat{\mathfrak{F}}$, then clearly $\mathfrak{F}$ is saturated as $\hat{\mathfrak{F}}$ is a locally defined formation.

Now suppose that $\mathfrak{F}$ is saturated. We want to show $\mathfrak{F} = \hat{\mathfrak{F}}$. Since $\mathfrak{F} \subseteq \hat{\mathfrak{F}}$, suppose that G is a group of minimal order such that $G \in \hat{\mathfrak{F}} \backslash \mathfrak{F}$. It follows that $\Phi(G) = \{1\}$. Otherwise $G/\Phi(G) \in \mathfrak{F}$ and as $\mathfrak{F}$ is saturated, $G \in \mathfrak{F}$. Similarly, G has a unique minimal normal subgroup N (if N, M were distinct minimal normal subgroups, then $G \in \mathfrak{F}$ since G/N, $G/M \in \mathfrak{F}$ and $N \cap M = \{1\}$). Applying Lemma 2.3 of Appendix C yields $N = \text{Fit}(G)$. Consequently (Theorem 2.6 of Appendix C) $C_G(N) = N$.

Now since $\mathfrak{F}$ is saturated, $G \notin \mathfrak{F}$ and $G/N \in \mathfrak{F}$ imply that N has a complement H and G. Let $|N|$ be a power of p and consider the formation locally befined by letting for each prime q

$$\mathcal{K}(q) = \begin{cases} \mathfrak{F} & \text{if } q \neq p, \\ \mathfrak{R} & \text{if } q = p, \end{cases}$$

where $\mathfrak{R} = \text{Form}(\{H/C_H(K/L): K/L$ is a p-chief factor of $H\})$.

We will derive the contradiction $H \in \mathfrak{R}$ and $H \notin \mathfrak{R}$. First $G \in \hat{\mathfrak{F}}$ implies $G/C_G(N) \in \mathcal{W}(p) \subseteq \mathcal{K}(p) = \mathfrak{R}$. But $G/C_G(N) = G/N \cong H$. Thus $H \in \mathfrak{R}$. On the other hand let m be the solvable length of H. Then for all p-chief factors K/L of H, $C_H(K/L)$ contains $\text{Fit}(H)$ (Theorem 2.5 of Appendix C), hence contains all Abelian normal subgroups of H, hence contains the last nontrivial subgroup in the derived series, $H^{(m-1)}$. Consequently the solvable length of $H/C_H(K/L)$ is less than or equal to that of $H/H^{(m-1)}$, which is $m - 1$. It is easily seen that the class of all groups of solvable length $\leq m - 1$ forms a formation, $\mathfrak{B}$, which contains each $H/C_H(K/L)$, hence contains $\mathfrak{R}$. Then $H \notin \mathfrak{B}$ implies $H \notin \mathfrak{R}$.

(iii) Parts (i) and (ii) imply that any saturated formation is locally defined. Hence any saturated formation containing $\mathfrak{F}$ must contain $\hat{\mathfrak{F}}$.

Now we have shown that the saturated formations are precisely the locally defined formations. Later we will consider the different possible systems of formations $\{\mathscr{F}(p)\}$ which locally define a formation $\mathscr{F}$.

However, before considering local formations in more detail we want to introduce the concept of an $\mathscr{F}$-projector. This is the concept (orginally called an $\mathscr{F}$-covering subgroup in [Gaschütz 1963]) which enables us to generalize   Carter's result about the existence and conjugacy of self-normalizing nilpotent subgroups of solvable groups. For we will prove that if $\mathscr{F}$ is a saturated formation, then any finite solvable group contains an $\mathscr{F}$-projector and all the $\mathscr{F}$-projectors are conjugate. Also we will show that for the saturated formation of nilpotent groups, $\mathscr{N}$, the $\mathscr{N}$-projectors are precisely the self-normalizing nilpotent subgroups.

DEFINITION   Let $\mathscr{F}$ be a formation and let G be a finite solvable group. A subgroup F of G is called an $\mathscr{F}$-**projector** provided the following conditions are satisfied:
(i) $F \in \mathscr{F}$
(ii) if $F \leq H \leq G$ and $N \lhd H$ so that $H/N \in \mathscr{F}$, then $H = FN$.

Remark   Let $\mathscr{F}$ be a nonempty formation. Then for any finite group G, there exists a normal subgroup $\rho_{\mathscr{F}}{}^*(G)$ of G so that $G/\rho_{\mathscr{F}}{}^*(G) \in \mathscr{F}$ and if $G/N \in \mathscr{F}$, then $\rho_{\mathscr{F}}{}^*(G) \leq N$. $\rho_{\mathscr{F}}{}^*(G)$ is called the $\mathscr{F}$-**residual** of G and it is defined as $\bigcap \{B : B \lhd G$ and $G/B \in \mathscr{F}\}$. It follows that a subgroup F of G so that $F \in \mathscr{F}$ is an $\mathscr{F}$-projector if and only if $F \leq H \leq G$ implies $H = F\rho_{\mathscr{F}}{}^*(H)$.

First we need a few lemmas. We let $\mathscr{F}$ denote a formation.

LEMMA 1.4   *Let F be an $\mathscr{F}$-projector of G. Then if $F \leq H \leq G$, we must have that F is an $\mathscr{F}$-projector of H.*
Proof: Clear from the definition.

LEMMA 1.5   *Let F be an $\mathscr{F}$-projector of G and suppose that $N \lhd G$. Then NF/N is an $\mathscr{F}$-projector of G/N.*
Proof: First we note that $NF/N \in \mathscr{F}$ as $NF/N \cong F/N \cap F \in \mathscr{F}$ since $\mathscr{F}$ is a formation.

Now suppose that $NF/N \leq H/N \leq G/N$ and $(H/N)/(M/N) \in \mathscr{F}$. Then $H/M \in \mathscr{F}$ and so as $F \leq H \leq G$ we get that $H = MF$ from the fact that F is an $\mathscr{F}$-projector of G. Hence it follows that $H/N = MF/N = (M/N)(FN/N)$ and consequently FN/N is an $\mathscr{F}$-projector of G/N.

The next lemma allows us to pull an $\mathscr{F}$-projector of a factor group of G back to an $\mathscr{F}$-projector of G.

LEMMA 1.6 *Let $N \lhd G$ and suppose that $\bar{F}/N$ is an $\mathcal{F}$-projector of G/N. Let F be an $\mathcal{F}$-projector of $\bar{F}$. Then F is an $\mathcal{F}$-projector of G.*

Proof: Clearly $F \in \mathcal{F}$. Now suppose that $F \leq H \leq G$ and $H/M \in \mathcal{F}$. Since F is an $\mathcal{F}$-projector of $\bar{F}$ and $\bar{F}/N \in \mathcal{F}$ it follows that $\bar{F} = NF$.

Since $\bar{F}/N$ is an $\mathcal{F}$-projector of G/N, it follows without difficulty that $H \cap \bar{F}/H \cap N$ is an $\mathcal{F}$-projector of $H/H \cap N$. By Lemma 1.4 F is an $\mathcal{F}$-projector of $H \cap \bar{F}$. Now if $H < G$, it would follow by induction that F is an $\mathcal{F}$-projector of H. As $H/M \in \mathcal{F}$, this would give that $H = MF$ as required.

Thus we suppose that $H = G$. Now $\bar{F}/N$ is an $\mathcal{F}$-projector of G/N. And $(G/N)/(MN/N) \cong G/MN \cong (G/M)/(MN/M) = (H/M)/(MN/M) \in \mathcal{F}$, and so $G/N = (\bar{F}/N)(MN/N)$. Thus $G = \bar{F}(MN) = \bar{F}M$.

Now $G/M = \bar{F}M/M \cong \bar{F} \cap M \in \mathcal{F}$ so $\bar{F} = F(\bar{F} \cap M)$. Thus $G = \bar{F}M = F(\bar{F} \cap M)M = FM$ as required.

Now we are ready to apply the above lemmas, as was originally done in [Gaschütz 1963], to show that every finite solvable group contains an $\mathcal{F}$-projector and that all of its $\mathcal{F}$-projectors are conjugate.

THEOREM 1.7 *Let $\mathcal{F}$ be a saturated formation. Then every solvable group contains an $\mathcal{F}$-projector and all of its $\mathcal{F}$-projectors are conjugate (in the group).*

Proof: Let G be a finite solvable group. If $G \in \mathcal{F}$, the result trivially holds. Thus we suppose that $G \notin \mathcal{F}$. Let M be a minimal normal subgroup of G.

Case 1: $G/M \in \mathcal{F}$. In this case, since $\mathcal{F}$ is saturated and $G \notin \mathcal{F}$, there exists a complement F of M and all the complements of M are conjugate. We will show that any complement of M is an $\mathcal{F}$-projector. (The converse incidentally—any $\mathcal{F}$-projector must be a complement of M—is also true.) Now if $F \leq H \leq G$, then $H = F(H \cap M)$ and as $G = MF$, we find that $G = MH$. Now $M \cap H$ is normal in G because it is normal in M (M being Abelian) and also normal in H. Since M is a minimal normal subgroup of G, $M \cap H = \{1\}$ or M so $H = F$ or $H = G$.

If $H = F$, we are okay (i.e., proving that $H = NF$ for any $N \lhd H$ with $H/N \in \mathcal{F}$ becomes a triviality). Thus suppose that $H = G$. Now if $G/N \in \mathcal{F}$ for some $N \lhd G$, it follows that $N \geq M$ (else $N \cap M = \{1\}$, G/N, $G/M \in \mathcal{F}$, would imply that $G \in \mathcal{F}$). Hence $G = FM = FN$ as required. Thus any complement of M is an $\mathcal{F}$-projector. The result follows.

Case 2: $G/M \notin \mathcal{F}$. We let by induction $\bar{F}/M$ be an $\mathcal{F}$-projector of G/M. Since $G/M \notin \mathcal{F}$, $\bar{F} < G$. It follows, again by induction, that there is an $\mathcal{F}$-projector F of $\bar{F}$. Hence, by Lemma 1.6, F is an $\mathcal{F}$-projector of G.

It only remains to show that any two $\mathcal{F}$-projectors of G must be conjugate in G. Suppose that F_1 and F_2 are $\mathcal{F}$-projectors of G. By lemma 1.5 MF_1/M and MF_2/M are $\mathcal{F}$-projectors of G/M. It follows by induction that there exists a $g \in G$ so that $(MF_1)^g = MF_2$ and hence that $F_1^g \leq MF_2$. But then F_1^g and F_2 are $\mathfrak{F}$-projectors of MF_2. Now $MF_2 < G$ (since $G/M \notin \mathfrak{F}$), so by induction F_1^g and F_2 (hence F_1 and F_2) must be conjugate in G.

Before going any further we want to refer again to the various examples of formations we looked at previously. We will describe the $\mathfrak{F}$-projectors in each particular case.

EXAMPLES

(i) We saw that $\mathcal{C}$, the class of all Abelian groups, did not form a saturated formation. It is easy to see that the quaternion group Q of order 8 has no $\mathcal{C}$-projectors. For if F were an $\mathcal{C}$-projector, then as $Q/Z(Q) \in \mathcal{C}$, it would follow that $Q = Z(Q)F = F$ as $Z(Q)$ is contained in every subgroup of Q. On the other hand, S_3 is an example of a group which does have $\mathcal{C}$-projectors; these are the subgroups of order 2.

(ii) The class $\mathfrak{N}$ of finite nilpotent groups is a saturated formation. We will show that the $\mathfrak{N}$-projectors of a solvable group G are precisely the self-normalizing nilpotent subgroups of G.

Suppose H is an $\mathfrak{N}$-projector of G. First, we will show that H is self-normalizing. Suppose $x \in N_G(H)$, then we have $H \leq H\langle x \rangle \leq G$ and $H\langle x \rangle / H \cong \langle x \rangle / H \cap \langle x \rangle \in \mathfrak{N}$. Hence $H\langle x \rangle = H$ and $x \in H$. Thus H is a self-normalizing nilpotent subgroup of G.

Conversely, suppose F is a self-normalizing nilpotent subgroup of G and $F \leq H \leq G$ with $H/N \in \mathfrak{N}$. Then FN/N is a self-normalizing subgroup of H/N (if $Nx \in N_{H/N}(FN/N)$, then $x^{-1}Fx \leq FN$: thus F and $x^{-1}Fx$ are self-normalizing nilpotent subgroups of FN. Since we may assume $|FN| < |G|$, F and $x^{-1}Fx$ are both $\mathfrak{N}$-projectors of FN by induction. Hence by Theorem 1.7 F and $x^{-1}Fx$ are conjugate in FN. Thus $x^{-1}Fx = aFa^{-1}$ for some $a \in N$ and so $xa \in N_G(F) = F$ implying $Nx \in FN/N$.) But since H/N is nilpotent, its only self-normalizing subgroup is H/N itself. Thus $H = FN$ as required.

The N-projectors of a finite solvable group are often referred to as **Carter subgroups**.

Remark The proof that the $\mathfrak{N}$-projectors of a finite solvable group are self-normalizing is actually very general. An examination of the proof reveals that if $\mathfrak{F}$ is a formation which contains $\mathbb{Z}_p$ for every prime p, then the $\mathfrak{F}$-projectors of a finite solvable group must be self-normalizing.

This implies that if a group G contains an $\mathcal{C}$-projector then it must be self-normalizing; i.e., it is an Abelian self-normalizing subgroup. Thus a solvable group G contains an $\mathcal{C}$-projector if and only if its Carter subgroups are Abelian.

We next note that a saturated formation $\mathcal{F}$ contains all the cyclic groups $\mathbb{Z}_p$ for all primes p if and only if it contains all finite nilpotent groups. For if it contains $\mathbb{Z}_p$, it must contain all elementary Abelian p-groups and, hence, as it is saturated, all p-groups. Thus a formation which contains all the cyclic groups $\mathbb{Z}_p$ for all primes p must contain $\mathcal{N}$.

Further, if $\mathcal{F}$ is locally defined by $\{\mathcal{F}(p)\}$, then $\mathcal{F}$ contains all $\mathbb{Z}_p$ for all primes p if and only if $\mathcal{F}(p) \neq \emptyset$ for all primes p. Hence, if $\mathcal{F}$ is a formation locally defined by $\{\mathcal{F}(p)\}$ and for all primes p $\mathcal{F}(p) \neq \emptyset$, then $\mathcal{F} \supseteq \mathcal{N}$.

EXAMPLES (continued)

(iii) Recall that $\mathcal{G}_\pi$ is the formation of π-groups. It follows that the $\mathcal{G}_\pi$-projectors of a group G are the Hall π-subgroups of G.

(iv) Now $\mathcal{F}_{q'}$ consisted of the groups with a normal Sylow q-subgroup. It is easily verified that the $\mathcal{F}_{q'}$-projectors of a group G are precisely the normalizers of the Sylow q-subgroups.

(v) Recall that $\mathcal{SS}$ is the formation of supersolvable groups. We will show that the $\mathcal{SS}$-projectors of a finite solvable group G are precisely the supersolvable subgroups F of G so that for every pair of subgroups H, K of G so that $F \leq H \leq K \leq G$, the index $[K:H]$ is not a prime.

First we suppose that F is an $\mathcal{SS}$-projector of G and that $F \leq H \leq K \leq G$. Suppose that $[K:H] = p$, a prime. Then H is a maximal subgroup of K. Let N be the largest normal subgroup of K which is contained in H and let A/N be a maximal Abelian normal subgroup of K/N. Then $A \cap H \lhd K$ so $A \cap H \leq N$. Hence, as $N \leq A \cap H$, we have $N = A \cap H$. Now as H is a maximal subgroup of K, $K = AH$. Thus $|K| = |A| \cdot |H|/|A \cap H| = |A| \cdot |H|/|N|$. It follows that $[A:N] = [K:H] = p$. It follows that AF/N is supersolvable $((AF/N)/(A/N) \cong AF/A \cong F/A \cap F$ is supersolvable, so AF/N is supersolvable by Theorem 1.2 of Chapter 1.) Hence $AF = NF$, a contradiction. This establishes that all $\mathcal{SS}$-projectors satisfy the above condition.

Now suppose F is a supersolvable subgroup satisfying the given condition. Let $F \leq H \leq G$ and suppose $H/N \in \mathcal{SS}$. Now if $NF \neq H$, there would exist a maximal subgroup containing NF whose index in H would be a prime, contrary to assumption.

(vi) If $\mathcal{C}$ is a class of groups, then the formation $\mathrm{Form}(\mathcal{C})$ does not have to be saturated. For instance $\mathrm{Form}(\mathbb{Z}_2)$ would be the formation of all elementary Abelian 2-groups.

Thus we see that every finite solvable group contains for each saturated formation $\mathcal{F}$ a class of conjugate subgroups, the $\mathcal{F}$-projectors. Later we will discuss another conjugacy class of subgroups of a group G related to the saturated formation $\mathcal{F}$; the so-called $\mathcal{F}$-normalizers. Before proceding with their discussion we consider in some detail how a locally defined formation

$\mathcal{F}$ is related to the system of formations $\{\dot{\mathcal{F}}(p)\}$ which locally defines it.

If $\mathcal{K}$ and $\mathcal{F}$ are classes of groups, then define $\mathcal{K} * \mathcal{F}$ to be the class $\{G : \text{there exists } N \in \mathcal{K}, N \triangleleft G \text{ such that } G/N \in \mathcal{F}\}$. Note that if $\mathcal{F}$ is a formation and $\mathcal{K}$ is a formation which is closed under normal subgroups, then $\mathcal{K} * \mathcal{F}$ is a formation. (Suppose $G \in \mathcal{K} * \mathcal{F}$ and $N \triangleleft G$. Since $G \in \mathcal{K} * \mathcal{F}$, there is some $M \in \mathcal{K}$ such that $M \triangleleft G$ and $G/M \in \mathcal{F}$. Then $MN/N \cong M/M \cap N \in \mathcal{K}$ and also $(G/N)/(MN/N) \cong G/MN \cong (G/M)/(MN/M) \in \mathcal{F}$. Hence $G/N \in \mathcal{K} * \mathcal{F}$. Secondly, if G/M and $G/N \in \mathcal{K} * \mathcal{F}$, then there exist subgroups J, K of G such that $J/M \triangleleft G/M, J/M \in \mathcal{K}, G/J \in \mathcal{F}, K/N \triangleleft G/N, K/N \in \mathcal{K}$, and $G/K \in \mathcal{F}$. It follows that $G/J \cap K \in \mathcal{F}$. Also $M(J \cap K)/M \cong J \cap K/(J \cap K) \cap M = J \cap K/M \cap K$. But since $\mathcal{K}$ is closed under normal subgroups and $M(J \cap K)/M \triangleleft J/M \in \mathcal{K}$, we have $J \cap K/M \cap K \in \mathcal{K}$. Similarly $J \cap K/J \cap N \in \mathcal{K}$ so $J \cap K/(M \cap K) \cap (J \cap N) = J \cap K/M \cap N \in \mathcal{K}$. It follows that $G/M \cap N \in \mathcal{K} * \mathcal{F}$.)

In particular, letting $\mathcal{K} = \mathcal{G}_p$, the formation of all finite p-groups, we have established that $\mathcal{G}_p * \mathcal{F}$ is a formation whenever $\mathcal{F}$ is.

DEFINITIONS Let $\{\mathcal{F}(p)\}$ be a system of formations which locally defines the formation $\mathcal{F}$.

(i) We say that the system $\{\mathcal{F}(p)\}$ is **integrated** provided for every prime $p, \mathcal{F}(p) \subseteq \mathcal{F}$.

(ii) We say that the system $\{\mathcal{F}(p)\}$ is **full** provided for every $p, \mathcal{G}_p * \mathcal{F}(p) = \mathcal{F}(p)$.

We will establish the following existence and uniqueness results, both essentially from [Carter and Hawkes 1967]: First, every locally defined formation can be locally defined by an integrated full system of formations. Secondly, if $\{\mathcal{F}_1(p)\}$ and $\{\mathcal{F}_2(p)\}$ are two integrated full systems which locally define the same formation, then $\mathcal{F}_1(p) = \mathcal{F}_2(p)$ for all primes p.

Note that uniqueness fails if two defining systems for a formation are not required to be full. For example, if we let

$$\mathcal{F}_1(p) = \begin{cases} \mathcal{G}_2 & \text{if } p = 2 \\ \varnothing & \text{if } p \neq 2 \end{cases}$$

and

$$\mathcal{F}_2(p) = \begin{cases} \text{class of elementary Abelian 2-groups} & \text{if } p = 2 \\ \varnothing & \text{if } p \neq 2 \end{cases}$$

then clearly $\{\mathcal{F}_1(p)\}$ and $\{\mathcal{F}_2(p)\}$ both locally define $\mathcal{G}_2$.

LEMMA 1.8 *Let $\{\mathcal{F}(p)\}$ be a system of formations which locally defines the formation $\mathcal{F}$. Then $\{\mathcal{F}(p) \cap \mathcal{F}\}$ will be an integrated system of formations which also locally defines $\mathcal{F}$.*

Proof: Given that a group G belongs to $\mathcal{F}$ if and only if $\mathrm{Aut}_G(H/K) \in \mathcal{F}(p)$ for all p-chief factors H/K of G, it immediately follows that $G \in \mathcal{F}$ if and only if $\mathrm{Aut}_G(H/K) \in \mathcal{F}(p) \cap \mathcal{F}$ for all p-chief factors H/K of G. Thus $\{\mathcal{F}(p) \cap \mathcal{F}\}$ locally defines $\mathcal{F}$ and is clearly integrated.

LEMMA 1.9 *Let $\{\mathcal{F}(p)\}$ be a system of formations which locally defines the formation $\mathcal{F}$. Then $\{\mathcal{G}_p * \mathcal{F}(p)\}$ is a full system of formations which locally defines $\mathcal{F}$. Further, $\{\mathcal{G}_p * \mathcal{F}(p)\}$ is integrated if and only if $\{\mathcal{F}(p)\}$ is.*

Proof: If H/K is a p-chief factor of a group G, then $\mathrm{Aut}_G(H/K) \cong G/C_G(H/K)$ has no nontrivial normal p-subgroups by Corollary 6.4 of Appendix C. We see then that $\mathrm{Aut}_G(H/K) \in \mathcal{F}(p)$ if and only if $\mathrm{Aut}_G(H/K) \in \mathcal{G}_p * \mathcal{F}(p)$. The result follows.

Further, if $\{\mathcal{G}_p * \mathcal{F}(p)\}$ is integrated, it clearly follows that $\{\mathcal{F}(p)\}$ is integrated. Now if we suppose that $\{\mathcal{F}(p)\}$ is integrated, then if $G \in \mathcal{G}_p * \mathcal{F}(p)$, we get a normal p-subgroup N of G so that $G/N \in \mathcal{F}(p)$. We seek to show $G \in \mathcal{F}$; thus it suffices to show $\mathrm{Aut}_G(H/K) \in \mathcal{F}(q)$ for all primes q and all q-chief factors H/K of G.

If $q \neq p$, then $K = NK \cap H$. Consequently $NH/NK = (NK)H/NK \cong H/NK \cap H \cong H/K$, and moreover $(NH/N)/(NK/N)$ is a q-chief factor of G/N. Now $G/N \in \mathcal{F}(p) \subseteq \mathcal{F}$ because $\{\mathcal{F}(p)\}$ is integrated. Thus $\mathrm{Aut}_{G/N}((NH/N)/(NK/N)) \in \mathcal{F}(q)$. Hence, by Theorems 2 and 3 of Appendix B, $\mathrm{Aut}_G(H/K) \cong \mathrm{Aut}_G(NH/NK) \cong \mathrm{Aut}_{G/N}((NH/N)/(NK/N)) \in \mathcal{F}(q)$.

Assume now that $q = p$. Then $N \leq O_{p',p}(G) \leq C_G(H/K)$ (the last inclusion by Theorem 3.2 of Appendix C). Hence $G/N \in \mathcal{F}(p)$ implies $\mathrm{Aut}_G(H/K) \cong G/C_G(H/K) \in \mathcal{F}(p)$ since $\mathcal{F}(p)$ is a formation.

THEOREM 1.10 *Let $\{\mathcal{F}_1(p)\}$ and $\{\mathcal{F}_2(p)\}$ be two integrated systems of formations which both locally define the formation $\mathcal{F}$. Then $\mathcal{G}_p * \mathcal{F}_1(p) = \mathcal{G}_p * \mathcal{F}_2(p)$ for each prime p. Consequently, for each saturated formation $\mathcal{F}$, there is a unique integrated full system of formations which locally defines it.*

Proof: From Lemma 1.9 it suffices to assume that $\{\mathcal{F}_1(p)\}$ and $\{\mathcal{F}_2(p)\}$ are full and then to show that $\mathcal{F}_1(p) = \mathcal{F}_2(p)$ for each prime p.

We assume $\mathcal{F}_1(p)$ is not contained in $\mathcal{F}_2(p)$ and seek a contradiction. Let $F \in \mathcal{F}_1(p)\backslash\mathcal{F}_2(p)$ have minimal order. Let H be the $\mathcal{F}_2(p)$-residual of F (i.e., $H = \bigcap\{J : J \lhd F \text{ and } F/J \in \mathcal{F}_2(p)\}$). Consequently H is a minimal normal subgroup of F for if $K \lhd F$ with $\{1\} < K < H$, then $F/K \in \mathcal{F}_1(p)$ would imply $F/K \in \mathcal{F}_2(p)$ by choice of F. Furthermore, and again by choice of F, H is the unique minimal normal subgroup of F. Let $|H| = q^b$ for some prime q. Then $q \neq p$, for otherwise $F \in \mathcal{G}_p * \mathcal{F}_2(p) = \mathcal{F}_2(p)$.

Let W be the standard wreath product $W = \mathbb{Z}_p \,\mathrm{wr}\, F$. We recall that W has an elementary Abelian normal p-subgroup B so that $W = BF$ and $B \cap F = \{1\}$. Now H cannot centralize all the chief factors of W below B; for if it did, BH would be nilpotent (if $\gamma_\infty(BH) \neq \{1\}$, then there is a chief factor $\gamma_\infty(BH)/S$ of W below B; hence $\gamma_\infty(BH)/S \leq Z(BH/S)$ because H centralizes $\gamma_\infty(BH)/S$; hence BH/S is nilpotent; hence $\gamma_\infty(BH) \leq S$). But nilpotency of BH implies $BH = B \oplus H$ since $(|B|, |H|) = 1$. This would be contrary to the fact that H operates nontrivially on B in the wreath product.

We next claim that $W \in \mathfrak{F}$. For any p-chief factor R/S of W, we have $B \leq \mathrm{Fit}(W) \leq C_W(R/S)$ (by Theorem 2.5 of Appendix C) so that $\mathrm{Aut}_W(R/S) \cong W/C_W(R/S) \in \mathfrak{F}_1(p)$ because $W/B \cong F \in \mathfrak{F}_1(p)$. Moreover, for any r-chief factor R/S of W with $r \neq p$, we have BR/BS is r-chief in W and $BS \cap R = S$ and hence $(BR/B)/(BS/B)$ is r-chief in W/B. Using Theorems 1 and 2 of Appendix B we conclude $\mathrm{Aut}_W(R/S) \cong \mathrm{Aut}_{W/B}((BR/B)/(BS/B))$ which belongs to $\mathfrak{F}_1(r)$ because $W/B \in \mathfrak{F}$ ($W/B \cong F \in \mathfrak{F}_1(p) \subseteq \mathfrak{F}$). This establishes the claim.

In the next-to-last paragraph we have produced a chief factor J/L of W with $J \leq B$ and H not contained in $C_W(J/L)$. Since B is Abelian, $B \leq C_W(J/L)$. If this inclusion were proper, then $C_W(J/L) \cap F$ would be a proper normal subgroup of F and would consequently contain the unique minimal normal subgroup H, a contradiction. Thus $B = C_W(J/L)$. Then $W \in \mathfrak{F}$ implies $\mathfrak{F}_2(p)$ contains $W/C_W(J/L) = W/B \cong F$, again a contradiction.

This establishes the first part of the theorem. The second is an immediate consequence of Theorem 1.3, Lemmas 1.8 and 1.9, and the first part.

Before continuing, we will need to know a few facts about the structure of maximal subgroups of finite solvable groups. The next lemma is part of Theorem 6.1.4 of [Gorenstein 1968].

LEMMA 1.11 *Suppose that M is a maximal subgroup of the solvable group G and let $L = \mathrm{Core}_G(M)$. Also let R/L denote $\mathrm{Fit}(G/L)$. Then*
(i) *R/L is the unique minimal normal subgroup of G/L.*
(ii) *R/L is complemented by M/L in G/L.*
(iii) *$[G:M]$ is a power of some prime p and M/L has no nontrivial normal p-subgroups.*
Proof: As $R/L \neq \{1\}$, $Z(R/L) \neq \{1\}$, and so $Z(R/L)$ has a nontrivial p-component for some prime p. Let P/L be the subgroup generated by all elements of $Z(R/L)$ of order p. Then $\{1\} \neq P/L \lhd G/L$ and P/L is elementary Abelian. We will show that $G/L = (P/L)(M/L)$. (Note this will show that $[G:M]$ is a power of p.) Now since M/L is a maximal subgroup of G/L and, since $(P/L)(M/L)$ is a subgroup of G/L, the desired assertion will follow if we show that P/L is not contained in M/L. In fact,

we shall show that M/L possesses no nontrivial normal subgroups of G/L. Indeed, if N/L is such a subgroup, then $N \lhd G$ and so $N \leq \bigcap_{x \in G} M^x = L$. This is a contradiction. It follows that $G/L = (P/L)(M/L)$.

Now let $C/L = C_{G/L}(P/L)$ and note that $R/L \leq C/L$. Moreover, $C/L \lhd G/L$, and so $(C/L) \cap (M/L) \lhd M/L$. But as P/L centralizes $(C/L) \cap (M/L)$ and $G/L = (P/L)(M/L)$, it follows that we must have $(C/L) \cap (M/L) = \{1\}$, by the above paragraph. Since $P/L \leq R/L \leq C/L$, we have $G/L = (C/L)(M/L)$, but then by order considerations $P = R = C$.

Now it follows that R/L is a minimal normal subgroup of G/L. Uniqueness is shown as follows: suppose B/L is some other minimal normal subgroup. Then $R/L \cap B/L = \{1\}$ and so $(R/L)(B/L) = R/L \oplus B/L$, hence $B/L \leq C_{G/L}(R/L) = R/L$, a contradiction.

All that remains to show is that M/L has no nontrivial normal p-subgroups, and since $M/L \cong G/R$, it suffices to show this for G/R. But if J/R were a p-group, $J/R \lhd G/R$, then J/L would also be a p-group, $J/L \lhd G/L$. Thus $J/L \leq R/L$ by definition of Fitting subgroup.

We are now ready for the following

DEFINITIONS Let $\{\mathcal{F}(P)\}$ be a system of formations which locally defines the formation $\mathcal{F}$.

(i) A p-chief factor H/K of G is called $\mathcal{F}$-**central** provided $\mathrm{Aut}_G(H/K) \in \mathcal{F}(p)$. Otherwise, it is called $\mathcal{F}$-**eccentric**.

(ii) A maximal subgroup M of G is called $\mathcal{F}$-**normal** provided $M/\mathrm{Core}_G(M) \in \mathcal{F}(p)$ where p is the prime dividing $[G : M]$. Otherwise, it is is called $\mathcal{F}$-**abnormal**.

Remarks

(1) It is clear that the definitions of $\mathcal{F}$-central and $\mathcal{F}$-normal reduce to the usual concepts of central and normal in the case that $\mathcal{F}(p)$ is the formation of trivial groups for each prime p (i.e., in the case that $\mathcal{F} = \mathfrak{N}$).

(2) In general, the concepts of $\mathcal{F}$-central and $\mathcal{F}$-normal depend upon the particular system of formations used to locally define $\mathcal{F}$. However, if the system $\{\mathcal{F}(p)\}$ is integrated, then the concepts of $\mathcal{F}$-central and $\mathcal{F}$-normal depend only upon $\mathcal{F}$ and not upon $\{\mathcal{F}(p)\}$.

For if $\{\mathcal{F}_1(p)\}$ and $\{\mathcal{F}_2(p)\}$ are two systems of integrated formations both of which locally define $\mathcal{F}$, we know that $\mathcal{G}_p * \mathcal{F}_1(p) = \mathcal{G}_p * \mathcal{F}_2(p)$ for each prime p by Theorem 1.10. But then as $\mathrm{Aut}_G(H/K)$ has no normal p-subgroups, when H/K is a p-chief factor of G (by Corollary 6.4 of Appendix C) it follows that $\mathrm{Aut}_G(H/K) \in \mathcal{F}_1(p)$ if and only if $\mathrm{Aut}_G(H/K) \in \mathcal{G}_p * \mathcal{F}_1(p)$ if and only if $\mathrm{Aut}_G(H/K) \in \mathcal{G}_p * \mathcal{F}_2(p)$ if and only if $\mathrm{Aut}_G(H/K) \in \mathcal{F}_2(p)$.

Similarly, if M is a maximal subgroup of G and $[G:M]$ is a power of p, then $M/\mathrm{Core}_G(M)$ has no normal p-subgroups (Lemma 1.11(iii)) and so, as above, $M/\mathrm{Core}_G(M) \in \mathcal{F}_1(p)$ if and only if $M/\mathrm{Core}_G(M) \in \mathcal{F}_2(p)$, as required.

Thus, in what follows, unless we explicitly state to the contrary, we will assume that $\mathcal{F}$ is a formation locally defined by an integrated system of formations $\{\mathcal{F}(p)\}$.

The next lemma shows the relationship between $\mathcal{F}$-normal maximal subgroups and $\mathcal{F}$-central chief factors. We will say that a maximal subgroup M of G **complements** a chief factor H/K if $G = HM$ and $H \cap M = K$.

LEMMA 1.12 *A maximal subgroup of a group G is $\mathcal{F}$-normal if and only if it complements an $\mathcal{F}$-central chief factor of G.*

Proof: Suppose that M is an $\mathcal{F}$-normal maximal subgroup of G with $[G:M]$ a power of p. Let $K = \mathrm{Core}_G(M)$. By Lemma 1.11 we have that G/K has a unique minimal normal subgroup H/K which is complemented by M/K. Thus $G/H \cong (G/K)/(H/K) \cong M/K \in \mathcal{F}(p)$.

Now since H/K is its own centralizer in G/K we obtain

$$\mathrm{Aut}_G(H/K) \cong G/C_G(H/K) \cong (G/K)/(C_G(H/K)/K)$$
$$= (G/K)/C_{G/K}(H/K) = (G/K)/(H/K)$$
$$\cong G/H \in \mathcal{F}(p)$$

as required.

Conversely, if M is a maximal subgroup of G which complements the $\mathcal{F}$-central p-chief factor H/K of G, then $\mathrm{Core}_G(M) = M \cap C$ where $C = C_G(H/K)$. (Recall $G/K = (M/K)(H/K)$, H/K is a minimal normal subgroup of G/K.)

Since $H \leq C$, we obtain $G = CM$. Thus $M/\mathrm{Core}_G(M) = M/(M \cap C) \cong MC/C = G/C \in \mathcal{F}(p)$. Therefore M is $\mathcal{F}$-normal in G as desired.

LEMMA 1.13 $G \in \mathcal{F}_p$ *if and only if every minimal normal subgroup of $G/\Phi_p(G)$ is $\mathcal{F}$-central. Similarly, $G \in \mathcal{F}$ if and only if every minimal normal subgroup of $G/\Phi(G)$ is $\mathcal{F}$-central.*

Proof: Recall that $\mathcal{F}_p$ is the class of groups G for which $\mathrm{Aut}_G(H/K) \in \mathcal{F}(p)$ for all p-chief factors H/K of G.

By Theorem 4.3 of Appendix C, $\mathrm{Fit}(G/\Phi_p(G))$ can be written $N_1/\Phi_p(G) \times \ldots \times N_r/\Phi_p(G)$ where the $N_i/\Phi_p(G)$ are minimal normal subgroups of $G/\Phi_p(G)$. Let C_i be defined by $C_i/\Phi_p(G) = C_{G/\Phi_p(G)}(N_i/\Phi_p(G))$ for each i. Then, since G is solvable and $\mathrm{Fit}(G/\Phi_p(G))$ is Abelian we obtain (Theorem 2.6 of Appendix C)

$$\mathrm{Fit}(G/\Phi_p(G)) = C_{G/\Phi_p(G)}(\mathrm{Fit}(G/\Phi_p(G))) = \bigcap_{i=1}^{r} C_i/\Phi_p(G).$$

Now if we assume that all the minimal normal subgroups of $G/\Phi_p(G)$ are $\mathcal{F}$-central, we get that $G/C_i \cong (G/\Phi_p(G))/(C_i/\Phi_p(G)) \in \mathcal{F}(p)$; and hence, since $\mathcal{F}(p)$ is a formation, $(G/\Phi_p(G))/\mathrm{Fit}(G/\Phi_p(G)) \in \mathcal{F}(p)$. It follows from Theorem 4.3 of Appendix C that $G/O_{p',p}(G) \in \mathcal{F}(p)$ and, hence, $G \in \mathcal{F}_p$.

Conversely, if $G \in \mathcal{F}_p$, we get that $G/O_{p',p}(G) \in \mathcal{F}(p)$. But as $\mathrm{Fit}(G/\Phi_p(G)) \le C_{G/\Phi_p(G)}(N_i/\Phi_p(G))$ we get that $(G/\Phi_p(G))$ $/C_{G/\Phi_p(G)}(N_i/\Phi_p(G))$ is a quotient group of $(G/\Phi_p(G))/\mathrm{Fit}(G/\Phi_p(G))$ $\cong G/O_{p',p}(G) \in \mathcal{F}(p)$. Hence, the $N_i/\Phi_p(G)$ are $\mathcal{F}$-central. This proves the first part of the lemma.

Clearly, if $G \in \mathcal{F}$, then every minimal normal subgroup of $G/\Phi(G)$ is $\mathcal{F}$-central (as all the chief factors are). Again, as in the above proof, we write $\mathrm{Fit}(G/\Phi(G)) = \mathrm{Fit}(G)/\Phi(G) = N_1/\Phi(G) \times \ldots \times N_r/\Phi(G)$ where $N_i/\Phi(G)$ are minimal normal subgroups of $G/\Phi(G)$. Also, we let $C_i/\Phi(G) = C_{G/\Phi(G)}(N_i/\Phi(G))$ for each i. Then, since each $N_i/\Phi(G)$ is $\mathcal{F}$-central we have $(G/\Phi(G))/(C_i/\Phi(G)) \in \mathcal{F}(p)$ when $N_i/\Phi(G)$ is a p-group. But as we are assuming the system $\{\mathcal{F}(p)\}$ to be integrated we get for each i, that $(G/\Phi(G))/(C_i/\Phi(G)) \in \mathcal{F}$. But as $\mathrm{Fit}(G/\Phi(G)) = \bigcap_{i=1}^r C_i/\Phi(G)$, we can conclude that $(G/\Phi(G))/(\mathrm{Fit}(G)/\Phi(G)) \in \mathcal{F}$. Hence each chief factor of $(G/\Phi(G))/(\mathrm{Fit}(G)/\Phi(G))$ is $\mathcal{F}$-central. Now by our hypothesis the chief factors of $G/\Phi(G)$ below $\mathrm{Fit}(G)/\Phi(G)$ are $\mathcal{F}$-central. It follows that $G/\Phi(G) \in \mathcal{F}$ (given any p-chief factor $(H/\Phi(G))/(K/\Phi(G))$ of $G/\Phi(G)$, Theorem 4 of Appendix B says that either

$$\left[(H\,\mathrm{Fit}(G)/\Phi(G))/(\mathrm{Fit}(G)/\Phi(G)) \right] / \left[(K\,\mathrm{Fit}(G)/\Phi(G))/(\mathrm{Fit}(G)/\Phi(G)) \right]$$

is a p-chief factor of $(G/\Phi(G))/(\mathrm{Fit}(G)/\Phi(G))$ or

$$\left[(H \cap \mathrm{Fit}(G))/\Phi(G) \right] / \left[(K \cap \mathrm{Fit}(G))/\Phi(G) \right]$$

is a p-chief factor of $G/\Phi(G)$). As $\mathcal{F}$ is saturated, $G \in \mathcal{F}$ as desired.

DEFINITIONS (i) A maximal subgroup M of G whose index is a power of the prime p will be called a p-**maximal subgroup**.

(ii) An $\mathcal{F}$-abnormal p-maximal subgroup M of G will be called an $\mathcal{F}(p)$-**critical** maximal subgroup provided $O_{p',p}(G)M = G$.

(iii) A complemented $\mathcal{F}$-eccentric p-chief factor H/K of G is called $\mathcal{F}(p)$-**critical** provided every p-chief factor below K is either $\mathcal{F}$-central or not complemented.

Remark G has $\mathcal{F}(p)$-critical maximal subgroups if and only if $G \notin \mathcal{F}_p$.

For if $G \notin \mathcal{F}_p$, then by Lemma 1.13 we see that G must have an $\mathcal{F}$-eccentric chief factor $N_i/\Phi_p(G)$ where $N_i \le O_{p',p}(G)$. Since $N_i > \Phi_p(G)$, N_i is not contained in all the p-maximal subgroups of G. Hence, as $N_i \le O_{p',p}(G)$, $N_i/\Phi_p(G)$ is complemented and the complement must be $\mathcal{F}(p)$-critical (pick as the complement for $N_i/\Phi_p(G)$ any p-maximal subgroup H not containing N_i. Then $G = HN_i$ and, as $N_i/\Phi_p(G)$ is a minimal normal subgroup of $G/\Phi_p(G)$, $H \cap N_i = \Phi_p(G)$. H is $\mathfrak{F}$-abnormal, since, otherwise, $N_i/\Phi_p(G)$ would have to be $\mathfrak{F}$-central.)

Conversely, if M is an $\mathfrak{F}$-abnormal p-maximal subgroup such that $O_{p',p}(G)M = G$, then

$$G/O_{p',p}(G) = O_{p',p}(G)M/O_{p',p}(G) \cong M/M \cap O_{p',p}(G) \notin \mathfrak{F}(p),$$

as $M/\mathrm{Core}_G(M) \cong (M/M \cap O_{p',p}(G))/(\mathrm{Core}_G(M)/M \cap O_{p',p}) \notin \mathfrak{F}(p)$. Hence, $G \notin \mathfrak{F}(p)$ as desired.

See Lemma 3.1 of Chapter 6 for a related result.

LEMMA 1.14 *Suppose $G \notin \mathfrak{F}_p$ and let H/K be a p-chief factor of G covered by an $\mathfrak{F}(p)$-critical maximal subgroup M of G ("covering" is defined in Appendix B). Then*
(i) *$H \cap M/K \cap M$ is a p-chief factor of M.*
(ii) *$\mathrm{Aut}_G(H/K) \cong \mathrm{Aut}_M(H \cap M/K \cap M)$.*
Proof: Consider an arbitrary chief series of G passing through H and K

$$\{1\} = K_0 < K_1 < \ldots < K_r = K < H = H_0 < H_1 \ldots < H_t = G.$$

Case 1: K is not contained in M. In this case there is an integer i so that $K_{i-1} \leq M$ but K_i is not contained in M. To prove (i), suppose $T \lhd M$ with $K \cap M \leq T \leq H \cap M$. It is easily seen then that $KT \lhd G$ with $K \leq KT \leq H$. Hence $KT = K$ or H implying $T = K \cap M$ or $H \cap M$.

As for (ii), we have

$$\mathrm{Aut}_M(H \cap M/K \cap M) \cong \mathrm{Aut}_{M/K_{i-1}}((H \cap M/K_{i-1})/(K \cap M/K_{i-1}))$$

(Theorem 2 of Appendix B) which is isomorphic to

$$(M/K_{i-1})/(C/K_{i-1}),$$

where C/K_{i-1} denotes $C_{M/K_{i-1}}((H \cap M/K_i)/(K \cap M/K_i))$, and hence to M/C. On the other hand, $\mathrm{Aut}_G(H/K) \cong G/C_G(H/K)$ which, since $K_{i-1} \leq C_G(H/K)$, is isomorphic to $(G/K_{i-1})/(C_G(H/K)/K_{i-1})$. But $G = C_G(H/K)M$ implies

$$G/K_{i-1} = (C_G(H/K)/K_{i-1})(M/K_{i-1})$$

so that

$$(G/K_{i-1})/(C_G(H/K)/K_{i-1}) \cong (M/K_{i-1})/((M \cap C_G(H/K))/K_{i-1})$$
$$\cong M/M \cap C_G(H/K).$$

Finally, $C = M \cap C_G(H/K)$, as is easily established since $H = K(H \cap M)$. Putting the pieces together yields

$$\mathrm{Aut}_M(H \cap M/K \cap M) \cong M/C = M/M \cap C_G(H/K) \cong \mathrm{Aut}_G(H/K).$$

Case 2: $K \leq M$. Now as M covers H/K, it follows that $H \leq M$ also. Since $O_{p',p}(G) \leq C_G(H/K)$ (by virtue of Theorem 3.2 of Appendix C) and $O_{p',p}(G)M = G$ (by hypothesis on M) we have $G = C_G(H/K)M$. This implies that for any T such that $K \leq T \leq H$; if $T \triangleleft M$, then $T \triangleleft G$ as well. Hence H/K is a chief factor of M, proving (i).

Furthermore we have

$$\mathrm{Aut}_G(H/K) \cong G/C_G(H/K) = C_G(H/K)M/C_G(H/K)$$
$$\cong M/M \cap C_G(H/K) = M/C_M(H/K)$$
$$\cong \mathrm{Aut}_M(H/K)$$

which proves (ii).

COROLLARY 1.15 *Under the hypotheses of the above theorem* $C_M(H \cap M/K \cap M) = M \cap C_G(H/K)$.
Proof: Clearly, $M \cap C_G(H/K) = C_M(H/K)$ is contained in $C_M(H \cap M/K \cap M)$. But as $G/C_G(H/K) = MC_G(H/K)/C_G(H/K) \cong M/M \cap C_G(H/K) = M/C_M(H/K)$, $C_M(H/K)$ has the same order as $C_M(H \cap M/K \cap M)$. Thus they are equal.

Remark If M is an $\mathfrak{F}(p)$-critical maximal subgroup of G and if $\{1\} = G_0 \leq G_1 \leq \ldots \leq G_n = G$ is a chief series of G, then M covers every chief factor of the series except for the $\mathfrak{F}$-eccentric factor which M complements. In particular $\{1\} \leq M \cap G_1 \leq \ldots \leq M \cap G_i \leq \ldots \leq M \cap G_n = M$ is a chief series of M. (Here we have omitted the factor avoided by M.). We prove this as follows:

Now M either covers of avoids each factor G_i/G_{i-1} (Theorem 5 of Appendix B). Suppose M avoids G_i/G_{i-1}. Let $L = \mathrm{Core}_G(M)$ and let $P/L = \mathrm{Fit}(G/L)$. Then $M \cap G_i = M \cap G_{i-1}$ and we must have $G_{i-1} \leq M$. Hence $G_{i-1} \leq L$. Thus $G_{i-1} \leq L \cap G_i \leq G_i$ so $G_{i-1} = L \cap G_i$ (since $G_i \leq L \leq M$ is contrary to assumption). Thus $G_i/G_{i-1} = G_i/L \cap G_i \cong G_iL/L$. Now G_iL/L is a minimal normal subgroup of G/L so $G_iL/L = P/L$ (recall from Lemma 1.11 that P/L is the unique minimal normal subgroup of G/L). Now $C_G(G_i/G_{i-1}) = C_G(P/L)$ so it follows that G_i/G_{i-1} is $\mathfrak{F}$-eccentric. Hence M covers all but the chief factor isomorphic to P/L. This and order considerations yield the chief series for M as given.

Now we are ready to define the $\mathfrak{F}$-normalizer of a solvable group G. This is a generalization of the notion of the normalizer of a Sylow system. The following ideas are basically from [Carter and Hawkes 1967].

In what follows we continue to let $\mathfrak{F}$ be a formation locally defined by the integrated system of formations $\{\mathfrak{F}(p)\}$.

DEFINITION AND NOTATION Let Σ be a Sylow system for the group G. For each $S \in \Sigma$, let S^p denote the p-complement in Σ. Let $C_p(G)$ denote the intersection of the centralizers of the $\mathfrak{F}$-central p-chief factors of G. Let $\rho_{\mathfrak{F}(p)}^*(G)$ denote the $\mathfrak{F}(p)$-residual of G. Let $T^p = S^p \cap C_p(G) \cap \rho_{\mathfrak{F}(p)}^*(G)$, which is equal to $S^p \cap \rho_{\mathcal{F}(p)}{}^*(G)$ as $\rho_{\mathcal{F}(p)}{}^*(G) \leq C_p(G)$ since $G/C_p(G)$ $\in \mathcal{F}(p)$. (Note that T^p is actually only a p-complement of $\rho_{\mathcal{F}(p)}{}^*(G)$.) Finally, $D = \bigcap_p N_G(T^p)$ will be called the $\mathcal{F}$-**normalizer** of G corresponding to Σ.

Remarks

(1) If $\mathcal{F} = \mathfrak{N}$ is the formation of nilpotent groups, then $T^p = S^p$ for each p. Thus D is just the set of elements of G which normalize all the S^p. As the Sylow subgroups in a Sylow system are formed by intersecting certain of the $S^{p'}$s, it follows that D is simply a system normalizer.

(2) Since any two Sylow systems of a group G are conjugate, it follows very easily that the $\mathcal{F}$-normalizers of G form a single conjugacy class of subgroups of G.

(3) If $G \in \mathcal{F}$, then $G/O_{p', p}(G) \in \mathcal{F}(p)$. Thus $\rho_{\mathcal{F}(p)}{}^*(G) \leq O_{p', p}(G)$. Thus the p-complement of $\rho_{\mathcal{F}(p)}{}^*(G)$ is $O_{p'}(G) \cap \rho_{\mathcal{F}(p)}{}^*(G) \lhd G$ for each prime p. Hence $D = G$.

(4) If the system of formations which locally defines $\mathcal{F}$ is not assumed to be integrated, the class of $\mathcal{F}$-normalizers obtained would not necessarily be the same as in the integrated case. Later we will give an example to show that some of our results fail if the sytsem $\{\mathcal{F}(p)\}$ is not assumed to be integrated.

(5) Suppose $\{1\} = G_0 \leq G_1 \leq \ldots \leq G_n = G$ is a chief series of G. Then $C_p(G) = \bigcap\{C_G(G_j/G_{j-1}) : G_j/G_{j-1}$ is an $\mathcal{F}$-central p-chief factor of $G\}$. In other words to compute $C_p(G)$ we can restrict our attention to the $\mathcal{F}$-central p-chief factors of any given chief series of G. We prove this as follows:
Suppose

$$x \in \bigcap\{C_G(G_j/G_{j-1}) : G_j/G_{j-1} \text{ is an } \mathcal{F}\text{-central } p\text{-chief factor of } G\}.$$

Let R/S be an $\mathcal{F}$-central p-chief factor of G. Pick H/K so that $C_G(H/K) = C_G(R/S)$ and among all the $\mathcal{F}$-central p-chief factors H/K satisfying $C_G(H/K) = C_G(R/S)$, $H \leq G_i$ with i minimal. Then H is not contained in G_{i-1} and it follows that $G_i = HG_{i-1}$. Two cases arise.
First suppose $K \leq G_{i-1}$. Then $K \leq H \cap G_{i-1} \leq H$ so $H \cap G_{i-1} = K$ and $G_i/G_{i-1} = HG_{i-1}/G_{i-1} \cong H/H \cap G_{i-1} = H/K$. Now

$$x \in C_G(G_i/G_{i-1}) = C_G(H/K) = C_G(R/S)$$

as required.

Now suppose K is not contained in G_{i-1}. Then $KG_{i-1} = G_i$. Hence $H \leq KG_{i-1}$. It follows that $H = K(H \cap G_{i-1})$ and

$$H/K = K(H \cap G_{i-1})/K \cong H \cap G_{i-1}/K \cap G_{i-1}.$$

It is easy to see that $H \cap G_{i-1}/K \cap G_{i-1}$ is an $\mathcal{F}$-central p-chief factor of G so that $C_G(H \cap G_{i-1}/K \cap G_{i-1}) = C_G(H/K)$ contrary to assumption (as $H \cap G_{i-1} \leq G_{i-1}$).

Thus $\bigcap\{C_G(G_j/G_{j-1}): G_j/G_{j-1}$ is an $\mathcal{F}$-central p-chief factor of $G\}$ $\leq C_p(G)$. This establishes the result, the other inclusion being clear.

Our next goal will be to prove that an $\mathcal{F}$-projector of a finite solvable group G must contain an $\mathcal{F}$-normalizer. Before proving that result we need to consider the structure of $\mathcal{F}$-normalizers in some detail.

LEMMA 1.16 *Suppose M is an $\mathcal{F}$-abnormal maximal subgroup containing S^p. Then $N_G(T^p) \leq M$.*

Proof: Let $K = \text{Core}_G(M)$ and suppose that H/K is the unique minimal normal subgroup of G/K (Lemma 1.11). Since M is $\mathcal{F}$-abnormal, $G/H \cong M/K \notin \mathcal{F}(p)$, hence $\rho_{\mathcal{F}(p)}{}^*(G)K > K$. Thus $H < \rho_{\mathcal{F}(p)}{}^*(G)K$.

Let L/H be a minimal normal subgroup of G/H such that $L \leq \rho_{\mathcal{F}(p)}{}^*(G)K$. Since $C_G(H/K) = H$ (recall that $G = MH$; as $C_G(H/K) \geq H$ we can conclude that $M \cap C_G(H/K) \triangleleft G$; so $M \cap C_G(H/K) \leq K$; thus $C_G(H/K) = H$), we know that G/H has no nontrivial normal p-subgroups (Corollary 6.4 of Appendix C). Thus L/H is a q-chief factor of G for some prime $q \neq p$. Now $(L \cap M)/K$ is a q-chief factor of M/K. Thus we see that $(L \cap M)/K$ is a p-complement of L/K ($L/L \cap M \cong LM/M = G/M$ and $[G:M]$ is a power of p because $S^p \leq M$). Now KT^p/K is a p-complement of $\rho_{\mathcal{F}(p)}{}^*(G)K/K$ and therefore $(L \cap KT^p)/K$ is a p-complement of L/K contained in $(L \cap M)/K$. Hence $L \cap KT^p = L \cap M$. Now $L \cap M/K \triangleleft M/K$ but not in G/K since $K = \text{Core}_G(M)$. Therefore, $M = N_G(L \cap M)$. But $N_G(T^p)$ normalizes L and KT^p, so it normalizes $L \cap KT^p = L \cap M$. Thus $N_G(T^p) \leq M$ as required.

We need the following corollary of the above result.

COROLLARY 1.17 *In the notation of the above lemma, if $\bar{T}^p = \rho_{\mathcal{F}(p)}{}^*(M) \cap S^p$, then*

(i) $\bar{T}^p \leq T^p$

(ii) $N_G(T^p) \leq N_M(\bar{T}^p)$.

Proof: Let $C = \rho_{\mathcal{F}(p)}{}^*(G)$ and $\bar{C} = \rho_{\mathcal{F}(p)}{}^*(M)$. Since M is $\mathcal{F}$-abnormal, $G = CM$. Thus $M/M \cap C \cong CM/C = G/C \in \mathcal{F}(p)$ and so $M \cap C \geq \bar{C}$.

Now, $\bar{T}^p = T^p \cap \bar{C}$ ($T^p \cap \bar{C} = S^p \cap C \cap \bar{C} = (S^p \cap C \cap \bar{C}) \cap M = (M \cap S^p) \cap \bar{C}$ (as $M \cap C \geq \bar{C}$) $= S^p \cap C = \bar{T}^p$). This proves (i).

Now from the lemma $N_G(T^p) = N_M(T^p)$. But clearly $N_M(T^p) \leq N_M(T^p \cap \bar{C}) = N_M(\bar{T}^p)$. Thus $N_G(T^p) \leq N_M(\bar{T}^p)$ proving (ii).

LEMMA 1.18 *In the above notation $N_G(T^p)$ covers the $\mathcal{F}$-central p-chief factors and avoids the $\mathcal{F}$-eccentric p-chief factors of G.*

Proof: The proof will be by induction on the order of G. First we note if $G \in \mathcal{F}_p$, then all the p-chief factors of G are $\mathcal{F}$-central. Also $\rho_{\mathcal{F}(p)}{}^*(G)$ is trivial, hence so too is T^p. Thus $N_G(T^p) = G$ which clearly covers each $\mathcal{F}$-central p-chief factor.

Thus we may assume $G \notin \mathcal{F}_p$. Now by the remark preceding Lemma 1.14, we know that G must have an $\mathcal{F}(p)$-critical maximal subgroup, M, containing S^p.

Now $T^p = S^p \cap \rho_{\mathcal{F}(p)}{}^*(G) = (S^p \cap M) \cap \rho_{\mathcal{F}(p)}{}^*(G) \geq (S^p \cap M) \cap \rho_{\mathcal{F}(p)}{}^*(M) = \overline{T}^p$. Now from Corollary 1.17 $N_G(T^p) \leq N_M(\overline{T}^p)$. Now by induction $N_M(\overline{T}^p)$ and hence $N_G(T^p)$ avoids the $\mathcal{F}$-central p-chief factors of M. It then follows from Lemma 1.14 and Theorem 5 of Appendix B that $N_G(T^p)$ avoids the $\mathcal{F}$-eccentric p-chief factors of G.

Now in order to show that $N_G(T^p)$ covers the $\mathcal{F}$-central p-chief factors we will actually prove a little more. We will show that if $F^p \leq S^p \cap C_p(G)$ (recall that $C_p(G)$ is the intersection of the centralizers of the $\mathcal{F}$-central p-chief factors), then $N_G(F^p)$ must cover all the $\mathcal{F}$-central p-chief factors. As $T^p = S^p \cap \rho_{\mathcal{F}(p)}{}^*(G) \leq S^p \cap C_p(G)$, the result will follow.

Now $M \cap C_p(G) = C_p(M)$ (this follows from Corollary 1.15, Lemma 1.14, the remark after Corollary 1.15, and Remark (5) preceding Lemma 1.16). Therefore, $F^p \leq (M \cap S^p) \cap C_p(G) = (M \cap S^p) \cap C_p(M)$. Since $F^p \leq M$, it follows by induction that $N_M(F^p)$ must cover all the $\mathcal{F}$-central p-chief factors of M. It follows that $N_G(F^p)$ covers the $\mathcal{F}$-central p-chief factors of M. But now Lemma 1.14 allows us to conclude that $N_G(F^p)$ must cover all the p-chief factors of G. This gives the result.

Now for some precise information about $\mathcal{F}$-normalizers.

THEOREM 1.19 *Let D be an $\mathcal{F}$-normalizer of G. Then D covers each $\mathcal{F}$-central chief factor of G and avoids each $\mathcal{F}$-eccentric chief factor. The order of D is the product of the orders of the $\mathcal{F}$-central factors in some chief series for G. The intersection of D with a chief series of G gives a chief series of D, and $\operatorname{Aut}_D(D \cap H / D \cap K) \cong \operatorname{Aut}_G(H/K)$ for all chief factors H/K of G.*
Proof: Since $D = \bigcap_p N_G(T^p)$, D must avoid all the p-chief factors that $N_G(T^p)$ avoids. Hence by Lemma 1.18, D must aviod each $\mathcal{F}$-eccentric chief factor.

Now for each p $[G : N_G(T^p)]$ is a power of p (as $T^p \lhd S^p$) and thus D is an intersection of subgroups with relatively prime indices (pairwise). It follows that $[G : D] = \prod [G : N_G(T^p)]$. Thus $|D|$ is the product of the orders of the $\mathcal{F}$-central factors in a chief series and so D must cover all the $\mathcal{F}$-central chief factors (let $\{1\} = H_0 \leq H_1 \leq \ldots \leq H_n = G$ be a chief series for G. Then $|D| = |H_0 \cap D| \cdot |H_1 \cap D / H_0 \cap D| \cdot \ldots \cdot |H_{i+1} \cap D / H_i \cap D| \cdot \ldots \cdot |H_n \cap D / H_{n-1} \cap D|$, where $|H_{i+1} \cap D / H_i \cap D| = 1$ provided D avoids H_{i+1}/H_i, and in any case $|H_{i+1} \cap D / H_i \cap D| = |H_{i+1} \cap D / (H_{i+1} \cap D) \cap H_i| = |(H_{i+1} \cap D)H_i / H_i| \leq |H_{i+1}/H_i|$. Now knowing that $|D|$ is the product of the orders of the $\mathcal{F}$-central chief factors and D avoids all the $\mathfrak{F}$-eccentric chief factors gives that for each $\mathfrak{F}$-central chief factor $|H_{i+1} \cap D/H_i \cap D| = |H_{i+1}/H_i|$. This gives $(H_{i+1} \cap D)H_i = H_{i+1}$ so D covers H_{i+1}/H_i for all such $\mathfrak{F}$-central chief factors.)

Let H/K be a chief factor of G. If H/K is $\mathfrak{F}$-eccentric, $H \cap D = K \cap D$. We will show that if H/K is $\mathfrak{F}$-central, then $H \cap D/K \cap D$ is a chief factor of D and $\mathrm{Aut}_G(H/K) \cong \mathrm{Aut}_G(H \cap D/K \cap D)$.

Let $C = C_G(H/K)$. Now if H/K is a p-group we have that $G/C \in \mathfrak{F}(p)$ and thus $G/C \in \mathfrak{F}$ as we are assuming $\{\mathfrak{F}(p)\}$ to be integrated. Hence, all the chief factors of G/C are $\mathfrak{F}$-central. Thus D covers G/C. This says that $G = CD$.

Now $\mathrm{Aut}_G(H/K) \cong G/C = CD/C \cong D/D \cap C$ while $\mathrm{Aut}_D(D \cap H/D \cap K) \cong D/C_D(D \cap H/D \cap K)$ so it suffices to show $C_D(D \cap H/D \cap K) = D \cap C$. Let $x \in C_D(D \cap H/D \cap K)$; then $x \in D$. Moreover, for all $h \in H$, $h \in KD$ (because D covers H/K) so $h = kd$ where $k \in K$, $d \in D$. Then $(x, h) = (x, d)d^{-1}(x, k)d$; so $(x, h) \in K$ ($(x, d) \in K$ becasue $x \in C_D(D \cap H/D \cap K)$ and $d^{-1}(x, k)d \in K$ because $k \in K$ $\triangleleft G$). Thus $x \in C$ which shows $C_D(D \cap H/D \cap K) \leq D \cap C$. The reverse inclusion is immediate, hence we conclude $\mathrm{Aut}_G(H/K) \cong \mathrm{Aut}_D(D \cap H/D \cap K)$.

That $H \cap D/K \cap D$ is a chief factor of D follows easily from the facts that H/K is a chief factor of G and $G = CD$. This completes the proof.

COROLLARY 1.20 *The $\mathfrak{F}$-normalizers of G are in $\mathfrak{F}$.*
Proof: From the above result all the chief factors of an $\mathfrak{F}$-normalizer are $\mathfrak{F}$-central. Note that in the part of the proof we are using we needed to know that $\{\mathfrak{F}(p)\}$ was integrated.

We will give an example to show that Corollary 1.20 is not necessarily true if the system of formations $\{\mathfrak{F}(p)\}$ used to define $\mathfrak{F}$ is not assumed to be integrated.

EXAMPLE Let $G = \mathbb{Z}_5 \wr S_4$, the wreath product of a cyclic group of order 5 with the symmetric group of degree 4 (in its natural representation as a permutation group). Thus $|G| = 5^4 \cdot 24$.
We let

$$\mathfrak{F}(p) = \begin{cases} \mathfrak{SS} & \text{if } p = 2 \\ \mathrm{Form}(S_4) & \text{if } p = 5 \\ \{1\} & \text{if } p \neq 2 \text{ or } 5. \end{cases}$$

The system $\{\mathfrak{F}(p)\}$ is not integrated because $S_4 \in \mathfrak{F}(5)$, but $S_4 \notin \mathfrak{F}$, the formation locally defined by $\{\mathfrak{F}(p)\}$ (S_4 has a noncentral 3-chief factor).

It is easy to see that all the chief factors of G are $\mathfrak{F}$-central except for the one of order 3 ($\mathbb{Z}_5^4 A_4/\mathbb{Z}_5^4 \mathbb{Z}_2^2$). It follows from the proof of the first part of

Theorem 1.19 (this didn't use the fact that $\{\mathcal{F}(p)\}$ was integrated) that the $\mathcal{F}$-normalizers of G have order $2^3 5^4$. Such an $\mathcal{F}$-normalizer has a chief factor of order 5^3 on which it induces an automorphism group which is dihedral of order 8. However, the only 2-groups in Form(S_4) are elementary Abelian. Thus the $\mathcal{F}$-normalizers of G are not in $\mathcal{F}$.

COROLLARY 1.21 *If N is normal in G and D is an $\mathcal{F}$-normalizer of G, then ND/N is an $\mathcal{F}$-normalizer of G/N.*

Proof: In the above notation $D = \bigcap_p N_G(S^p \cap \rho_{\mathcal{F}(p)}{}^*(G))$. Now $N\rho_{\mathcal{F}(p)}{}^*(G)/N$ is the $\mathcal{F}(p)$-residual of G/N and $N(S^p \cap \rho_{\mathcal{F}(p)}{}^*(G))/N$ is a p-complement of $N\rho_{\mathcal{F}(p)}{}^*(G)/N$. Now D normalizes $N(S^p \cap \rho_{\mathcal{F}(p)}{}^*(G))$ for each p; hence ND/N is contained in the $\mathcal{F}$-normalizer $\bigcap_p N_{G/N}(N(S^p \cap \rho_{\mathcal{F}(p)}{}^*(G))/N)$ of G/N. But $|ND/N|$ is the product of the orders of the $\mathcal{F}$-central chief factors of a chief series of G/N by Theorem 1.19 and this is of the order of an $\mathcal{F}$-normalizer of G/N (let $\{1\} = H_0 \leq H_1 \leq \ldots H_i \leq N \leq H_{i+1} \leq \ldots \leq H_n = G$ be a chief series passing through N. Then $N/N \leq H_{i+1}/N \leq \ldots \leq H_n/N = G/N$ is a chief series of G/N with isomorphic factors. Now

$$|ND/N| = |D/N \cap D|$$
$$= |H_{i+1} \cap D/N \cap D| \cdot |H_{i+2} \cap D/H_{i+1} \cap D| \cdot$$
$$\ldots \cdot |H_n \cap D/H_{n-1} \cap D|.$$

Since D is an $\mathcal{F}$-normalizer of G we know that D avoids the $\mathcal{F}$-eccentric and covers the $\mathcal{F}$-central chief factors of G. This implies $|ND/N|$ is the product of the orders of the $\mathcal{F}$-central chief factors of G/N).

Thus ND/N is an $\mathcal{F}$-normalizer of G/N.

The next result gives us a relationship between the $\mathcal{F}$-normalizers of G and the $\mathcal{F}$-abnormal maximal subgroups of G.

THEOREM 1.22 *A maximal subgroup M of G contains some $\mathcal{F}$-normalizer of G if and only if M is $\mathcal{F}$-abnormal in G.*

Proof: Suppose M is $\mathcal{F}$-abnormal in G and $[G:M]$ is a power of p. Let S^p be a p-complement of G contained in M and $T^p = S^p \cap \rho_{\mathcal{F}(p)}{}^*(G)$. By Lemma 1.16, $N_G(T^p) \leq M$. Thus M contains an $\mathcal{F}$-normalizer of G.

Conversely, if M is $\mathcal{F}$-normal in G, then M complements some $\mathcal{F}$-central chief factor (Lemma 1.12) of G. If M contained an $\mathcal{F}$-normalizer D of G, D would avoid H/K contrary to Theorem 1.19.

Our proof that every $\mathcal{F}$-projector contains an $\mathcal{F}$-normalizer is due to Avino'am Mann ([Mann 1968]). We are, again, assuming that $\mathcal{F}$ is a formation locally defined by an integrated system of formations $\{\mathcal{F}(p)\}$.

THEOREM 1.23 *Let G be a finite solvable group. Then each $\mathscr{F}$-projector of G contains an $\mathscr{F}$-normalizer of G.*

Proof: Let C be an $\mathscr{F}$-projector of G. We shall prove by induction on the order of G that if $\{S^p\}$ reduces to C, (recall from the remark following Theorem 4 of Appendix A that it is always possible to pick such a system $\{S^p\}$) then $D = \bigcap_p N_G(T^p)$ is contained in C.

If $G = C$, there is nothing to prove, so we assume that $G \neq C$; hence $G \notin \mathscr{F}$. Let M be a minimal normal subgroup of G. Then DM/M is an $\mathscr{F}$-normalizer of G/M by Corollary 1.21 and CM/M is an $\mathscr{F}$-projector of G/M by Lemma 1.5. Hence, by induction, $DM/M \leq CM/M$. It follows that $D \leq CM$.

Case 1: $CM \neq G$. Since $\{S^p\}$ reduces to C and $M \lhd G$ it follows that $\{S^p\}$ reduces to CM. (Let $|M| = q^n$. If $q = p$, the result is clear. If $q \neq p$, $M \leq S^p$ and $S^p \cap CM = (S^p \cap C)M$ is a p-complement of CM, by order consideration.)

Let $U^p = S^p \cap CM$ and $Q^p = \rho_{\mathscr{F}(p)}{}^*(CM)$. Since C is an $\mathscr{F}$-projector, as $\mathscr{F}(p) \subseteq \mathscr{F}$ and $G/\rho_{\mathscr{F}(p)}{}^*(G) \in \mathscr{F}(p)$, we obtain the fact that $G = C\rho_{\mathscr{F}(p)}{}^*(G)$.

Hence $CM/CM \cap \rho_{\mathscr{F}(p)}{}^*(G) \cong CM\rho_{\mathscr{F}(p)}{}^*(G)/\rho_{\mathscr{F}(p)}{}^*(G) = G/\rho_{\mathscr{F}(p)}{}^*(G)$ $\in \mathscr{F}(p)$. It follows that $Q^p \leq CM \cap \rho_{\mathscr{F}(p)}{}^*(G)$. Now for each prime p, $T^p \cap Q^p = S^p \cap \rho_{\mathscr{F}(p)}{}^*(G) \cap Q^p = S^p \cap Q^p = (S^p \cap CM) \cap Q^p = U^p \cap Q^p$. This gives that $\{U^p \cap Q^p\}$ reduces to C. ($U^p \cap C = S^p \cap C$ is a p-complement for C.) Because C is an $\mathscr{F}$-projector of $CM < G$, by induction, $\bigcap_p N_{CM}(T^p \cap Q^p) \leq C$. But obviously $D \leq N_{CM}(T^p \cap Q^p)$ for all p. Thus $D \leq C$ as required.

Case 2: $G = CM$. Since M is a minimal normal subgroup of G, M is an elementary Abelian group of order q^n for some prime q. It follows that $C \cap M = \{1\}$ and C is a maximal subgroup of G. Now $G \notin \mathscr{F}$ but $G/M \cong C \in \mathscr{F}$. As M is a minimal normal subgroup of G, it follows that $\rho_{\mathscr{F}}{}^*(G) = M$.

We claim that without loss of generality M is the unique minimal normal subgroup of G. For if N is another minimal normal subgroup of G, either the theorem holds by Case 1 with M replaced by N or else $G = CN$. But if $G = CN$, as above, $\rho_{\mathscr{F}}{}^*(G) = N$ which also equals M.

Now suppose that $R/M = O_q(G/M)$ and let V_q be a p-complement of R. Now we have that $V_q \leq C$ and $R \cap C = V_q$. Hence $V_q \lhd C$. Thus $N_G(V_q) = C$ ($N_G(V_q) \geq C$, but V_q is not normal since it doesn't contain M).

Now since $[G : C] = q^n$ and $S^q \cap C$ is a q-complement of C, $S^q \leq C$. Hence $T^q = S^q \cap \rho_{\mathscr{F}(q)}{}^*(G) \leq C$.

Now as $G/M \in \mathscr{F}$ we must have that $\rho_{\mathscr{F}(q)}{}^*(G) \leq S$ where $S/M = O_{q',q}(G/M)$. ($G/S \cong (G/M)/(S/M) \in \mathscr{F}$. Consequently, it follows that $(G/M)/(S/M) \cong G/S \in \mathscr{F}(q)$.)

Therefore $T^q = S^q \cap \rho_{\mathscr{F}(q)}{}^*(G) \leq R$ where $R/M = O_q(G/M)$.

As $T^q \leq C$, T^q normalizes V_q. Therefore $T^q \leq V_q$ (V_q was a q-complement of R, the result follows as $T^q \leq R$). Now $T^q = V_q \cap (S^q \cap \rho_{\mathcal{F}(q)}*(G)) = V_q \cap \rho_{\mathcal{F}(q)}*(G)$. Hence $T^q \lhd C$. Now since C is maximal, either $N_G(T^q) = C$ or $N_G(T^q) = G$.

If $N_G(T^q) = G$, then $T^q \lhd G$, contrary to the fact that T^q does not contain M.

Hence $C = N_G(T^q)$ and thus it follows that $D \leq C$ as required.

Remark Because any two $\mathcal{F}$-normalizers are conjugate, Theorem 1.23 also shows that each $\mathcal{F}$-normalizer of a finite solvable group is contained in an $\mathcal{F}$-projector.

Now we want to consider the formation of supersolvable groups $\mathcal{SS}$. We have already shown that $\mathcal{F}$ is an $\mathcal{SS}$-projector of a finite solvable group G if and only if F is a supersolvable subgroup of G such that if $F \leq H \leq K \leq G$, then $[K:H]$ is not a prime.

From previous results (concerning $\mathcal{F}$-normalizers) we see that the supersolvable normalizers of a finite solvable group G have the following properties.

THEOREM 1.24 (i) *A supersolvable normalizer covers each cyclic chief factor and avoids each noncyclic chief factor.*

(ii) *The order of the supersolvable normalizers of G is equal to the product of the orders of the cyclic factors in any chief series for G.*

(iii) *A maximal subgroup M of G contains a supersolvable normalizer of G if and only if the index of M in G is not prime.*

(iv) *Each supersolvable normalizer of G is contained in some $\mathcal{SS}$-projector of G.*

Proof: (i) This follows from Theorem 1.19, as a cyclic chief factor must have prime order and hence must be $\mathcal{SS}$-central. Similarly, the $\mathcal{SS}$-eccentric chief factors are precisely the noncyclic chief factors by Theorem 1.4 of Chapter 1.

(ii) This follows from comments in (i) and Theorem 1.19.

(iii) If M is a maximal subgroup of G and $[G:M] = p$, a prime, we have that $|\text{Fit}(G/\text{Core}_G(M))| = [G:M]$ (Lemma 1.11). Thus we see that $M/\text{Core}_G(M)$ can be embedded in $\text{Aut}(\text{Fit}(G/\text{Core}_G(M)))$ and so $M/\text{Core}_G(M)$ is a cyclic group of exponent dividing $p - 1$. Thus $M/\text{Core}_G(M) \in \mathcal{F}(p)$. It follows that a maximal subgroup M of G is $\mathcal{SS}$-abnormal if and only if $[G:M]$ is not a prime. The result now follows from Theorem 1.22.

(iv) By the remark just after Theorem 1.23.

We remark that Theorem 1.24 provides an alternate proof of Huppert's characterization of supersolvable groups (Theorem 3.1 of Chapter 1). For if all the maximal subgroups of G have prime index, then an $\mathfrak{SS}$-normalizer cannot be a proper subgroup. But this means that $G \in \mathfrak{SS}$ (as G must be the $\mathfrak{SS}$-normalizer) by Corollary 1.20.

For the last result in this section we will establish the fact that if a formation consists entirely of nilpotent groups, then it must be subgroup-closed. This result is due to Peter Neumann and follows from the following two lemmas. These results are found in [Bryant, Bryce, and Hartley 1970].

LEMMA 1.25 *If Y is a subgroup of the nilpotent group X whose class is smaller than the class of X, then the class of $\langle Y^X \rangle$ is also smaller than the class of X.*

Proof: We will first show by induction on d that

$$(*) \qquad \gamma_d(\langle Y^X \rangle) \le \gamma_{d+1}(X)\gamma_d(Y)$$

for all d. If $d = 0$, we have, since $y^x = x^{-1}yx = (x, y^{-1})y$, that

$$\gamma_0(\langle Y^X \rangle) = \langle Y^X \rangle \le \gamma_1(X)Y = \gamma_1(X)\gamma_0(Y).$$

Thus induction begins. Now suppose $d \ge 1$. By standard commutator identities

$$\gamma_{d+1}(\langle Y^X \rangle) = \left[\gamma_d(\langle Y^X \rangle), \langle Y^X \rangle \right]$$

$$\le \left[\gamma_{d+1}(X)\gamma_d(Y), \langle Y^X \rangle \right]$$

$$= \left[\gamma_{d+1}(X), \langle Y^X \rangle \right]\left[\gamma_d(Y), \langle Y^X \rangle \right]$$

$$\le \gamma_{d+2}(X)\left[\gamma_d(Y), \gamma_1(X)Y \right]$$

(by use of the $d = 0$ step and because $Y \le X$)

$$= \gamma_{d+2}(X)\left[\gamma_d(Y), Y \right]\left[\gamma_d(Y), \gamma_1(X) \right]$$

$$\le \gamma_{d+2}(X)\gamma_{d+1}(Y)$$

(because $[\gamma_d(X), \gamma_1(X)] \le \gamma_{d+2}(X)$—an immediate consequence of the 3-subgroup lemma).

This completes the induction proof of $(*)$. Letting $d = \text{class}(X) - 1$ in $(*)$ gives the desired conclusion.

With the aid of Lemma 1.25 we can now show that every supplement of $\text{Fit}(G)$ must belong to $\text{Form}(G)$.

LEMMA 1.26 *If G is a finite solvable group, $T \triangleleft G, T$ nilpotent, and $G = TS$ for some subgroup S, then $S \in \text{Form}(G)$.*

Proof: We induce on the class of T.

Consider the following subgroups of $G \times G \times G$; $K = \{(s, s, s) : s \in S\}$, $D_1 = \{(t, t, 1) : t \in T\}$, and $D_2 = \{(1, t, t) : t \in T\}$. Let H be the subgroup they generate in $G \times G \times G$. Note that D_1 and D_2 are each normalized by K.

Also it is clear that H is a subdirect product ($H \leq G \times G \times G$ and clearly the three projection maps are onto.) It follows that $H \in \text{Form}(G)$.

Suppose T is Abelian. Then D_1 and D_2 centralize each other. Also we have the $D_1 D_2 \cap K = \{1\}$. (An arbitrary element of $D_1 D_2$ looks like (t, tu, u) for some $t, u \in T$. This would belong to K if and only if $t = tu = u$ and so $t = u = 1$.) Consequently when T is Abelian $S \cong K \cong H/D_1 D_2 \in \text{Form}(G)$ as required.

Now suppose that the class c of T is greater that 1. It is easily seen that $Z(D_1)$ and $Z(D_2)$ are normal in H. Thus $M = Z(D_1)Z(D_2)$ is also normal in H. Furthermore, as above, $M \cap K = \{1\}$. Let $C = \langle MD_1, MD_2 \rangle$. Since $C \leq T \times T \times T$, C is nilpotent of class at most c; hence so too is C/M. However, $\text{class}(C/M) > c - 1$ because

$$\left[\gamma_{c-1}(D_1), D_2 \right] = \{ (1, w, 1) \quad : w \in \gamma_c(T) \}$$

is not contained in M so $\gamma_{c-1}(C/M) \neq \{1\}$. Thus we have $\text{class}(C/M) = c$. On the other hand, MD_2/M has class $c - 1$. Thus Lemma 1.25 yields that the normal closure B/M of MD_2/M in C/M also has class $c - 1$. Since it is seen without difficulty that K normalizes B, we have $B/M \lhd H/M$. Thus

$$MD_1 K/M \in \text{Form}(H/M)$$

by induction. Next we note that $MD_1/M \lhd MD_1 K/M$ and $\text{class}(MD_1/M) = c - 1$ so, again induction,

$$MK/M \in \text{Form}(MD_1 K/M).$$

The last two displays, together with $H/M \in \text{Form}(G)$ (because $H \in \text{Form}(G)$) yield $MK/M \in \text{Form}(G)$. Finally $S \cong K \cong MK/M$ so $S \in \text{Form}(G)$.

But the result of Peter Neumann is a direct consequence of this.

THEOREM 1.27 *A formation consisting of nilpotent groups must be subgroup-closed.*
Proof: Let $G \in \mathcal{F}$, where $\mathcal{F}$ is a formation consisting of nilpotent groups. Then every subgroup of G is a supplement to $\text{Fit}(G) = G$, and so, by Lemma 1.26, every subgroup of G belongs to $\text{Form}(G) \subseteq \mathcal{F}$.

Theorem 1.27 has the following corollary.

COROLLARY 1.28 *Let $\mathcal{F}$ be a formation consisting of nilpotent groups. Let $\mathcal{F}^* = \{G : \text{all nilpotent subgroups of } G \text{ belong to } \mathcal{F}\}$. Then $\mathcal{F}^*$ is a formation which is subgroup-closed. Furthermore $\mathcal{F}^* = \{G : \text{all Sylow subgroups of } G \text{ belong to } \mathcal{F}\}$.*
Proof: We show $\mathcal{F}^*$ is a formation. First if $G \in \mathcal{F}^*$ and $N \lhd G$, we need to show that $G/N \in \mathcal{F}^*$. It is clearly sufficient to assume that N is a mininal normal subgroup of G. Let H/N be a nilpotent subgroup of G/N.

Let H_p/N be a Sylow p-subgroup of H/N. Now either H_p is a Sylow p-subgroup of G and $H_p \in \mathcal{F}$ or $H_p = NL_p$ where L_p is a Sylow p-subgroup of G and $L_p \in \mathcal{F}$. In either case $H_p/N \in \mathcal{F}$. Since H/N is nilpotent and all of its Sylow subgroups belong to $\mathcal{F}$ we get that $H/N \in \mathcal{F}$. Hence $G/N \in \mathcal{F}^*$.

Now suppose G/N_1, $G/N_2 \in \mathcal{F}^*$ where $N_1 \cap N_2 = \{1\}$. We must show $G \in \mathcal{F}^*$. Let H be a nilpotent subgroup of G. HN_1/N_1 is a nilpotent subgroup of G/N_1 so $HN_1/N_1 \in \mathcal{F}$ and hence $H/H \cap N_1 \in \mathcal{F}$. Similarly $H/H \cap N_2 \in \mathcal{F}$. Hence $H \in \mathcal{F}$ and so $G \in \mathcal{F}^*$.

Clearly if $H \leq G$, every nilpotent subgroup of H is contained in G so if $G \in \mathcal{F}^*$, then $H \in \mathcal{F}^*$. So $\mathcal{F}^*$ is subgroup-closed.

Suppose all the Sylow subgroups of G belong to $\mathcal{F}$ and let H be a nilpotent subgroup of G. Let H_p be a Sylow p-subgroup of H. Then $H_p \leq G_p$ where G_p is some Sylow p-subgroup of G. Hence $G_p \in \mathcal{F}$ and, by Theorem 1.27, $H_p \in \mathcal{F}$. It follows that $H \in \mathcal{F}$ and hence $G \in \mathcal{F}^*$. The other inclusion being obvious, the result follows.

In the next section we will see how Fischer dualized the concept of a formation.

2. Fitting Classes

In this section we will consider classes which are, in some sense, dual to the formations. These classes, referred to as Fitting classes, were introduced by B. Fischer in his dissertation. This introduction to Fitting classes comes, for most part, from the paper [Fischer, Gaschütz, and Hartley 1967].

DEFINITION A class of finite solvable groups $\mathcal{F}$ will be called a **Fitting class** provided the following two conditions are satisfied:
(1) if $G \in \mathcal{F}$ and $N \lhd G$, then $N \in \mathcal{F}$.
(2) if N_1, $N_2 \lhd G$ and N_1, $N_2 \in \mathcal{F}$, then $N_1 N_2 \in \mathcal{F}$.

Comparing the definition of a Fitting class with that of a formation, we see in what sense the two concepts are the duals of each other. "Closed under epimorphic images" has changed to "closed under normal subgroups" and "closed under subdirect products" has changed to "closed under normal products."

It is convenient to introduce the following terminology.

DEFINITION Let $\mathcal{R}$ be any class of groups. Suppose H is a subgroup of some group G. Then we will say that H is an $\mathscr{R}$-**maximal** subgroup of G provided $H \in \mathscr{R}$ and if $H < K \leq G$, then $K \notin \mathscr{R}$.

Condition (2) in the definition of a Fitting class $\mathscr{F}$ says that there is a unique, maximal, normal $\mathscr{F}$-subgroup of G which we will call the $\mathscr{F}$-radical of G and denote by $\rho_{\mathscr{F}}(G)$. Since $\rho_{\mathscr{F}}(G)$ is a characteristic subgroup of G, if $H \lhd G$ it follows that $\rho_{\mathscr{F}}(H) \leq \rho_{\mathscr{F}}(G)$. Hence $\rho_{\mathscr{F}}(G)$ contains every subnormal $\mathscr{F}$-subgroup of G.

And now dual to the notion of an $\mathscr{F}$-projector, we define the notion of an $\mathscr{R}$-injector, for $\mathscr{R}$ some class of finite solvable groups.

DEFINITION Let $\mathscr{R}$ be a class of groups. A group V will be called an $\mathscr{R}$-**injector** of G provided:
(i) $V \leq G$
(ii) If N is a subnormal subgroup of G, then $N \cap V$ is an $\mathscr{R}$-maximal subgroup of N.

It follows that if V is an $\mathscr{R}$-injector and N is any subnormal subgroup of G, then $N \cap V$ is an $\mathscr{R}$-injector of N, and conversely.

As an application of the definition we note that if V is an $\mathscr{F}$-injector of G (for some Fitting class $\mathscr{F}$), then $N_G(V)$ is self-normalizing. for suppose g normalizes $N_G(V)$. Then V, $V^g \lhd N_G(V)$ so by property (2) of the definition of Fitting class we see that $VV^g \in \mathscr{F}$. However, since V is an $\mathscr{F}$-injector, V is a maximal $\mathscr{F}$- subgroup. Thus $V = V^g$; hence $g \in N_G(V)$.

We also note at this time that the definition of an $\mathscr{F}$-injector shows that an $\mathscr{F}$-injector is $\mathscr{F}$-maximal and contains $\rho_{\mathscr{F}}(G)$.

Just as we showed that every finite solvable group contains an $\mathscr{F}$-projector and that all the $\mathscr{F}$-projectors are conjugate where $\mathscr{F}$ is a saturated formation, we will show that if $\mathscr{F}$ is a Fitting class, then every finite solvable group contains an $\mathscr{F}$-injector and all the $\mathscr{F}$-injectors are conjugate.

First we need a lemma. Recall that if $\mathscr{N}$ is the formation of all finite nilpotent groups, we call the $\mathscr{N}$-projectors Carter subgroups and that the Carter subgroups of a finite solvable group G are precisely the self-normalizing nilpotent subgroups of G.

LEMMA 2.1 *Let $\mathscr{F}$ be a Fitting class and let G be a finite solvable group with $N \lhd G$ such that G/N is nilpotent. Further, suppose that W is $\mathscr{F}$-maximal in N and V_1 and V_2 are $\mathscr{F}$-maximal in G with $W \leq V_1 \cap V_2$. Then V_1 and V_2 are conjugate in G.*
Proof: Since $\mathscr{F}$ is a Fitting class and we have that $N \cap V_1 \lhd V_1$ and $N \cap V_2 \lhd V_2$, it follows that $N \cap V_1 \in \mathscr{F}$ and $N \cap V_2 \in \mathscr{F}$. But as $W \leq N \cap (V_1 \cap V_2)$ we have that $W \leq N \cap V_1$ and $W \leq N \cap V_2$. As W was $\mathscr{F}$-maximal in N, it follows that $W = N \cap V_1 = N \cap V_2$. Now V_1, V_2

$\leq N_G(W)$ and hence it will suffice to assume without loss of generality that $N_G(W) = G$.(Replace N by $N \cap N_G(W)$.)

We may further assume that $W = \{1\}$; for if W is nontrivial, then all of the hypotheses are met for G/W, N/W, V_1/W, and V_2/W, so V_1/W and V_2/W are conjugate by induction. This forces V_1 to be conjugate to V_2.

First we will prove that for $i = 1, 2$, V_i is normal in some Carter subgroup C_i of G. Let $M_i = N_G(V_i)$ and let C_i be any Carter subgroup of M_i. Then $M_i/M_i \cap N$ is nilpotent, being isomorphic to $M_iN/N \leq G/N$. Thus, since $V_i \leq M_i$,

$$[\, V_i, \underbrace{M_i, \ldots, M_i}_{r}\,] \leq M_i \cap N$$

for sufficiently large r. Moreover, since $V_i \triangleleft M_i$,

$$[\, V_i, \underbrace{M_i, \ldots, M_i}_{r}\,] \leq V_i \cap M_i \cap N = V_i \cap N = W = \{1\}.$$

This implies that $V_i \leq Z^*(M_i)$ and so $V_i \leq C_i$ (it follows easily from Theorem 6.2 of Appendix C that the hypercenter of any group M is contained in all Carter subgroups of M). Then since $V_i \triangleleft C_i$, all that remains to establish the claim is to show C_i is self-normalizing in G.

Thus we suppose that for some $x \in G$, $C_i^x = C_i$. Then $V_i^x \triangleleft C_i$. Then the definition of a Fitting class implies that $V_iV_i^x \in \mathfrak{F}$. But now the fact that V_i is $\mathfrak{F}$-maximal in G is invoked to yield that $V_i = V_i^x$ and, hence, $x \in M_i$. But as C_i is a Carter subgroup of M_i we must have $x \in C_i$. It follows that C_i is a Carter subgroup of G.

Now we complete the proof. We know there exists a $y \in G$ so that $C_1^y = C_2$ and hence V_1^y and V_2 are both normal subgroups of C_2. Again, as above, by the definition of Fitting class and $\mathfrak{F}$-maximality of V_2 we get $V_1^y = V_2$, as required.

Now we are ready to prove the existence and conjugacy of $\mathfrak{F}$-injectors in finite solvable groups whenever $\mathfrak{F}$ is a Fitting class.

THEOREM 2.2 *Let $\mathfrak{F}$ be a Fitting class. Then every finite solvable group contains an $\mathfrak{F}$-injector and all the $\mathfrak{F}$-injectors are conjugate.*
Proof: The proof proceeds by induction on the order of G.

Let M be a proper normal subgroup of G such that G/M is nilpotent. By induction we know that there exists an $\mathfrak{F}$-injector R of M. Let V be $\mathfrak{F}$-maximal in G with $R \leq V$. We will show V is an $\mathfrak{F}$-injector of G.

To show this it will suffice to show that $G^* \cap V$ is an $\mathfrak{F}$-injector in every maximal normal subgroup G^* of G. Now let $N = G^* \cap M$. It follows that G/N is nilpotent (as G/G^* and G/M are). By induction we know that G^* contains an $\mathfrak{F}$-injector, call one such V^*. Then $N \cap R$ and $N \cap V^*$ are $\mathfrak{F}$-injectors of N. Moreover, $N \cap R = N \cap V$ ($N \cap V \triangleleft V \in \mathfrak{F}$ implies

$N \cap V \in \mathcal{F}$; but $N \cap R \le N \cap V \le N$ and $N \cap R$ is an $\mathcal{F}$-injector of N; thus $N \cap R = N \cap V$ by $\mathcal{F}$-maximality) so we have $N \cap V$ and $N \cap V^*$ are $\mathcal{F}$-injectors of N. By induction then $N \cap V = (N \cap V^*)^t$ for some $t \in N$. Let $\overline{V}$ be $\mathcal{F}$-maximal with $(V^*)^t \le \overline{V}$ and denote $N \cap V$ by W. We then have $W \le V \cap \overline{V}$. Lemma 2.1 can now be applied to yield $V^x = \overline{V}$ for some $x \in G$. Hence $(V \cap G^*)^x = V^x \cap G^* = \overline{V} \cap G^*$.

Now it is easy to see that conjugates of $\mathcal{F}$-injectors are also $\mathcal{F}$-injectors. In particular, $(V^*)^t$ is an $\mathcal{F}$-injector of G^* because V^* is. Then from $(V^*)^t \le \overline{V} \cap G^*$ and $\overline{V} \cap G \in \mathcal{F}$ (because $\overline{V} \cap G^* \vartriangleleft \overline{V} \in \mathcal{F}$) we conclude that $(V^*)^t = \overline{V} \cap G^*$. Combining this with the equations at the end of the previous paragraph yields $(V \cap G^*)^x = (V^*)^t$. Consequently, $V \cap G^*$ is an $\mathcal{F}$-injector of G^* because $(V^*)^t$ is.

To show the conjugacy of two $\mathcal{F}$-injectors V_1 and V_2 of G, let G^* be any maximal normal subgroup of G. Then $G^* \cap V_1$ and $G^* \cap V_2$ are $\mathcal{F}$-injectors of G^* so by induction $(G^* \cap V_1)^x = G^* \cap V_2$ for some $x \in G$. Now G/G^* is nilpotent and V_1^x and V_2 are $\mathcal{F}$-maximal in G. It follows from Lemma 2.1 that V_1 and V_2 are conjugate in G.

From the proof of the theorem we obtain the following corollary.

COROLLARY 2.3 *Let $\mathcal{F}$ be a Fitting class and suppose that $\{1\} = G_0 \vartriangleleft G_1 \vartriangleleft \ldots \vartriangleleft G_n = G$ with G_{i+1}/G_i nilpotent for all i. Then V is an $\mathcal{F}$-injector of G provided $G_i \cap V$ is $\mathcal{F}$-maximal in G_i for all i.*

Proof: By induction $G_{n-1} \cap V$ is an $\mathcal{F}$-injector of G_{n-1}. Let W be an $\mathcal{F}$-injector of G. Then $G_{n-1} \cap W$ is an $\mathcal{F}$-injector of G_{n-1}. Now it follows, as in the above proof, that V and W are conjugate in G and, hence, that V is an $\mathcal{F}$-injector of G.

Also, the next result follows easily.

THEOREM 2.4 *Let $\mathcal{F}$ be a Fitting class and suppose that V is an $\mathcal{F}$-injector of the group G and that $V \le H \le G$. Then V is an $\mathcal{F}$-injector of H.*
Proof: Let $\{1\} = G_0 \vartriangleleft G_1 \vartriangleleft \ldots \vartriangleleft G_n = G$ with G_{i+1}/G_i nilpotent (as in Corollary 2.3). Letting $H_i = G_i \cap H$ for all i, we have $\{1\} = H_0 \vartriangleleft H_1 \vartriangleleft \ldots \vartriangleleft H_n = H$ and H_{i+1}/H_i is nilpotent for each $i = 0, \ldots, n-1$

Then $H_i \cap V = G_i \cap H \cap V = G_i \cap V$ is $\mathcal{F}$-maximal in $H_i = G_i \cap H$. Thus, by Corollary 2.3, V is an $\mathcal{F}$-injector of H.

The next lemma, which is from [Hartley 1969], discusses a few properties of $\mathcal{F}$-injectors. We say that a subgroup H of a group G is **pronormal** provided for each $x \in G$, H and H^x are conjugate in $\langle H, H^x \rangle$.

LEMMA 2.5 *Let $\mathcal{F}$ be a Fitting class and suppose that V is an $\mathcal{F}$-injector of the group G. Then,*

(i) V is a pronormal subgroup of G,

(ii) if $H \lhd G$, then $V \cap H$ is an $\mathcal{F}$-injector of H and all the $\mathcal{F}$-injectors of H have this form; also $G = HN_G(V \cap H)$,

(iii) V covers or avoids each chief factor of G.

Proof: (i) Both V and V^x are $\mathcal{F}$-injectors of $\langle V, V^x \rangle$ by Theorem 2.4. It follows that V, V^x are conjugate in $\langle V, V^x \rangle$ by Theorem 2.2.

(ii) We have previously observed the first statement, i.e., $V \cap H$ is an $\mathcal{F}$-injector of H. Since all the $\mathcal{F}$-injectors are conjugate, it follows that they are all of the form $(V \cap H)^x = V^x \cap H$ for some $x \in G$. Also, if $x \in G$, then $V \cap H$ and $(V \cap H)^x$ are conjugate in H, so by a simple modification of the Frattini argument, $G = HN_G(V \cap H)$.

(iii) Let H/K be a chief factor of G. It follows from part (ii) that $G = HN_G(V \cap H)$. Hence $K(V \cap H) \lhd G$. Thus either $K(V \cap H) = H$ or $K(V \cap H) = K$, as required.

We are going to look at some particular Fitting classes and their injectors. In several of the examples we leave many of the details to the reader.

EXAMPLES

(i) Let $\mathfrak{N}$ be the class of all finite nilpotent groups. It follows from standard theorems that $\mathfrak{N}$ is a Fitting class. This was shown by Fischer ([Fischer 1966]) who characterized the $\mathfrak{N}$-injectors as those nilpotent maximal subgroups which contain the Fitting subgroup (i.e., the $\mathfrak{N}$-radical) of G.

Remarks after the definition of an injector show that an $\mathfrak{N}$-injector of G must be a maximal nilpotent subgroup of G containing $\rho_{\mathfrak{N}}(G)$. So by Theorem 2.2 all we need to show in order to prove Fischer's characterization is that any two maximal nilpotent subgroups of G containing $F = \text{Fit}(G)$ must be conjugate. Our proof of this is essentially from [Mann 1971].

For a prime p dividing $|F|$ we let $F_p{'}$ denote the p-complement of F, C_p denote $C_G(F_p{'})$, and S_p be any Sylow p-subgroup of C_p. First we note that if p, q are distinct primes dividing $|F|$, then $C_p, C_q \lhd G$ so $[C_p, C_q] \le C_p \cap C_q = C_G(F) \le F$. Thus C_p normalizes $S_q F = S_q \oplus F_q{'}$. This implies that C_p normalizes S_q. Hence, for $p \ne q$, each of S_p, S_q normalizes the other so $[S_p, S_q] = \{1\}$. Now let $S = \langle S_p : p$ is a prime dividing $|F| \rangle$. Thus S is a nilpotent subgroup of G.

We will show that if V is a maximal nilpotent subgroup containing F, then V is conjugate to S. This will give the result. Let V_p be the Sylow p-subgroup of V and $V_p{'}$ be the p-complement of V. Now as $V \ge F$, $V_p \le C_G(V_p{'}) \le C_G(F_p{'}) = C_p$. So for every prime p dividing $|F|$ there is some $a_p \in C_p$ so that $V_p \le S_p^{a_p}$. Now a_p normalizes S_q for $q \ne p$. Let a be the product (in any order) of the elements a_p over all primes p dividing $|F|$. Then $V \le \prod S_p^{a_p} = (\prod S_p)^a = S^a$. As S is nilpotent we get $V = S^a$.

Before looking at futher examples we show how Fitting classes may be combined to create other Fitting classes.

Remark Suppose $\mathscr{F}$, $\mathscr{G}$ are Fitting classes and that $\mathscr{G}$ is closed under quotients. Then $\mathscr{F} * \mathscr{G} = \{\, H : \text{there exists } N \in \mathscr{F}, \; N \vartriangleleft H \text{ such that } H/N \in \mathscr{G} \,\}$ is a Fitting class. (Note that since $\mathscr{G}$ is closed under quotients, $\mathscr{F} * \mathscr{G} = \{ H : H/\rho_{\mathscr{F}}(H) \in \mathscr{G} \}$.) We will verify the two conditions for a Fitting class.

(1) Suppose $H \in \mathscr{F} * \mathscr{G}$ and $M \vartriangleleft H$. Now $M \cap \rho_{\mathscr{F}}(H) = \rho_{\mathscr{F}}(M)$ so $M/\rho_{\mathscr{F}}(M) = M/M \cap \rho_{\mathscr{F}}(H) \cong M\rho_{\mathscr{F}}(H)/\rho_{\mathscr{F}}(H) \vartriangleleft H/\rho_{\mathscr{F}}(H) \in \mathscr{G}$. Thus $M \in \mathscr{F} * \mathscr{G}$ as required.

(2) Suppose $M_1, M_2 \vartriangleleft G$ with $M_1, M_2 \in \mathscr{F} * \mathscr{G}$. We need to show that $M_1 M_2/\rho_{\mathscr{F}}(M_1 M_2) \in \mathscr{G}$. Since $\mathscr{G}$ is closed under quotients and $\rho_{\mathscr{F}}(M_1 M_2) \geq \rho_{\mathscr{F}}(M_1)\rho_{\mathscr{F}}(M_2)$, it suffices to show that $M_1 M_2/\rho_{\mathscr{F}}(M_1)\rho_{\mathscr{F}}(M_2) \in \mathscr{G}$. Now $M_i \cap \rho_{\mathscr{F}}(M_1)\rho_{\mathscr{F}}(M_2) = \rho_{\mathscr{F}}(M_i)$ $(i = 1, 2)$, so $M_1\rho_{\mathscr{F}}(M_2)/\rho_{\mathscr{F}}(M_1)\rho_{\mathscr{F}}(M_2) \cong M_1/M_1 \cap \rho_{\mathscr{F}}(M_1)\rho_{\mathscr{F}}(M_2) = M_1/\rho_{\mathscr{F}}(M_1) \in \mathscr{G}$ and similarly we get that $M_2\rho_{\mathscr{F}}(M_1)/\rho_{\mathscr{F}}(M_1)\rho_{\mathscr{F}}(M_2) \in \mathscr{G}$. Hence $M_1 M_2/\rho_{\mathscr{F}}(M_1)\rho_{\mathscr{F}}(M_2) = (M_1\rho_{\mathscr{F}}(M_2)/\rho_{\mathscr{F}}(M_1)\rho_{\mathscr{F}}(M_2))(M_2\rho_{\mathscr{F}}(M_1)/\rho_{\mathscr{F}}(M_1)\rho_{\mathscr{F}}(M_2)) \in \mathscr{G}$.

EXAMPLES (continued)

(ii) The class $\mathfrak{N}_k$ of groups of nilpotent length $\leq k$ form a Fitting class. It is easy to see that $\mathfrak{N}_{k+1} = \mathfrak{N}_k * \mathfrak{N}$ and that V is an $\mathfrak{N}_{k+1}$-injector of G if and only if $V/\rho_{\mathfrak{N}_k}(G)$ is an $\mathfrak{N}$-injector of $G/\rho_{\mathfrak{N}_k}(G)$.

(iii) For π a set of primes we let $\mathscr{G}_\pi$ be the class of finite π-groups. It is clear that $\mathscr{G}_\pi$ is a Fitting class and that the $\mathscr{G}_\pi$-injectors of a group are its Hall π-subgroups.

(iv) Suppose $\mathscr{F} = \mathscr{G}_\pi * \mathscr{G}_{\pi'}$. The $\mathscr{F}$-injectors of G are those subgroups V so that $V/O_\pi(G)$ is a Hall π'-subgroup of $G/O_\pi(G)$. To see this we first note that if V is an $\mathscr{F}$-injector of G, then $O_\pi(V) = O_\pi(G)$. For since V is an $\mathscr{F}$-injector, $V \geq O_{\pi, \pi'}(G)$. This shows that $[O_\pi(V), O_{\pi, \pi'}(G)] \leq O_\pi(V) \cap O_{\pi, \pi'}(G) = O_\pi(G)$. It follows that $O_\pi(V)/O_\pi(G) \leq C_{G/O_\pi(G)}(O_{\pi, \pi'}(G)/O_\pi(G)) \leq O_{\pi, \pi'}(G)/O_\pi(G)$ (for this last inclusion see [Gorenstein 1968, Theorem 6.3.2]). Hence $O_\pi(V) \leq O_{\pi, \pi'}(G)$. It follows that $O_\pi(V) \leq O_\pi(G)$. The other inclusion is clear, so $O_\pi(V) = O_\pi(G)$. Properties of Hall subgroups now imply that V is an $\mathscr{F}$-injector if and only if $V/O_\pi(G)$ is a Hall π'-subgroup of $G/O_\pi(G)$.

(v) Let $\mathscr{S}\mathscr{T}\mathscr{P}$ be the class of groups having the Sylow tower property. Theorem 3.2 of Chapter 4 essentially shows that $\mathscr{S}\mathscr{T}\mathscr{P}$ is a Fitting class. By way of comparison, recall that the example preceding Theorem 1.13 of Chapter 1 shows that the class of supersolvable groups does not form a Fitting class.

We will give another example after the next definition.

For the remainder of this section we will be concerned only with a certain type of Fitting class, defined below. The results are mainly from [Blessenohl and Gaschütz 1970].

DEFINITION   A Fitting class $\mathcal{F}$ will be called **normal** provided for each finite solvable group G, the $\mathcal{F}$-injectors of G are normal subgroups of G.

Note that for a normal Fitting class $\mathcal{F}$ every finite solvable group G must contain a unique $\mathcal{F}$-injector which must contain every normal $\mathcal{F}$-subgroup.

We are about to give a general procedure for constructing normal Fitting classes.

EXAMPLES (continued)

(vi) Let A be an Abelian group (possibly infinite) and suppose $B \leq A$. Futher, suppose that for every group G we have a homomorphism $f_G : G \to A$ so that if $N \triangleleft G$, then f_G restricted to N is just f_N. Let $\mathcal{K} = \{ H : f_H(H) \leq B \}$. Then $\mathcal{K}$ is a normal Fitting class and $\rho_{\mathcal{K}}(G) = f_G^{-1}(B)$ for all G. We show that as follows:

First suppose $N \triangleleft H \in \mathcal{K}$. Then $f_N(N) = f_H(N) \leq f_H(H) \leq B$, so $N \in \mathcal{K}$. Next suppose $H = N_1 N_2$ with $N_1, N_2 \triangleleft H$ and $N_1, N_2 \in \mathcal{K}$. Then for any $h \in H$, $h = n_1 n_2$ with $n_1 \in N_1$, $n_2 \in N_2$, so that $f_H(h) = f_H(n_1 n_2) = f_H(n_1) f_H(n_2) = f_{N_1}(n_1) f_{N_2}(n_2) \in B$. Hence $H \in \mathcal{K}$.

Thus $\mathcal{K}$ is a Fitting class. Now A is Abelian so $f_G^{-1}(B) \triangleleft G$ and thus $f_G^{-1}(B) \in \mathcal{K}$. However, if $V \geq f_G^{-1}(B)$ and $V \in \mathcal{K}$, then $V \triangleleft G$ (as $V \geq G'$) and $f_G(V) = f_V(V) \leq B$ so $V \leq f_G^{-1}(B)$. Thus $V = f_G^{-1}(B)$ must be the unique $\mathcal{K}$-injector of G.

Now we state a few properties of the standard wreath product that we will need. Also, we will establish some notation that will be used in the subsequent proofs.

Recall that the standard wreath product, $G \operatorname{wr} H$, of G by H is the semi-direct product of a group $G^* = \prod_{\alpha \in H} G_\alpha$ where for each α, $G_\alpha \cong G$ with the action of H on G^* being given by the regular permutation representation of H (acting on the subscripts). See [Scott 1964] for more details.

If $L \leq G$ we denote by L^* the naturally associated subgroup of G^*, $\prod_{\alpha \in H} L_\alpha$ with $L_\alpha \cong L$ under the same isomorphism used between G_α and G.

The following three properties can be established by the reader without difficulty:

(1) If $L \leq G$, then $L^* H \cong L \operatorname{wr} H$.

(2) If $L \triangleleft G$, then $L^* \triangleleft G \operatorname{wr} H$ and $(G \operatorname{wr} H)/L^* \cong (G/L) \operatorname{wr} H$.

(3) If $J \leq H$, then $G^* J \cong \underbrace{(G \times G \times \ldots \times G)}_{[H : J] \text{ factors}} \operatorname{wr} J$.

Also, we recall the following fact from Chapter 2 (Chapter 2, Theorem 2.2)

(4) Every extension of G by H can be embedded in $G \operatorname{wr} H$.

We begin by determining a few properties of normal Fitting classes. The next lemma will be very important in subsequent proofs.

LEMMA 2.6 *Let $\mathcal{F}$ be a normal Fitting class, $\{1\} \neq G \in \mathcal{F}$, and let p be a prime. Then there exists a number m such that for all n,*

$$(\underbrace{G \times G \times \ldots \times G}_{mn \text{ factors}}) \operatorname{wr} \mathbb{Z}_p \in \mathcal{F}.$$

Proof: Suppose that G is a minimal counter-example. Let H be any subgroup of G containing G' so that $[G : H] = q$, a prime. Let Y be any group of order $p^a q$, where a can be any natural number in which all minimal normal subgroups are p-groups (e.g., choose $Y = \mathbb{Z}_p \operatorname{wr} \mathbb{Z}_q$). Let X be a subgroup of Y so that $|X| = q$.

Now if $H \neq \{1\}$, then we have by induction there exists m_1 so that

$$(\underbrace{H \times H \times \ldots \times H}_{nm_1 \text{ factors}}) \operatorname{wr} X \in \mathcal{F}$$

for every natural number n. If $H = \{1\}$ then

$$G \cong X \cong (\underbrace{H \times H \times \ldots H}_{n \text{ factors}}) \operatorname{wr} X \in \mathcal{F}$$

for all n. Here we take $m_1 = 1$.

Now we let

$$K = (\underbrace{G \times G \times \ldots \times G}_{nm_1 \text{ factors}}) \operatorname{wr} Y$$

for some natural number n. We set

$$D = (\underbrace{G \times G \times \ldots \times G}_{nm_1 \text{ factors}})^*$$

and let

$$D_1 = (\underbrace{H \times H \times \ldots \times H}_{nm_1 \text{ factors}})^*$$

Since $G \in \mathcal{F}$, it follows that $D \in \mathcal{F}$. Also, we see from properties (1) and (3) which preceded this lemma that

$$D_1 X \cong \underbrace{(H \times \ldots \times H)}_{nm_1[Y:X] \text{ factors}} \operatorname{wr} X$$

which belongs to $\mathcal{F}$ from above.

Now note that $D_1 X \lhd DX$. It follows that $DX = D(D_1 X) \in \mathcal{F}$. Therefore, the $\mathcal{F}$-injector of K must be a normal subgroup which properly contains D.

Now let M/D be a minimal normal subgroup of K/D contained in the $\mathcal{F}$-injector of K. Then since $K/D \cong Y$, M/D is a p-group by our assumption about Y. We have $M \in \mathcal{F}$ and $M = D(M \cap Y)$. Then $M \cap Y$ is Abelian so if we let Z be a subgroup of order p contained in $M \cap Y$ we have that $DZ \lhd M$ so $DZ \in \mathcal{F}$.

But (from property (3) preceding this lemma)

$$DZ \cong \underbrace{(G \times G \times \ldots \times G)}_{m_1 n[Y:X] \text{ factors}} \operatorname{wr} Z.$$

This is contrary to the fact that G is a counterexample. Thus the result has been established.

We have seen that for $\mathcal{N}$, the Fitting class of finite nilpotent groups, the $\mathcal{N}$-injectors of a group G are maximal nilpotent subgroups containing $\operatorname{Fit}(G)$. Thus $\mathcal{N}$ is not a normal Fitting class (the $\mathcal{N}$-injectors of S_4 are the Sylow 2-subgroups of S_4). However, the next result (attributed by Blessenohl and Gaschütz to J. Cossey) establishes a containment.

THEOREM 2.7 *The Fitting class of finite nilpotent groups is contained in every nontrivial normal Fitting class.*

Proof: Let $\mathcal{F}$ be a normal Fitting class and let p be any prime. From Lemma 2.6 there is a group H and subgroups D, Z_1 such that $H \in \mathcal{F}$, $D \lhd H$, $Z_1 \cong \mathbb{Z}_p$, $H = DZ_1$, and $H \cap Z_1 = \{1\}$.

Let $R = H \oplus Z$ where $Z \cong \mathbb{Z}_p$. Let y, x generate Z_1, Z respectively. Note that $D \lhd R$ and let $H_1 = D\langle xy \rangle$. Then $H_1 \lhd R$, $H_1 \cong H$ (both are isomorphic to R/Z), and $R = HH_1$. It follows that $R \in \mathcal{F}$, hence $Z \in \mathcal{F}$ as well since $Z \lhd R$.

Now we have shown that $\mathcal{F}$ must contain the cyclic group of order p for all primes p.

But if a Fitting class $\mathcal{F}$ contains a p-group for some prime p, it must contain all p-groups. For if not, there would exist a p-group G of smallest order p^n so that $G \notin \mathcal{F}$. Now since $n > 1$, it would follow that G must be cyclic. (All the other groups of order p^n would be the product of two normal subgroups of order p^{n-1}, i.e., two normal subgroups which would belong to $\mathcal{F}$.) But then (by property (4) preceding Lemma 2.6) G is isomorphic to a subgroup of $H = X \operatorname{wr} Z$ where X is cyclic of order p^{n-1} and $Z \cong \mathbb{Z}_p$. However, as $X, Z \in \mathcal{F}$, it follows that $H \in \mathcal{F}$ (H is a p-group, hence nilpotent, so every subgroup of H is subnormal; an $\mathcal{F}$-injector of H then must contain X^* and Z). Then the subgroup of H isomorphic to G, being subnormal in H, must belong to $\mathcal{F}$. Hence $G \in \mathcal{F}$, contrary to assumption.

Thus $\mathcal{F}$ contains all p-groups. Taking products completes the proof.

THEOREM 2.8 *Let $\mathcal{F}$ be a nontrivial normal Fitting class. Let H be any finite solvable group. Then there exists a group $G \in \mathcal{F}$ such that $H \leq G$.*
Proof: For $H = \{1\}$, the result is trivial. Let K be a normal subgroup of H such that $H/K \cong \mathbb{Z}_p$ for some prime p. If $K = \{1\}$, then $H \in \mathcal{F}$ by Theorem 2.7. If $K \neq \{1\}$, then by induction on the order of H, there exists $G_1 \in \mathcal{F}$ so that $K \leq G_1$. Now $G_1 \neq \{1\}$ so by Lemma 2.6

$$G = \underbrace{(G_1 \times \ldots \times G_1)}_{m \text{ factors}} \operatorname{wr} \mathbb{Z}_p \in \mathcal{F}$$

for some m.

But then $K \operatorname{wr} \mathbb{Z}_p \leq G$ and $K \operatorname{wr} \mathbb{Z}_p$ contains a subgroup isomorphic to H (H is an extension of K by $\mathbb{Z}_p$ so H can be embedded in $K \operatorname{wr} \mathbb{Z}_p$ by property (4) preceding Lemma 2.6.)

We need the following two lemmas to establish a characterization of normal Fitting classes.

LEMMA 2.9 *Let $\mathcal{F}$ be a Fitting class and suppose that G is a group with normal subgroups $N_1, \ldots, N_r$ such that $G = N_1 N_2 \ldots N_r$. Then*

$$\rho_{\mathcal{F}}(G)/\rho_{\mathcal{F}}(N_1)\rho_{\mathcal{F}}(N_2) \ldots \rho_{\mathcal{F}}(N_r) \leq Z(G/\rho_{\mathcal{F}}(N_1)\rho_{\mathcal{F}}(N_2) \ldots \rho_{\mathcal{F}}(N_r)).$$

Proof: Clearly $\rho_{\mathcal{F}}(G) \cap N_i = \rho_{\mathcal{F}}(N_i)$. It follows that $[\rho_{\mathcal{F}}(G), G] \leq [\rho_{\mathcal{F}}(G), N_1] \ldots [\rho_{\mathcal{F}}(G), N_r] \leq \rho_{\mathcal{F}}(N_1) \ldots \rho_{\mathcal{F}}(N_r)$, as required.

LEMMA 2.10 *Let X and Y be groups. Suppose that $N \triangleleft X \operatorname{wr} Y$ with N Abelian, $N \leq X^*$. Further, suppose that there exists an element $k \notin X^*$ so that $(Nx^*)^k = Nx^*$ for every $x^* \in X^*$. Then X is Abelian.*
Proof: Now $k \in X^* y$ for some $1 \neq y \in Y$. Also, $(X_h)^y = X_{hy} = X_{h_1}$ where $h_1 = hy \neq h$. But then $(X_h)^k = X_{h_1}$. However, by our assumption $(NX_h)^k = NX_h$ so $NX_h = NX_{h_1}$. (It follows that $X_{h_1} X_h \leq NXh_1$.) Thus $X \cong X_h \cong X_h X_{h_1}/X_{h_1} \leq NX_{h_1}/X_{h_1} \cong N/N \cap X_{h_1}$ is Abelian, as required.

Now we are ready to characterize the normal Fitting classes.

THEOREM 2.11 *Let $\mathcal{F}$ be a nontrivial Fitting class. Then $\mathcal{F}$ is a normal Fitting class if and only if $\rho_{\mathcal{F}}(G) \geq G'$ for all groups G.*

Proof: If it is true for every group G that $\rho_{\mathcal{F}}(G) \geq G'$, then clearly all the $\mathcal{F}$-injectors must be normal (as all the $\mathcal{F}$-injectors must contain $\rho_{\mathcal{F}}(G)$) and hence $\mathcal{F}$ is a normal Fitting class.

In the other direction we let G be a minimal counterexample to the result. That is, we suppose that $G/\rho_{\mathcal{F}}(G)$ is not Abelian, but $H/\rho_{\mathcal{F}}(H)$ is Abelian for every proper subgroup H of G. Now if $\rho_{\mathcal{F}}(G) \leq H \leq G$, then $\rho_{\mathcal{F}}(G)$ is, also, the $\mathcal{F}$-injector of H (Theorem 2.4) and therefore $\rho_{\mathcal{F}}(G) = \rho_{\mathcal{F}}(H)$. Hence $H/\rho_{\mathcal{F}}(G)$ is Abelian. It follows that $G/\rho_{\mathcal{F}}(G)$ is a minimal non-Abelian group. Now by a theorem of Miller and Moreno (see [Scott 1964, 6.5.8]) we can conclude there are two possible structures for $G/\rho_{\mathcal{F}}(G)$. In each of these cases we can conclude that the center $Z(G/\rho_{\mathcal{F}}(G))$ is a p-group.

Lemma 2.6 shows that there exists a natural number m so that

$$\underbrace{(\rho_{\mathcal{F}}(G) \times \ldots \times \rho_{\mathcal{F}}(G))}_{m \text{ factors}} \operatorname{wr} \mathbb{Z}_p \in \mathcal{F}.$$

Let

$$K = \underbrace{G \times \ldots \times G}_{m \text{ factors}} \operatorname{wr} \mathbb{Z}_p$$

and let

$$D = \underbrace{(G \times \ldots \times G}_{m \text{ factors}})^*.$$

Then D contains the normal subgroup

$$D_1 = \underbrace{(\rho_{\mathcal{F}}(G) \times \ldots \times \rho_{\mathcal{F}}(G))}_{m \text{ factors}}^*$$

and $D_1 \triangleleft K$. Let $K_1 = D_1 \mathbb{Z}_p$. Then

$$K_1 \cong \underbrace{(\rho_{\mathcal{F}}(G) \times \ldots \times \rho_{\mathcal{F}}(G))}_{m \text{ factors}} \operatorname{wr} \mathbb{Z}_p$$

so $K_1 \in \mathcal{F}$.

Hence $\rho_{\mathcal{F}}(K) \geq D_1$. Suppose that $\rho_{\mathcal{F}}(K) \leq D$. Then we must have that $\rho_{\mathcal{F}}(K) = \rho_{\mathcal{F}}(D)$.

Now Lemma 2.9 yields that $\rho_{\mathcal{F}}(K)/D_1 = \rho_{\mathcal{F}}(D)/D_1 \leq Z(D/D_1)$ and we can note that $Z(D/D_1) \cong Z(G/\rho_{\mathcal{F}}(G)) \times \ldots Z(G/\rho_{\mathcal{F}}(G))$ must be a p-group. It follows that $K_1 \rho_{\mathcal{F}}(K)/D_1$ must be a p-group. It follows that K_1 is a subnormal subgroup of $K_1 \rho_{\mathcal{F}}(K)$. But this would imply that $\rho_{\mathcal{F}}(K) \geq K_1$, contrary to the fact that $\rho_{\mathcal{F}}(K) = \rho_{\mathcal{F}}(D)$. Therefore, we assume that $\rho_{\mathcal{F}}(K)$ is not contained in D.

Now we let $D_1 x \in \rho_{\mathscr{F}}(K)/D_1 \backslash D/D_1$. As $[\rho_{\mathscr{F}}(K), D] \leq \rho_{\mathscr{F}}(K) \cap D$ $= \rho_{\mathscr{F}}(D)$, it follows that $D_1 x$ induces the identity automorphism on $D/\rho_{\mathscr{F}}(D)$. On the other hand it follows from Lemma 2.9 that $\rho_{\mathscr{F}}(D)/D_1$ $\leq Z(D/D_1)$. Hence $\rho_{\mathscr{F}}(D)/D_1$ is an Abelian normal subgroup of K/D_1. From our remarks about wreath products

$$K/D_1 \cong \underbrace{(G/\rho_{\mathscr{F}}(G) \times \ldots \times G/\rho_{\mathscr{F}}(G))}_{m \text{ factors}} \text{wr } \mathbb{Z}_p$$

where

$$D/D_1 \cong \underbrace{(G/\rho_{\mathscr{F}}(G) \times \ldots \times G/\rho_{\mathscr{F}}(G))}_{m \text{ factors}}{}^*.$$

Lemma 2.10 now says that $G/\rho_{\mathscr{F}}(G) \times \ldots \times G/\rho_{\mathscr{F}}(G)$ is Abelian. But this implies $G/\rho_{\mathscr{F}}(G)$ is Abelian, contrary to our assumption.

We finish this section with two corollaries of the above result.

COROLLARY 2.12 *Let $\mathcal{C}$ be a family of nontrivial normal Fitting classes. Then $\bigcap \mathcal{C}$ is a nontrivial normal Fitting class. In particular, there is a smallest nontrivial normal Fitting class.*
Proof: Clearly, the intersection of Fitting classes is a Fitting class. From Theorem 2.7 the intersection is nontrivial. Clearly, the $\bigcap \mathcal{C}$-radical of G is the intersection of the $\mathscr{F}$-radicals of G for all $\mathscr{F} \in \mathcal{C}$. It follows that the $\bigcap \mathcal{C}$-radical of G contains G' for every finite solvable group G. Hence $\bigcap \mathcal{C}$ is a normal Fitting class.

Hans Lausch ([Lausch 1973]) actually describes this smallest nontrivial normal Fitting class. He shows that a group G belongs to this class if and only if there exists a group R so that $G \lhd R$ and $G \leq \langle [N, \mathrm{Aut}(N)] : N \lhd R \rangle$.

The next corollary concerns the structure of the factor group $G/\rho_{\mathscr{F}}(G)$ where $\mathscr{F}$ is not a normal Fitting class.

COROLLARY 2.13 *Let $\mathscr{F}$ be a nontrivial nonnormal Fitting class and let H be any group. Then there is a group G so that $G/\rho_{\mathscr{F}}(G)$ contains a subgroup isomorphic to H.*
Proof: Since $\mathscr{F}$ is nontrivial and nonnormal, there exists a group A so that $A/\rho_{\mathscr{F}}(A)$ is non-Abelian. Let G be A wr H. Then $(\rho_{\mathscr{F}}(A))^* \leq \rho_{\mathscr{F}}(G) \leq A^*$. (For if $\rho_{\mathscr{F}}(G)$ is not contained in A^*, then it would follow as in the above theorem using Lemmas 2.9 and 2.10 that $A/\rho_{\mathscr{F}}(A)$ is Abelian.) The group $G/\rho_{\mathscr{F}}(G)$ contains $\rho_{\mathscr{F}}(G)H/\rho_{\mathscr{F}}(G) \cong H$ as a subgroup. This concludes the proof.

3. *Homomorphs and Normal Homomorphs*

We have seen that if $\mathscr{F}$ is a saturated formation, then every finite solvable group G contains an $\mathscr{F}$-projector and all the $\mathscr{F}$-projectors form a conjugacy class of subgroups of G. It was Wielandt who localized the concept of a formation by studying $Q(G)$, the epimorphic images of a single group G. He showed (in [Wielandt 1967/1968] that it was only the properties of the collection $\mathscr{F} \cap Q(G)$ which were relevant to the existence and conjugacy of $\mathscr{F}$-projectors of G. In this section we will see that we can construct $\mathscr{F}$-projectors of a group G for any class of factors closed under taking of epimorphic images within G. The results are basically due to Joseph A. Troccolo ([Troccolo 1975]). The subgroups obtained in this manner coincide with those of Schunk if $\mathscr{F}$ is a saturated homomorph (see [Schunk 1967]).

DEFINITION For any group G we let $Q(G) = \{ H/H_0 : \ H_0 \lhd H, \ H \le G \}$.

DEFINITIONS For some group G, let $\mathscr{F}$ be any subset of $Q(G)$.
(i) A subgroup E of G is said to have the $\mathscr{F}$-**covering property** in G provided whenever $E \le F \le G$ and $F/F_0 \in \mathscr{F}$, then $EF_0 = F$.
(ii) A subgroup E of G with the $\mathscr{F}$-covering property in G is said to be an $\mathscr{F}$-**projector** of G if no proper subgroup of E has the $\mathscr{F}$-covering property in E.

Note that if $\mathscr{F}^*$ is a saturated formation and we let $\mathscr{F} = \mathscr{F}^* \cap Q(G)$, the above definition coincides with the usual one.
Troccolo has determined the precise conditions on $\mathscr{F}$ necessary to ensure that $\mathscr{F}$-projectors exist and are all conjugate. Thus we make the following definition.

DEFINITION Let $\mathscr{F}$ be a subset of $Q(G)$ for some finite solvable group G. We will say that $\mathscr{F}$ is a **homomorph** provided
(i) If $A_0 \lhd A$, $B_0 \lhd B$, $A_0 B = A$, and $A_0 \cap B = B_0$, then $A/A_0 \in \mathscr{F}$ if and only if $B/B_0 \in \mathscr{F}$.
(ii) If $R/R_0 \in \mathscr{F}$ and $R_0 \le R_1 \lhd R$, then $R/R_1 \in \mathscr{F}$.
Furthermore, we will say homomorph $\mathscr{F}$ is **normal** provided it satisfies the following additional condition
(iii) If $R/S \in \mathscr{F}$ and $g \in G$, then $R^g/S^g \in \mathscr{F}$.

These conditions may seem strange but it turns out that they are just what is needed to show the existence and conjugacy of $\mathscr{F}$-projectors. Also, it is clear that all the above conditions are satisfied if we assume that $\mathscr{F}$ is closed with respect to the taking of epimorphic images within G.

For the rest of this section $\mathcal{F}$ will denote a normal homomorph of G. If $H \leq G$, then $\mathcal{F} \cap Q(H)$ will be a normal homomorph of H. Also, if $N \triangleleft G$, and we identify the factor $(L/N)/(L_0/N)$ of G/N with the factor L/L_0 of G, then the set $\mathcal{F} \cap Q(G/N)$ is a normal homomorph of G/N. We refer to the $\mathcal{F} \cap Q(H)$-projectors of H and the $\mathcal{F} \cap Q(G/N)$-projectors of G/N as $\mathcal{F}$-projectors of G and G/N respectively.

Now we are ready to begin to show that G has an $\mathcal{F}$-projector and all the $\mathcal{F}$-projectors of G are conjugate. Again this is from [Troccolo 1975]. Note the similarity of these proofs to the proofs in the case of a saturated formation.

LEMMA 3.1 *Let E be an $\mathcal{F}$-projector of G. Then*
(i) *If $E \leq H \leq G$, then E is an $\mathcal{F}$-projector of H.*
(ii) *If $N \triangleleft G$, then EN/N is an $\mathcal{F}$-projector of G/N.*
Proof: (i) This is clear from the definition.

(ii) Suppose that $EN/N \leq F/N$ and that $N \leq F_0 \triangleleft F$ with $F/F_0 \in \mathcal{F}$. Now since $E \leq F$ and E is an $\mathcal{F}$-projector of G, we can conclude that $EF_0 = F$ (by defintion of the $\mathcal{F}$-covering property). Hence $(EN/N)(F_0/N) = ENF_0/N = F/N$, as required.

Now suppose that E_1/N has the $\mathcal{F}$-covering property in EN/N. Now if $E/E_0 \in \mathcal{F}$ for some E_0, then $EN/E_0N \in \mathcal{F}$ (for since $E/E_0 \in \mathcal{F}$ and $E_0 \leq E_0N \cap E \triangleleft E$, we can conclude that $E/E_0N \cap E \in \mathcal{F}$. But then from the facts that $E_0N \cap E \triangleleft E$ and $E_0N \triangleleft EN$ we can conclude, since $E/E_0N \cap E \in \mathcal{F}$, that $EN/E_0N \in \mathcal{F}$).

Thus $E_1(E_0N) = EN$. Thus, since $E_1 \geq N$, we get $E_1E_0 = EN$ and so $E = (E \cap E_1)E_0$, $(E_0 \leq E \leq E_0E_1)$. Thus $E \cap E_1$ has the $\mathcal{F}$-covering property in E. It follows that $E \cap E_1 = E$. Therefore $EN = E_1$ as required. Thus EN/N is an $\mathcal{F}$-projector of G/N.

LEMMA 3.2 *Let E/N be an $\mathcal{F}$-projector of G/N and let D be an $\mathcal{F}$-projector of E. Then D is an $\mathcal{F}$-projector of G.*
Proof: By part (ii) of Lemma 3.1 DN/N is an $\mathcal{F}$-projector of E/N. Therefore $DN = E$ (since the only subgroup of E/N which has the $\mathcal{F}$-covering property in E/N must be E/N).

Now we will show that D is an $\mathcal{F}$-projector for G. First, suppose that $D \leq F$ and $F/F_0 \in \mathcal{F}$. Then $E/N = DN/N \leq FN/N$ and we can conclude that $E(F_0N) = FN$. (As in the above lemma from the fact that $F/F_0 \in \mathcal{F}$, we can conclude that $FN/F_0N \in \mathcal{F}$.) Hence we have that $EF_0 = FN$. So $F_0 \leq F \leq F_0E$ from which it follows that $F = (E \cap F)F_0$ and hence that $E \cap F/E \cap F_0 \in \mathcal{F}$ (the fact that $E \cap F/E \cap F_0 \in \mathcal{F}$ follows from $F_0 \triangleleft F$ and $E \cap F_0 \triangleleft E \cap F$ since we know that $F/F_0 \in \mathcal{F}$). Now $D \leq E \cap F$ and therefore $D(E \cap F_0) = E \cap F$. Thus (since $F = (E \cap F)F_0$) $F = D(E \cap F_0)F_0$. That is, $DF_0 = F$. Hence D has the $\mathcal{F}$-covering property with respect to G. However, as D was an $\mathcal{F}$-projector of E, it is clear that no proper subgroup of D has the $\mathfrak{F}$-covering property with respect to D. Thus D is an $\mathfrak{F}$-projector of G and the proof is complete.

Finally we are ready for the main result of this section.

THEOREM 3.3 *If $\mathfrak{F}$ is a normal homomorph of G, then G has an $\mathfrak{F}$-projector and all $\mathfrak{F}$-projectors of G are conjugate.*
Proof: First we will prove existence. The proof proceeds by induction on the order of G.

Let N be a minimal normal subgroup of G. Then by induction G/N has an $\mathfrak{F}$-projector E/N. Now if $E < G$, then E has an $\mathfrak{F}$-projector D (by induction) and so by Lemma 3.2 G has an $\mathfrak{F}$-projector, namely D. Thus we assume that G/N is its own $\mathfrak{F}$-projector.

Now if G is its own $\mathfrak{F}$-projector, there is nothing to prove. So suppose that G is not its own $\mathfrak{F}$-projector. This means there must be a subgroup E of G which has the $\mathfrak{F}$-covering property in G. We will show that E is an $\mathfrak{F}$-projector of G. As E has the $\mathfrak{F}$-covering property in G, we need only to show that no proper subgroup of E has the $\mathfrak{F}$-covering property in E.

Thus we suppose that $E_1 \leq E$ and E_1 has the $\mathfrak{F}$-covering property in E. Now from the proof of Lemma 3.1(ii) we see that EN/N has the $\mathfrak{F}$-covering property in G/N and hence that $EN = G$ as G/N is its own $\mathfrak{F}$-projector. But then as N is a minimal normal subgroup we must have $E \cap N = \{1\}$.

But $E_1 N/N$ has the $\mathfrak{F}$-covering property in $EN/N = G/N$. So $E_1 N = G$ also. Therefore, since $E_1 \leq E$, we must have $E_1 = E$ and thus E is an $\mathfrak{F}$-projector of G.

To show that any two $\mathfrak{F}$-projectors of G are conjugate, we again proceed by induction on $|G|$. Thus we assume that E and F are $\mathfrak{F}$-projectors of G. Now by Lemma 3.1(ii) we can conclude that EN/N are FN/N are $\mathfrak{F}$-projectors of G/N where N is some minimal normal subgroup of G. Thus by induction $EN = (FN)^g = F^g N$ for some $g \in G$. Since F^g is also an $\mathfrak{F}$-projector of G, we may as well assume that $EN = FN$. Now if $EN < G$, then Lemma 3.1(i) shows that E, F are both $\mathfrak{F}$-projectors of EN and hence by induction E and F are conjugate in G.

Hence we can assume that $EN = FN = G$ for every minimal normal subgroup N of G. Then E and F are maximal subgroups of G, and $\text{Core}_G(E) = \text{Core}_G(F) = \{1\}$. It follows that $C_G(N) = \{1\}$ and hence that G has a self-centralizing, complemented, minimal normal subgroup. From Ore's theorem (see [Schunk 1967]) it follows that any two complements (namely E and F) of N must be conjugate. The result follows.

Summary on Closure Properties and Characterizations

Michael Weinstein

In this last chapter we summarize 8 of the classes which have been examined throughout the book. After restating the definition for each class, we address the questions of whether or not the class is subgroup-closed, quotient-closed, and product-closed. Finally we list alternate definitions or characterizations of the classes.

1 CLT-groups

DEFINITION The group G is **CLT** iff G has a subgroup of order n for every divisor n of $|G|$.

That the class $\mathcal{CLT}$ is neither subgroup- nor quotient-closed was shown in Theorem 1.4 of Chapter 3.

THEOREM 1.1 $\mathcal{CLT}$ *is product-closed.*
Proof: Let A and B be two CLT-groups of orders $p_1^{e_1} p^{e_2} \ldots p_k^{e_k}$ and $p_1^{d_1} p_2^{d_2} \ldots p_k^{d_k}$ respectively (we can use the same primes for both $|A|$ and $|B|$ by allowing some exponents to be zero). Then $|A \times B|$

$= p_1^{e_1+d_1} p_2^{e_2+d_2} \ldots p_k^{e_k+d_k}$. Let n divide $|A \times B|$ and seek a subgroup of order n. This divisor n thus has the form $n = p_1^{j_1} p_2^{j_2} \ldots p_k^{j_k}$ where $j_i \leq e_i + d_i$ for all i. Let $a_i = \min\{j_i, e_i\}$ and let $b_i = j_i - a_i$ for all i. We claim $b_i \leq d_i$. If, to the contrary, $d_i < b_i = j_i - a_i$, Then $d_i + a_i < j_i \leq e_i + d_i$ so $a_i < e_i$. But $a_i \neq e_i$ implies $a_i = j_i$ so $d_i < j_i - a_i = 0$, which can't happen. Thus $b_i \leq d_i$ for all i as claimed. Let $n' = p_1^{a_1} p_2^{a_2} \ldots p_k^{a_k}$ and let $n'' = p_1^{b_1} p_2^{b_2} \ldots p_k^{b_k}$. Then A has a subgroup H of order n' while B has a subgroup K of order n''. Thus $H \times K$ has order n.

THEOREM 1.2 *For a finite solvable group G, the following are equivalent:*
(1) *G is CLT.*
(2) *G has a subgroup of index p^n for all primes p and all natural numbers n such that p^n divides $|G|$.*
Proof: That (1) implies (2) is obvious. To see that (2) implies (1), let n be any divisor of $|G|$ and seek a subgroup of order n. Factor the natural number $|G|/n$ as $p_1^{e_1} p_2^{e_2} \ldots p_k^{e_k}$. Apply (2) to yield existence of B_i such that $[G : B_i] = p_i^{e_i}$ for all i. Then since indices of $B_1, B_2, \ldots, B_k$ are pairwise coprime, the subgroup $B = B_1 \cap B_2 \cap \ldots \cap B_k$ has index $p_1^{e_1} p_2^{e_2} \ldots p_k^{e_k}$, hence has order n.

2 QCLT-groups

DEFINITION The group G is **QCLT** iff all quotient groups of G are CLT-groups.

The group S_4 is QCLT but its subgroup A_4 is not. This dispenses with the question of subgroup-closure. Quotient-closure clearly holds by definition. Product-closure also holds: This will be shown following two lemmas.

LEMMA 2.1 *The class $\mathcal{CLT}$ is closed under central extensions.*
Proof: Suppose that H is a normal subgroup of a solvable group G such that $H \leq Z(G)$ and G/H is CLT. To show that G is CLT, let n divide $|G|$.
    Denote $|H|$ by m. Then m divides $|G|$, say $|G| = mr$. Let $d = (m, n)$. Thus $m = dm'$ and $n = dn'$ where $(m', n') = 1$. Now $n \mid mr$ implies $n' \mid m'r$ and so $n' \mid r$ because $(m', n') = 1$. Thus we have n' divides $|G/H|$, so that G/H has a subgroup A/H of order n'. Accordingly, $|A| = mn'$. Now H is central, hence Abelian, so H has a subgroup F of order d. Since $F \leq Z(G)$, F is normal in G.
    Consider the group A/F. It has order $|A|/|F| = mn'/d = dm'n'/d = m'n'$. Let B/F be a Hall subgroup of A/F of order n' (existence by Theorem 1 of Appendix A). Then $|B| = n' \cdot |F| = n'd = n$.

LEMMA 2.2 *If $G = J \oplus K$ and $N \lhd G$, then $JN \cap KN/N \leq Z(G/N)$.*
Proof: It suffices to show the commutator (x, g) belongs to N for all $x \in JN \cap KN$ and $g \in G$. For such x and g we have $g = jk$ for some $j \in J$, $k \in K$. Then $(x, g) = x^{-1}k^{-1}j^{-1}xjk = x^{-1}k^{-1}x(x, j)k$. But $(x, j) \in J$ (since $j \in J \lhd G$) so (x, j) commutes with k. Thus

$$(x, g) = (x, k)(x, j).$$

Now $x \in JN \cap KN$ so $x = j_2 n_1 = k_2 n_2$ for some $j_2 \in J$, $k_2 \in K$, and n_1, $n_2 \in N$. Then $(x, k) = n_1^{-1} j_2^{-1} k^{-1} j_2 n_1 k = (n_1, k)$ because j_2 and k commute. Similarly $(x, j) = (n_2, j)$. We now have $(x, k), (x, j) \in N$. This and the displayed factorization shows $(x, g) \in N$ as required.

THEOREM 2.3 $\mathcal{QCLT}$ *is product-closed.*
Proof: Let $G = J \oplus K$ where J, K are QCLT, let $N \lhd G$, and seek to show G/N is a CLT-group. Since G/JN and G/KN are quotient groups of $G/J \cong K$ and $G/K \cong J$ respectively, they are CLT-groups. Also $G/JN \cap KN = [JN/JN \cap KN] \oplus [KN/JN \cap KN] \cong G/KN \times G/JN$, so that $G/JN \cap KN$ is CLT by Theorem 1.1. Now note that G/N is a central extension by the group $G/JN \cap KN$ by Lemma 2.2. Therefore Lemma 2.1 applies to show that G/N is CLT.

3. *Nilpotent-by-Abelian Groups*

DEFINITION The group G is **nilpotent-by-Abelian** iff G' is nilpotent.

Thus a nilpotent-by-Abelian group is an extension of a nilpotent group by an Abelian group; whence the name. Conversely any extension of a nilpotent group by an Abelian group is easily seen to have a nilpotent commutator subgroup.

The class nilpotent-by-Abelian is subgroup-closed (since $A' \leq G'$ for any $A \leq G$), quotient-closed (since $(G/N)' = G'N/N \cong G'/G' \cap N$ for any quotient group G/N of G), and product-closed (since $(A \times B)' = A' \times B'$ for any groups A, B).

Before giving characterizations of the class nilpotent-by-Abelian, it will be convenient to record here a more general lemma which characterizes the groups in a locally defined formation. This lemma is related to, but not identical with, the remark following Lemma 1.13 in Chapter 5.

LEMMA 3.1 *Let $\mathcal{F}$ be a formation locally defined by $\{\mathcal{F}(p)\}$ (we do not assume $\{\mathcal{F}(p)\}$ to be either integrated or full). Then $G \in \mathcal{F}$ iff $M/\mathrm{Core}_G(M)$*

$\in \mathcal{F}(p)$ *for all p-maximal subgroups M of G (in other words iff all maximal subgroups of G are $\mathcal{F}$-normal).*

Proof: Suppose $G \in \mathcal{F}$. Let M be a p-maximal subgroup of G. Let K denote the subgroup $\text{Core}_G(M)$ and let R/K be $\text{Fit}(G/K)$. Then $C_{G/K}(R/K) \le R/K$ (Theorem 2.6 of Appendix C) and hence $C_G(R/K) \le R$. By Lemma 1.11 of Chapter 5, R/K is a p-chief factor of G. Then $G/C_G(R/K) \in \mathcal{F}(p)$ so $G/R \in \mathcal{F}(p)$. But (again by Lemma 1.11 of Chapter 5) $G/R \cong M/K$. Therefore $M/K \in \mathcal{F}(p)$.

Conversely suppose the condition on p-maximal subgroups holds. It is easily seen that the class of all groups that satisfy this condition is quotient-closed (note that $\text{Core}_G(M)/N = \text{Core}_{G/N}(M/N)$ for any group G and any subgroups $N \le M \le G$ with N normal). By virtue of induction then, we may consider that every proper quotient group of G belongs to $\mathcal{F}$.

Now by Lemmas 1.8 and 1.9 of Chapter 5 $\mathcal{F}$ may also be locally defined by the system $\{\mathcal{G}_p * (\mathcal{F}(p) \cap \mathcal{F})\}$; this new system being both integrated and full. Now for any maximal subgroup M (with, say, index p^a) again let $K = \text{Core}_G(M)$ and $R/K = \text{Fit}(G/K)$. Again by Lemma 1.11 of Chapter 5 we have $G/R \cong M/K$ and R/K is a p-group. Then $M/K \in \mathcal{F}(p)$ (by hypothesis) so $G/R \in \mathcal{F}(p)$. But $G/R \in \mathcal{F}$ as well, by induction. Hence $G/K \in \mathcal{G}_p * (\mathcal{F}(p) \cap \mathcal{F}) \subseteq \mathcal{F}$ which implies $\rho_{\mathcal{F}}*(G) \le K \le M$. As this was shown for all maximal subgroups M, we have $\rho_{\mathcal{F}}*(G) \le \Phi(G)$ so that $G/\Phi(G) \in \mathcal{F}$. But $\mathcal{F}$ is saturated (since it is locally defined). Therefore $G \in \mathcal{F}$.

THEOREM 3.2 *For a finite solvable group G, the following are equivalent:*
(1) *G is nilpotent-by-Abelian.*
(2) $G' \le \text{Fit}(G)$.
(3) $G'' \le \Phi(G)$.
(4) $\text{Aut}_G(H/K)$ *is Abelian for all p-chief factors H/K of G (in other words, the class of nilpotent-by-Abelian groups is the formation locally defined by $\{\mathcal{F}(p)\}$ where for each prime p, $\mathcal{F}(p) = \mathcal{Q}$).*
(5) $G/O_{p',p}(G)$ *is Abelian for all primes p.*
(6) $M/\text{Core}_G(M)$ *is Abelian for all maximal subgroups M of G.*
(7) $G' \cap M \vartriangleleft G$ *for all maximal subgroups M of G.*
(8) (Huppert and Inagaki) $H' \vartriangleleft G$ *for all Hall subgroups H of G.*

Proof: (1)$\Rightarrow$(2) This is immediate.

(2)$\Rightarrow$(3) If $G' \le \text{Fit}(G)$, then $G'' \le \text{Fit}(G)'$. But $\text{Fit}(G)$ is nilpotent so $\text{Fit}(G)' \le \Phi(\text{Fit}(G))$. Finally $\text{Fit}(G) \vartriangleleft G$ implies $\Phi(\text{Fit}(G)) \le \Phi(G)$ by Lemma 1.1 of Appendix C. Conclude $G'' \le \Phi(G)$.

(3)$\Rightarrow$(1) Let $N = \Phi(G)G'$. Then $N/\Phi(G) \cong G'/\Phi(G) \cap G'$. By hypothesis $G'' \le \Phi(G)$ hence $G'/\Phi(G) \cap G'$ is Abelian. Thus $N/\Phi(G)$ is Abelian. We may now apply Lemma 1.2 of Appendix C and conclude that N is nilpotent. Then $G' \le N$ implies G' is nilpotent.

$(1) \Rightarrow (5)$ If G' is nilpotent, then $G' = A \oplus B$ where A is a p-group and B is a p'-group; moreover, $A, B \lhd G$. Thus $B \leq O_{p'}(G)$ so $B \leq O_{p',p}(G)$. Also $A \cap O_{p'}(G) = \{1\}$ so $AO_{p'}(G)/O_{p'}(G) \cong A$, a p group. Consequently $AO_{p'}(G)/O_{p'}(G) \leq O_p(G/O_{p'}(G))$ so $A \leq O_{p',p}(G)$. We now have $G' = A \oplus B \leq O_{p',p}(G)$. Thus $G/O_{p',p}(G)$ is Abelian for all primes p.

(5) iff (4) By the remark following Theorem 1.2 of Chapter 5.

$(4) \Rightarrow (1)$ Let $\mathcal{C}$ denote the formation locally defined by $\{\mathcal{F}(p)\}$ where $\mathcal{F}(p) = \mathcal{Q}$ for each prime p. Let $G \in \mathcal{C}$. Let N be a minimal normal subgroup of G contained in G'. By hypothesis, $G/C_G(N)$ is Abelian so that $G' \leq C_G(N)$ and hence $N \leq Z(G')$. Now since $\mathcal{C}$ is a formation (Theorem 1.2 of Chapter 5), $G/N \in \mathcal{C}$ and so $(G/N)'$ is nilpotent by induction; i.e., G'/N is nilpotent. This and $N \leq Z(G')$ imply G' is nilpotent.

$(1) \Rightarrow (7)$ Let M be a maximal subgroup of G. If $G' \leq M$, then $G' \cap M = G'$ which is certainly normal, hence we assume G' is not contained in M. Thus $G = G'M$ and $G' \cap M < G'$. Now $G' \cap M \lhd M$ so $M \leq N_G(G' \cap M)$. Since $G' \cap M < G'$, $G' \cap M < N_{G'}(G' \cap M)$ because G' is nilpotent. Thus there is some $x \in N_{G'}(G' \cap M) \backslash G' \cap M$. Thus $x \in N_G(G' \cap M) \backslash M$, which together with $M \leq N_G(G' \cap M)$ imply $N_G(G' \cap M) = G$.

$(7) \Rightarrow (6)$ Let M be a maximal subgroup of G. Then $G' \cap M \lhd G$ by (7) so $G' \cap M \leq \mathrm{Core}_G(M)$. Now $M/G' \cap M \cong G'M/G' \leq G/G'$ and so $M/G' \cap M$ is Abelian. Hence so too is $M/\mathrm{Core}_G(M)$.

(4) iff (6) By Lemma 3.1.

$(1) \Rightarrow (8)$ Let H be a Hall π-subgroup of G. Then $G' \cap H$ is a Hall π-subgroup of G'. Let N be a Hall π'-subgroup of G'. Then G' nilpotent implies N is characteristic in G' which in turn implies $N \lhd G$. If N is trivial, then $G' = G' \cap H$ so $G' \leq H$ which implies $H \lhd G$. This and H' characteristic in H imply $H' \lhd G$. We may, therefore, assume $N \neq \{1\}$.

Now since H is a Hall subgroup of G, NH/N is a Hall subgroup of G/N. Also G/N is nilpotent-by-Abelian because G is. We may now apply induction to conclude that $(NH/N)' \lhd G/N$. Then, as it is easily seen that $(NH/N)' = NH'/N$, we get $NH'/N \lhd G/N$ which in turn implies $NH' \lhd G$. Note that $H' \leq G' \cap H$ so that H' is a π-group. Therefore $N \cap H' = \{1\}$, which yields $[NH' : H'] = [N : N \cap H'] = |N|$, a π'-number. This and H' is a π-group mean that H' is a Hall π-subgroup of NH'. But $NH' \leq G'$ so NH' is nilpotent and hence its Hall subgroups are characteristic. Finally H' characteristic in NH' and $NH' \lhd G$ imply $H' \lhd G$.

$(8) \Rightarrow (1)$ It is seen without difficulty that the class of groups satisfying condition (8) is quotient-closed (if G satisfies condition (8), $N \lhd G$, and A is a Hall π-subgroup of G, then $A' \lhd G$ and $NA'/N = (NA/N)'$ so $(NA/N)' \lhd G/N$; but for any Hall π-subgroup H/N of G/N, $(H/N)' = (NA/N)'$). Let G be a minimal counterexample. Then by quotient-closure and induction all proper quotient groups of G are nilpotent-by-Abelian. Thus G must have a unique minimal normal subgroup, call in N. Applying Theorem 1.1 of Chapter 5 (which can be done because the class of nilpotent-by-Abelian groups is a saturated formation by condition (4) above) produces a complement P for N. Thus $[G : P] = |N| = p^a$ for some prime p.

Let K be a Hall p'-subgroup of P, so that K is also a Hall p'-subgroup of G; therefore $K' \triangleleft G$. Hence K is Abelian (lest $N \leq K' \leq P$ by uniqueness of N). Let S be a Sylow p-subgroup of P.

Now NS is a Sylow p-subgroup of G. Since $S < NS$, there is some $w \in N_{NS}(S) \setminus S$. Then $w \notin P$ because $P \cap NS = S$. Since S' is characteristic in S, it is normal in $N_{NS}(S)$, i.e., $N_{NS}(S) \leqslant N_G(S')$. Thus $w \in N_G(S') \setminus P$. This and P maximal in G (as is easily seen) shows $S' \triangleleft G$. Therefore S is Abelian (lest $N \leq S' \leq P$ by uniqueness of N). This implies NS/N is Abelian, so that $(NS)' \leq N$. By condition (8), $(NS)' \triangleleft G$ so $(NS)' = \{1\}$ or N. Letting M be any maximal subgroup of NS which contains S, we get $(NS)' \leq M$ by nilpotency of NS. This rules out the possibility $(NS)' = N$ and we conclude $(NS)' = \{1\}$.

We now have $G = (NS)K$ where NS and K are both Abelian. Applying Ito's theorem ([Scott 1964, 13.3.2]) yields $G'' = \{1\}$, the desired contradiction.

4. The Sylow Tower Property

DEFINITION Let G be a group of order $p_r^{e_r} \ldots p_2^{e_2} p_1^{e_1}$ where $p_1 > p_2 > \ldots > p_r$. The G satisfies the **Sylow tower property** iff there exist S_1, $S_2, \ldots, S_r$ such that each S_i is a Sylow p_i-subgroup of G and $S_k \ldots S_2 S_1 \triangleleft G$ for $k = 1, 2, \ldots, r$.

To show subgroup-closure, let G satisfy the Sylow tower property by virtue of subgroups $S_1, S_2, \ldots, S_r$ as in the definition, and let H be a subgroup of G. Since for $1 \leq k \leq r$ each $S_k \ldots S_2 S_1$ is a normal Hall subgroup of G, it follows that $S_k \ldots S_2 S_1 \cap H$ is a normal Hall subgroup of H. Thus $S_1 \cap H$ is a normal Sylow p_1-subgroup of H and starts a Sylow tower. Now let B_2 be a Sylow p_2-subgroup of $S_2 S_1 \cap H$ and note that it is a Sylow p_2-subgroup of H as well. Moreover $S_1 \cap H \triangleleft H$ implies $(S_1 \cap H)B_2$ is a subgroup of $S_2 S_1 \cap H$ so by a cardinality consideration $S_2 S_1 \cap H = (S_1 \cap H)B_2$ and so $(S_1 \cap H)B_2 \triangleleft H$. Continuing the argument by letting B_3 be a Sylow p_3-subgroup of $S_3 S_2 S_1 \cap H$, etcetera, produces a Sylow tower for H.

To show quotient-closure, again let G satisfy the Sylow tower property by virtue of subgroups $S_1, S_2, \ldots, S_r$ as in the definition, and now let $H \triangleleft G$. Then each HS_i/H is a Sylow p_i-subgroup of G/H and for $k = 1, 2, \ldots, r$ we have $S_k \ldots S_2 S_1 \triangleleft G$ hence $HS_k \ldots S_2 S_1/H \triangleleft G/H$. Thus

$(HS_k/H) \ldots (HS_2/H)(HS_1/H) \lhd G/H$ for $k = 1, 2, \ldots, r$ so G/H satisfies the Sylow tower property.

Closure under normal products, hence certainly closure under products, was shown in Theorem 3.2 of Chapter 4.

Before giving characterizations, it will be useful to record the following extension lemma.

LEMMA 4.1 *Let G be a finite solvable group of order $p_r^{e_r} \ldots p_2^{e_2}p_1^{e_1}$ where $p_1 > p_2 > \ldots > p_r$. Suppose H is a normal p_1-subgroup of G such that G/H satisfies the Sylow tower property. Then G does also.*
Proof: We have subgroups $B_1/H, B_2/H, \ldots, B_r/H$ such that each B_i/H is a Sylow p_i-subgroup of G/H and for $k = 1, 2, \ldots, r$, $(B_k/H) \ldots (B_2/H)(B_1/H) \lhd G/H$. Consequently each $B_k \ldots B_2B_1$ is normal in G. Note that B_1 is a Sylow p_1-subgroup of G. For $k \geq 2$ let R_k be a Sylow p_k-subgroup of B_k. Letting p_1^a denote the order of H, we get $|B_k| = p_k^{e_k}p_1^a$ because $|B_k/H| = p_k^{e_k}$. Thus $|R_k| = p_k^{e_k}$ for $k \geq 2$, so that for such k, R_k is a Sylow p_k-subgroup of G. Moreover, by order considerations, $B_k = HR_k$. Thus for all $k \geq 2$ $B_k \ldots B_2B_1 = HB_kHB_{k-1} \ldots HB_2B_1$ $= R_kR_{k-1} \ldots R_2B_1$ so that $R_kR_{k-1} \ldots R_2B_1 \lhd G$. Thus G satisfies the Sylow tower property by virtue of Sylow subgroups $B_1, R_2, R_3, \ldots, R_r$.

THEOREM 4.2 *Let G be a finite solvable group of order $p_r^{e_r} \ldots p_2^{e_2}p_1^{e_1}$ where $p_1 > p_2 > \ldots > p_r$. Then the following are equivalent:*
(1) G satisfies the Sylow tower property.
(2) Every quotient group of G has a normal Sylow p-subgroup for the largest prime p dividing its order.
(3) Every subgroup of G is p-nilpotent (i.e., has a normal p-complement) for the smallest prime p dividing its order.
(4) If H/K is a p-chief factor of G, then $\mathrm{Aut}_G(H/K)$ is a π-group where every prime in π is less than p (in other words, G belongs to the formation locally defined by $\{\mathcal{F}(p)\}$ where for each prime p, $\mathcal{F}(p)$ is the class of π-groups where every prime in π is less than p).
(5) For all primes p, $G/O_{p',p}(G)$ is a π-group where every prime in π is less than p.
(6) For all p-maximal subgroups M of G, $M/\mathrm{Core}_G(M)$ is a π-group where every prime in π is less than p.
Proof:$(1) \Rightarrow (2)$ Since G satisfies the Sylow tower property so does any quotient group G/H. Hence G/H has a normal Sylow p-subgroup for the largest prime dividing its order.

$(2) \Rightarrow (1)$ By (2), G has a normal Sylow p_1-subgroup S_1. Clearly the quotient group G/S_1 also satisfies condition (2) so that G/S_1 satisfies the Sylow tower property by induction. Hence G satisfies the Sylow tower property by Lemma 4.1.

(1)$\Rightarrow$(3) Any group satisfying the Sylow tower property is clearly p-nilpotent for its smallest prime divisor (Sylow subgroups $S_1, S_2, \ldots, S_r$ as in the definition yield the normal p_r-complement $S_{r-1}, \ldots, S_2 S_1$). The conclusion then follows by subgroup-closure of the Sylow tower property.

(3)$\Rightarrow$(1) By (3), G has a normal p_r-complement, call it W. This subgroup W has order $p_{r-1}^{e_{r-1}} \ldots p_2^{e_2} p_1^{e_1}$ and it is seen to satisfy the Sylow tower property by induction. Thus it has Sylow subgroups $B_1, B_2, \ldots, B_{r-1}$ such that $B_k \ldots B_2 B_1 \lhd W$ for $k = 1, 2, \ldots, r-1$. Since each B_j has order $p_j^{e_j}$, each B_j is a Sylow p_j-subgroup of G as well as of W. Next note that any normal Hall subgroup of W is characteristic in W and hence normal in G so we get $B_k \ldots B_2 B_1 \lhd G$ for $k = 1, 2, \ldots, r-1$. Thus G satisfies the Sylow tower property by virtue of the Sylow subgroups $B_1, B_2, \ldots, B_{r-1}, S_r$.

(1)$\Rightarrow$(5) Let p divide $|G|$. Thus $p = p_t$ for some t, $1 \leq t \leq r$. By the Sylow tower property G has a normal subgroup $W = S_{t-1} \ldots S_2 S_1$ where each S_i, $1 \leq i \leq t-1$ is a Sylow p_i-subgroup of G (if $t = 1$, we simply have $W = \{1\}$). Now G/W has order $p_r^{e_r} \ldots p_t^{e_t}$ and it satisfies the Sylow tower property because G does. Hence G/W has a normal Sylow p_t-subgroup, call it A/W. Now $AO_{p'}(G)/O_{p'}(G) \cong A/A \cap O_{p'}(G)$ whose order is a power of p_t. Therefore $AO_{p'}(G)/O_{p'}(G) \leq O_p(G/O_{p'}(G))$, meaning that $A \leq O_{p',p}(G)$. Now $W \leq A \leq O_{p',p}(G)$ hence any divisor of $|G/O_{p',p}(G)|$ must also divide $|G/W| = p_r^{e_r} \ldots p_t^{e_t}$. Thus $p_r, p_{r-1}, \ldots, p_t$ are the only possible divisors of $|G/O_{p',p}(G)|$. We can further eliminate the prime p_t: If p_t divides $[G : O_{p',p}(G)]$, it would also divide $[G : A] = [G/W : A/W]$ contrary to the fact that A/W is a Sylow p_t-subgroup of G/W. Hence the only divisors of $|G/O_{p',p}(G)|$ are strictly smaller than p_t.

(5) iff (4) By the remark following Theorem 1.2 of Chapter 5.

(4)$\Rightarrow$(1) Let $\mathcal{C}$ denote the formation locally defined by $\{\mathcal{F}(p)\}$ in condition (4). We must show every $G \in \mathcal{C}$ satisfies the Sylow tower property. Let N be a minimal normal subgroup of G of order, say, p^a. Since $\mathcal{C}$ is a formation (Theorem 1.2 of Chapter 5), $G/N \in \mathcal{C}$ so G/N satisfies the Sylow tower property by induction. If $p = p_1$, then we are through by Lemma 4.1. Assume then that $p < p_1$.

Suppose $\Phi(G) \neq \{1\}$. Then $G/\Phi(G) \in \mathcal{C}$ so that $G/\Phi(G)$ satisfies the Sylow tower property by induction. Let $B/\Phi(G)$ be a Sylow p_1-subgroup of $G/\Phi(G)$. Then $B/\Phi(G) \lhd G/\Phi(G)$ so that $B \lhd G$. Also, B is nilpotent by Lemma 1.2 of Appendix C. Let T be a Sylow p_1-subgroup of B. Then T is characteristic in B (since B is nilpotent) hence $T \lhd G$. Now note that T is a Sylow p_1-subgroup of G. Moreover, $G/T \in \mathcal{C}$ implies that G/T satisfies the Sylow tower property by induction. Hence Lemma 4.1 applies to show G satisfies the Sylow tower property.

We may therefore assume $\Phi(G) = \{1\}$. Then N cannot be the only minimal normal subgroup of G for if it were, we would have $\mathrm{Fit}(G) = N$ by

Lemma 2.3 of Appendix C, and consequently $C_G(N) \leq N$ by Theorem 2.6 of Appendix C. But $G \in \mathcal{C}$ implies $G/C_G(N)$ is a π-group where every prime in π is less than p. Thus $p_1 \nmid |G/C_G(N)|$ so $p_1 \nmid |G/N|$, a contradiction. Conclude G must have some minimal normal subgroup $N^* \neq N$. Then $G/N, G/N^* \in \mathcal{C}$ imply $G/N, G/N^*$ both satisfy the Sylow tower proerty by induction. Consequently G also satisfies the Sylow tower property as it is isomorphic to a subgroup of $G/N \times G/N^*$.

(4) iff (6) By Lemma 3.1.

5. Supersolvable Groups

DEFINITION The group G is **supersolvable** iff all chief factors of G are cyclic.

The class of supersolvable groups is subgroup-closed, quotient-closed, and product closed; all of this was shown in Theorem 1.1 of Chapter 1.

THEOREM 5.1 *For a finite solvable group G the following are equivalent:*
(1) *G is supersolvable.*
(2) *$Aut_G(H/K)$ is Abelian of exponent dividing $p - 1$ for all p-chief factors H/K (in other words, G belongs to the formation locally defined by $\{\mathfrak{F}(p)\}$ where for each prime p, $\mathfrak{F}(p)$ is the class of Abelian groups of exponent dividing $p - 1$).*
(3) *G satisfies the Sylow tower property and $N_G(S)/C_G(S)$ is strictly p-closed for every Sylow p-subgroup S of G.*
(4) *G is equichained.*
(5) *All maximal subgroups of G have prime index.*
(6) *For all maximal subgroups M of G, either $\mathrm{Fit}(G) \leq M$ or $M \cap \mathrm{Fit}(G)$ is a maximal subgroup of $\mathrm{Fit}(G)$.*
(7) *For all maximal subgroups M of G and all $N \lhd G$, either $N \leq M$ or $M \cap N$ is maximal in N.*
(8) *All subgroups of G are CLT-groups.*
(9) *$G = \mathrm{genz}^*(G)$.*
(10) *$G = Q^*(G)$ (or equivalently $G = \mathrm{SE}(G)$ or $G = (WZ)^*(G)$ since $Q^*(A)$, $\mathrm{SE}(A)$, and $(WZ)^*(A)$ coincide for any finite solvable group A).*
(11) *All Sylow systems are weakly normal in G.*
(12) *All maximal subgroups of G are weakly normal in G.*
(13) *All proper subgroups of G are properly contained in their weak normalizers.*
(14) *All subgroups of G are weakly subnormal in G.*

(15) $G/O_{p',p}(G)$ is Abelian of exponent dividing $p-1$ for all primes p.

(16) $M/\text{Core}_G(M)$ is Abelian of exponent dividing $p-1$ for all p-maximal subgroups M of G.

(17) G satisfies the Sylow tower property and is QCLT.

(18) G satisfies the Sylow tower property and the permutizer condition.

(19) G is nilpotent-by-Abelian and $N_G(S)/C_G(S)$ is strictly p-closed for every Sylow p-subgroup S of G.

(20) G is nilpotent-by-Abelian and QCLT.

(21) G is nilpotent-by-Abelian and satisfies the permutizer condition.

(22) (McLain) For any two characteristic subgroups $A \le B$ of G, B/A is a CLT-group.

(23) (Baer) $x^{p-1}y = yx^{p-1}$ for all x, y such that $y \in G'$, $o(y)$ is a power of p, and $o(x)$ is coprime to p.

Proof: The equivalence of (1)–(14) was shown in Chapter 1 (more particularly; (1) iff (2) by Chapter 1, Corollary 1.5; (1) iff (3) by Chapter 1, Theorem 1.12; (1) iff (4) by Chapter 1, Theorem 2.4; (1) iff (5) by Chapter 1, Theorems 1.7 and 3.1; (1) iff (6) by Chapter 1, Theorem 3.3; (1) iff (7) by Chapter 1, Corollary 3.4; (1) iff (8) by Chapter 1, Theorem 4.1; (1) iff (9) by Chapter 1, Corollary 5.10; (1) iff (10) by Chapter 1, Theorem 7.10 and Corollaries 7.13 and 7.14; (1) iff (11) iff (12) by Chapter 1, Theorem 8.7; (1) iff (13) iff (14) by Chapter 1, Corollary 8.8).

(2) iff (15) By the remark following Theorem 1.2 of Chapter 5.

(2) iff (16) By Lemma 3.1.

(1)$\Rightarrow$(17) That supersolvable groups satisfy the Sylow tower property was shown in Chapter 1, Theorem 1.8; that they are QCLT follows from Chapter 1, Theorem 4.1 and the fact that $\mathcal{S}\mathcal{S}$ is quotient-closed.

(17)$\Rightarrow$(1) Suppose G satisfies (17). Since the class of groups satisfying (17) is quotient-closed, we may conclude by induction that all proper quotient groups of G are supersolvable. We may assume $\Phi(G)$ is trivial, for otherwise supersolvability of $G/\Phi(G)$ would imply supersolvability of G (Corollary 3.2 of Chapter 1). Therefore Fit(G) is Abelian by Lemma 2.3 of Appendix C. By the Sylow tower property G has a normal Sylow p-subgroup S for the largest prime p dividing $|G|$. Let $|S| = p^n$. Since $S \lhd G$, $S \le \text{Fit}(G)$ and so S is Abelian. Because G is QCLT, it has a subgroup B of index p^{n-1}. Since $[G:S]$ and $[G:B]$ are coprime, $G = SB$. The subgroup $S \cap B$ is normal in B and is also normal in S because S is Abelian. Thus $S \cap B \lhd G$. Moreover $|S \cap B| = p$ so $S \cap B$ is cyclic. Hence G is an extension of a cyclic group by a supersolvable group (G/N is supersolvable by induction) so G is supersolvable by Theorem 1.2 of Chapter 1.

(1)$\Rightarrow$(18) By Chapter 1, Theorem 1.8 and Chapter 1, Theorem 6.10.

(18)$\Rightarrow$(5) Suppose G satisfies (18). It is easily seen that the class of groups which satisfy the permutizer condition is quotient-closed. So is the class $\mathcal{S}\mathcal{T}\mathcal{P}$. Hence we conclude by induction that all proper quotient groups of G satisfy (5). We may therefore assume that $\Phi(G)$ is trivial for if not, then $G/\Phi(G)$ satisfies (5), hence so does G. Consequently Fit(G) is Abelian (by Lemma 2.3 of Appendix C). By the Sylow tower property G has a normal Sylow p-subgroup S for the largest prime p dividing $|G|$. Since $S \le$ Fit(G), S is Abelian.

Let M be any maximal subgroup and seek to show $[G:M]$ is a prime. If $\mathrm{Core}_G(M) \ne \{1\}$, then $M/\mathrm{Core}_G(M)$ is maximal in the group $G/\mathrm{Core}_G(M)$ which, as noted above, satisfies (5). Thus, since $[G:M] = [G/\mathrm{Core}_G(M) : M/\mathrm{Core}_G(M)]$, we get that $[G:M]$ is a prime.

We may consequently assume that $\mathrm{Core}_G(M) = \{1\}$. Hence S is not contained in M, so we have $G = SM$. Moreover $S \cap M$ is normal in M and in S (because S is Abelian) so $S \cap M \lhd G$ and thus $\mathrm{Core}_G(M) = \{1\}$ forces $S \cap M = \{1\}$.

By the permutizer condition $G = P_G(M)$ so there is some $x \in G$ such that $G = \langle x \rangle M$. Factor the order of x as $p^m d$ where p^m and d are coprime. Then $1 = id + jp^m$ for some integers i, j. Letting $a = x^{id}$ and $b = x^{jp^m}$, we have $x = ab = ba$. Now $[G:M] = [\langle x \rangle M : M] = [\langle x \rangle : \langle x \rangle \cap M]$ and so $|\langle x \rangle| = [G:M] \cdot |\langle x \rangle \cap M|$. Hence d divides $[G:M] \cdot |\langle x \rangle \cap M|$. But $[G:M] = [SM:M] = [S:S \cap M]$ is a power of p so we get d divides $|\langle x \rangle \cap M|$. Since $b^d = 1$, $o(b)$ divides d and so $o(b)$ divides $|\langle x \rangle \cap M|$. Thus $\langle x \rangle \cap M$ has a subgroup B of order $o(b)$. Since a cyclic group ($\langle x \rangle$ in this case) cannot have distinct subgroups of the same order, we conclude $B = \langle b \rangle$. Hence $b \in M$. This allows us to show that $G = \langle a \rangle M$ (given $g \in G = \langle x \rangle M$, $g = x^k y$ for some $y \in M$ and some integer k. Then since a, b commute, $g = a^k b^k y \in \langle a \rangle M$). Further, note that $o(a)$ is a power of p and S is the unique Sylow p-subgroup of G, so we have $a \in S$.

We now have $\langle a \rangle \le S$, $G = SM = \langle a \rangle M$ and $S \cap M = \{1\}$; all of which implies $\langle a \rangle = S$ (because $|\langle a \rangle| = |S|$). Hence $\langle a \rangle \lhd G$. Once again note that $G/\langle a \rangle$ satisfies condition (5) by induction. But conditions (5) and (1) are equivalent so we get $G/\langle a \rangle$ is supersolvable. Hence so too is G. Hence G satisfies (5).

(1)$\Rightarrow$(19) By Chapter 1, Theorem 1.6 and Chapter 1, Theorem 1.12.

(19)$\Rightarrow$(1) Suppose G satisfies condition (19). The class of groups which satisfy condition (19) is seen without difficulty to be quotient-closed. Hence we conclude by induction that all proper quotient groups of G are supersolvable. If $\Phi(G) \ne \{1\}$, then supersolvability of $G/\Phi(G)$ implies supersolvability of G, so we may assume $\Phi(G) = \{1\}$. We may also assume G has a unique minimal normal subgroup N. Applying Lemma 2.3 of Appendix C yields Fit(G) = N. Let $|N| = p^a$. Let S be a Sylow p-subgroup of G such that $N \le S$. Since G is nilpotent-by-Abelian, we have $G' \le$ Fit(G) $= N \le S$ and so $S \lhd G$. This further implies $S \le$ Fit(G) so that Fit(G) = N $= S$.

By condition (19) $G/C_G(S)$ is strictly p-closed. But $S = N$ implies S is Abelian so $S \le C_G(S)$. Therefore p does not divide $|G/C_G(S)|$ (because S is a Sylow p-subgroup of G). This and $G/C_G(S)$ being strictly p-closed imply $G/C_G(S)$ is Abelian of exponent dividing $p - 1$. Now apply Chapter 1, Theorem 1.4 to conclude that S is cyclic. This and G/S supersolvable (by induction) yield that G is supersolvable.

$(1) \Rightarrow (20)$ By Chapter 1, Theorem 1.6 and Chapter 1, Theorem 4.1 (and the fact that $\mathcal{SS}$ is quotient-closed).

$(20) \Rightarrow (1)$ Let G satisfy (20). Since the class groups which satisfy condition (20) is quotient-closed, we may conclude by induction that all proper quotient groups of G are supersolvable. As in the proof that $(19) \Rightarrow (1)$, we may assume that $\Phi(G) = \{1\}$, that G has a unique minimal normal subgroup N and that $G' \le \text{Fit}(G) = N = S$ for some Sylow p-subgroup S of G. Since G is QCLT, there exists a subgroup A of index p. It cannot happen that $N \le A$ because N is a Sylow p-subgroup. Consequently $G = NA$. We then have $N \cap A \lhd G$ because $N \cap A \lhd A$ and $N \cap A \lhd N$ (since N is Abelian). Minimality of N now shows $N \cap A = \{1\}$ or $N \cap A = N$. The second case would mean that $N \le A$ which we 've already noted cannot happen; conclude $N \cap A = \{1\}$ so that $|N| = [G:A] = p$. Thus N is cyclic. This and G/N Abelian $(G' \le \text{Fit}(G) = N)$ imply G is supersolvable.

$(1) \Rightarrow (21)$ By Chapter 1, Theorem 1.6 and Chapter 1, Theorem 6.10.

$(21) \Rightarrow (5)$ Suppose G satisfies (21). Since the class of groups which satisfy condition (21) is quotient-closed, we may conclude by induction that all proper quotient groups of G satisfy condition (5).

Let M be any maximal subgroup of G and seek to show $[G:M]$ is a prime. If $\text{Core}_G(M) \ne \{1\}$, then $M/\text{Core}_G(M)$ is maximal in the group $G/\text{Core}_G(M)$ which, as noted above, satisfies (5). Thus, since $[G:M] = [G/\text{Core}_G(M) : M/\text{Core}_G(M)]$, we get that $[G:M]$ is a prime.

We may consequently assume that $\text{Core}_G(M) = \{1\}$. Apply Lemma 1.11 of Chapter 5 to get $G = \text{Fit}(G)M$ and $\text{Fit}(G) \cap M = \{1\}$. Then since $[G:M]$ is a prime-power, so is $|\text{Fit}(G)|$, say $|\text{Fit}(G)| = p^a$. Let S be a Sylow p-subgroup of G containing $\text{Fit}(G)$. Then G nilpotent-by-Abelian implies $G' \le \text{Fit}(G) \le S$ so that $S \lhd G$. Therefore $S \le \text{Fit}(G)$. That is, $\text{Fit}(G) = S$.

As in the proof that $(18) \Rightarrow (5)$, the permutizer condition yields existence of $a \in S = \text{Fit}(G)$ such that $G = \langle a \rangle M$. Now $\langle a \rangle \le \text{Fit}(G)$, $G = \text{Fit}(G)M = \langle a \rangle M$, and $\text{Fit}(G) \cap M = \{1\}$ together yield that $\langle a \rangle = \text{Fit}(G)$ (because $|\langle a \rangle| = |\text{Fit}(G)|$). Hence $\langle a \rangle \lhd G$. Now note that $G/\langle a \rangle$ satisfies condition (5) by induction. But conditions (5) and (1) are equivalent so we get $G/\langle a \rangle$ is supersolvable. Hence so too is G. Hence G satisfies condition (5). Therefore $[G:M]$ is a prime, as was required to show.

$(1) \Rightarrow (22)$ Let G be supersolvable, let A, B be characteristic subgroups of G with $A \le B$. Then B is supersolvable, hence so too is B/A. Therefore B/A is a CLT-group (Theorem 4.1 of Chapter 1).

$(22) \Rightarrow (1)$ Let G satisfy (22). We begin by noting that if K is any characteristic subgroup of G, then both K and G/K also satisfy (22) (this follows easily from the facts that (i) J is characteristic in K and K is characteristic in G imply J is characteristic in G and (ii) K is characteristic in G and L/K is characteristic in G/K imply L is characteristic in G). Hence by induction, K and G/K are supersolvable for all proper nontrivial characteristic subgroups K. Now let q, p denote the smallest and respectively the largest primes dividing $|G|$. By hypothesis G has a subgroup H of index q. Then $H \lhd G$ so that $G' \leq H$ and consequently q divides $|G/G'|$. Factor the Abelian group G/G' as $G/G' = A/G' \oplus B/G'$ where A/G' is the Sylow q-subgroup and B/G' is the Hall q'-subgroup of G/G'. Then B is a proper characteristic subgroup of G (proper because q divides $|G/G'|$) and hence B is supersolvable. Apply Theorem 1.8 of Chapter 1 to obtain a normal Sylow p-subgroup S of B and since $[G : B]$ is a power of q, S is a Sylow p-subgroup of G as well. Furthermore $S \lhd G$ because S is characteristic in B and $B \lhd G$.

We may assume that $\Phi(G) = \{1\}$ otherwise (because $\Phi(G)$ is characteristic in G) $G/\Phi(G)$ would be supersolvable as noted above, hence so too would G. Therefore $\mathrm{Fit}(G)$ is Abelian (Lemma 2.3 of Appendix C). Then $S \leq \mathrm{Fit}(G)$ so S is Abelian.

By hypothesis G has a subgroup W of index p^{n-1} where $p^n = |S|$. Since $[G : S]$ and $[G : W]$ are coprime, $G = SW$. The subgroup $S \cap W$ is normal in W and is also normal in S because S is Abelian. Thus $S \cap W \lhd G$. Moreover, $|S \cap W| = p$ so $S \cap W$ is cyclic. We have at this stage produced a nontrivial cyclic normal subgroup of G. This implies $\mathrm{SE}(G) \neq \{1\}$. But $\mathrm{SE}(G)$ is characteristic in G so we may conclude $G/\mathrm{SE}(G)$ is supersolvable. This in turn implies that G is supersolvable (for instance by Lemma 7.1 of Chapter 1, $\mathrm{SE}(G/\mathrm{SE}(G)) = \{1\}$ so that $G/\mathrm{SE}(G)$ is supersolvable iff it is trivial iff $G = \mathrm{SE}(G)$ iff G is supersolvable).

$(1) \Rightarrow (23)$ Let x and y be elements as in the statement of (23). We consider 2 cases:

First suppose $O_p(G) = \{1\}$. Hence $O_{p',p}(G) = O_p(G)$ so by (15) (which we have already seen to be equivalent to (1)) We have $G/O_p(G)$ is Abelian of exponent dividing $p - 1$. Therefore $x^{p-1} \in O_p(G)$. But since $o(x)$ is coprime to p, so is $o(x^{p-1})$ and hence $x^{p-1} = 1$ so trivially x^{p-1} commutes with y.

Now suppose $O_p(G) \neq \{1\}$. For convenience let A denote $O_p(G)$. Let D be the Sylow p-subgroup of G'. Since G' is nilpotent (Chapter 1, Theorem 1.6), D is normal in G', hence characteristic in G', hence normal in G. We may assume that $D \neq \{1\}$ for $D = \{1\}$ implies $y = 1$ (as $y \in D$) and hence y trivially commutes with x^{p-1}.

We now have $A \neq \{1\}$, $D \neq \{1\}$, and $A \cap D = \{1\}$. Now $Ay \in G'A/A = (G/A)'$, $o(Ay)$ is a power of p, and $o(Ax)$ is coprime to p. Now G/A satisfies condition (23) by induction and so we have $(Ax)^{p-1}Ay = Ay(Ax)^{p-1}$ so that $(x^{p-1}, y) \in A$. Furthermore, $y \in D \lhd G$ implies $(x^{p-1}, y) \in D$ and so $(x^{p-1}, y) \in A \cap D = \{1\}$ showing that x^{p-1} and y commute.

(23)$\Rightarrow$(1) First we want to show that the class of groups which satisfy condition (23) is quotient-closed. So let F satisfy this condition and let $N \triangleleft F$. It is clearly sufficient to assume that N is a minimal normal subgroup of F. Thus $|N| = q^a$ for some prime q. Let Nx, $Ny \in F/N$ such that $Ny \in (F/N)'$, $o(Ny) = $ a power of p, and $o(Nx)$ is coprime to p.

Case one: $p = q$ Since $Ny \in (F/N)' = F'N/N$, we have $Ny = Nw$ where $w \in F'$. Hence $o(Nw) = o(Ny) = $ a power of p. This and N is a p-group imply $o(w)$ is a power of p.

Let $e = o(Nx)$. Since $x^e \in N$, $x^{ep} = 1$ since N is an elementary Abelian p-group. Hence $o(x^p)$ is coprime to p. We apply condition (23) to the situation $w \in F'$, $o(w)$ is a power of p, and $o(x^p)$ is coprime to p to get $(x^p)^{p-1}w = w(x^p)^{p-1}$. Therefore $(Nx)^{p(p-1)}Nw = Nw(Nx)^{p(p-1)}$ and so $(Nx)^{p(p-1)}$ commutes with $Nw = Ny$. Then since e and p are coprime we get $(Nx)^{p-1}$ commutes with Ny as was required.

Case two: $p \neq q$ Again $Ny = Nw$ where $w \in F'$. Now $o(Ny) = o(Nw)$ is a power of p, say p^i, so $w^{p^i q} = 1$ since N is an elementary Abelian q-group. Hence $w^q \in F'$ and $o(w^q)$ is a power of p.

Let $e = o(Nx)$. Since $x^e \in N$, $x^{eq} = 1$. Therefore $o(x)|eq$ and so $o(x)$ is coprime to p as both e and q are. We apply condition (23) to the situation $w^q \in F'$, $o(w^q)$ is a power of p, and $o(x)$ is coprime to p to get $x^{p-1}w^q = w^q x^{p-1}$. Therefore $(Nx)^{p-1}(Nw)^q = (Nw)^q(Nx)^{p-1}$, and hence $(Nw)^q$ commutes with $(Nx)^{p-1}$. Since q and p^i are coprime we have $1 = \lambda p^i + \mu q$ for some integers λ, μ. Thus $Nw = (Nw)^{\lambda p^i}(Nw)^{\mu q} = (Nw)^{\mu q}$ and so $Ny = Nw$ commutes with $(Nx)^{p-1}$. This completes the demonstration of quotient-closure.

The next step of the proof will be to show that if G satisfies condition (23), then G is nilpotent-by-Abelian. It suffices to show that if S_p and S_q are, respectively, a Sylow p-subgroup of G' and a Sylow q-subgroup of G', $p \neq q$, then S_p, S_q commute elementwise. Let $a \in S_p$, $\in S_q$. Without loss of generality, $p < q$. Hence $p - 1$ and q are coprime, so $p - 1$ and $o(b)$ are also. Consequently there exist integers i, j such that $1 = i(p - 1) + jo(b)$ and so $b = b^{i(p-1)}$. But by condition (23), b^{p-1} commutes with a. Hence so too does b, as required.

To complete the proof, let G satisfy condition (23). Then by induction all proper quotient groups of G are supersolvable. We may therefore assume that $\Phi(G) = \{1\}$ and that G has a unique minimal normal subgroup N. By Lemma 2.3 of Appendix C, $\text{Fit}(G) = N$. Let $|N| = p^a$ and let S be a Sylow p-subgroup of G such that $N \leq S$. Since G is nilpotent-by-Abelian, $G' \leq \text{Fit}(G) = N \leq S$ and so $S \triangleleft G$. This further implies $S \leq \text{Fit}(G)$ so that $\text{Fit}(G) = N = S$. Let K denote a Sylow p-complement of G so that $G = NK$.

Since $G' \leq N \leq C_G(N)$, $G/C_G(N)$ is Abelian. Moreover we claim $G/C_G(N)$ has exponent dividing $p - 1$. Let $C_G(N)g$ be a typical element. Then $g \in G = NK$ so $g = ak$ where $a \in N$, $k \in K$. By condition (23),

$k^{p-1}y = yk^{p-1}$ for any $y \in N$ (we are assuming $N \leq G'$; if not, then $G' = \{1\}$ since N is the unique minimal normal subgroup and the proof is over), which means that $k^{p-1} \in C_G(N)$. Now since $G/C_G(N)$ is Abelian, $(C_G(N)g)^{p-1} = (C_G(N)a)^{p-1}(C_G(N)k)^{p-1} = 1$. Thus $G/C_G(N)$ is Abelian of exponent dividing $p - 1$. Apply Theorem 1.4 of Chapter 1 to yield that N is cyclic. This and G/N supersolvable (by induction) completes the proof.

6. The Class $\mathcal{Y}$

DEFINITION $G \in \mathcal{Y}$ iff for all subgroups H of G, if $p \,|\, [G:H]$ where p is a prime, then G has a subgroup W such that $H \leq W$ and $[W : H] = p$.

Quotient-closure for this class is easily verified so we turn to the counterexamples.

A famous counterexample which shows the failure of product-closure is $G = S_3 \times \mathbb{Z}_3$. This group has the presentation $G = (x, y, z : x^3 = y^2 = (xy)^2 = 1, z^3 = 1, xz = zx, yz = zy)$. Consider the subgroup $R = \langle xz \rangle$ of order 3. Since the result of conjugating xz by y is $y^{-1}(xz)y = y^{-1}xyz = yxyz = x^{-1}z$ (by the relations), and since $x^{-1}z$ is equal to neither xz nor $(xz)^2$ ($x^{-1}z = xz$ would imply $x^2 = 1$; $x^{-1}z = (xz)^2$ would imply $1 = z$) we have $y^{-1}(xz)y \notin R$ so R is not normal in G. Since $[G : R] = 6$, to show $G \notin \mathcal{Y}$ it suffices to show there is no 6-element subgroup of G containing R. Assume to the contrary that U is such a subgroup. Let $T = \langle x, z \rangle$, the Sylow 3-subgroup of G. Then $[G : U]$ and $[G : T]$ are coprime so $G = UT$. But $R \triangleleft T$ since T is Abelian and $R \triangleleft U$ because U has a unique Sylow 3-subgroup by Sylow's theorem. Therefore $R \triangleleft G$, the desired contradiction showing $G \notin \mathcal{Y}$.

The next example—failure of subgroup closure—is from [McLain 1957]. Let $G = (a, b, c, d : a^3 = b^3 = c^3 = d^2 = 1, a^{-1}b^{-1}ab = c, ac = ca, bc = cb, dc = cd, dad = a^2, dbd = b^2)$. This group can be realized as a subgroup of $T(3,3)$, the group of triangular matrices over GF(3) by letting

$$a = \begin{pmatrix} 1 & 0 & 0 \\ 0 & 1 & 0 \\ 0 & 1 & 1 \end{pmatrix} \quad b = \begin{pmatrix} 1 & 0 & 0 \\ 1 & 1 & 0 \\ 0 & 0 & 1 \end{pmatrix} \quad c = \begin{pmatrix} 1 & 0 & 0 \\ 0 & 1 & 0 \\ 1 & 0 & 1 \end{pmatrix} \quad d = \begin{pmatrix} 2 & 0 & 0 \\ 0 & 1 & 0 \\ 0 & 0 & 2 \end{pmatrix}$$

Let U denote the group of lower triangular matrices, which has order 27. Then $U \leq G$ (since U is generated by the matrices a, b, c) and moreover $U \triangleleft G$ because $U \triangleleft T(3,3)$. We thus have $G = U\langle D \rangle$ and so $|G| = 54$.

We claim $G \in \mathfrak{Y}$. Since $U \lhd G$, and U is a nilpotent Hall subgroup and $G/U \cong \mathbb{Z}_2$ is nilpotent, in order to show $G \in \mathfrak{Y}$ it suffices, by Theorem 4.3 of Chapter 1, to show that $G = UN_G(H)$ for all subgroups H of U. So let $H \le U$. If $H \lhd G$, then $G = N_G(H)$, so we may assume that H is not normal in G. Now it happens that every 9-element subgroup of U is normal in G (Let R be a 9-element subgroup of U. Then R is maximal in U, so $U' \le R$. But $U' = \langle c \rangle$, hence $c \in R$. Then for any

$$r = \begin{pmatrix} 1 & 0 & 0 \\ i & 1 & 0 \\ j & k & 1 \end{pmatrix} \in R,$$

$d^{-1}rd = r^2 c^{-(j+ik)} \in R$. Hence $d \in N_G(R)$. But also $U \le N_G(R)$ since R is maximal in U. Thus $R \lhd G$). Consequently our nonnormal subgroup H must have order 3, and moreover $H \ne \langle c \rangle$ since $\langle c \rangle = U' \lhd G$. Let h generate H. Then

$$h = \begin{pmatrix} 1 & 0 & 0 \\ i & 1 & 0 \\ j & k & 1 \end{pmatrix} = b^i a^k c^j$$

where either $i \ne 0$ or $k \ne 0$ ($i = k = 0$ imply $h \in \langle c \rangle$). Let

$$w = \begin{cases} db^{-i} a^{j/i} c^j & \text{if } i \ne 0 \\ db^{-j/k} & \text{if } i = 0 \text{ and } k \ne 0. \end{cases}$$

Then $w^{-1}hw = h^2 \in H$ and $o(w) = 2$. Thus 2 divides $|N_G(H)|$, so 54 divides $|UN_G(H)|$, hence $UN_G(H) = G$. This completes the proof that $G \in \mathfrak{Y}$.

Since it has already been shown that $S_3 \times \mathbb{Z}_3 \notin \mathfrak{Y}$, to complete the counterexample it suffices to show G has a subgroup isomorphic to $S_3 \times \mathbb{Z}_3$. The subgroup $\langle a, d, c \rangle$ is easily seen to satisfy the relations $a^3 = d^2 = (ad)^2 = 1$, $c^3 = 1$, $ac = ca$, $dc = cd$, for $S_3 \times \mathbb{Z}_3$ and to have the same order as $S_3 \times \mathbb{Z}_3$ (namely 18). Hence $\langle a, d, c \rangle \cong S_3 \times \mathbb{Z}_3$.

THEOREM 6.1 *For a finite solvable group G, the following are equivalent:*
(1) *$G \in \mathfrak{Y}$.*
(2) *For all subgroups H of G and all divisors d of $[G:H]$, G has a subgroup W such that $H \le W$ and $[W:H] = d$.*
(3) *G has a normal Hall subgroup N such that N and G/N are both nilpotent and $G = NN_G(H)$ for all $H \le N$.*
(4) *Every subgroup of G can be written as an intersection of subgroups of pairwise coprime prime-power indices.*

(5) (McLain) *For any two subnormal subgroups A, B of G with $A \leq B$ and for any divisor d of $[B:A]$ there exists a subgroup C such that $A \leq C \leq B$ and $[C:A] = d$.*

Proof: The equivalence of (1), (2), and (3) was shown in Chapter 1 (see Chapter 1, Theorem 4.3 and the discussion preceding it) and the equivalence of (1) and (4) was shown in Chapter 4, Theorem 5.2.

(4)$\Rightarrow$(5) First we show that condition (4) is closed under normal subgroups. Suppose a group T satisfies condition (4) and $N \lhd T$. Let $A \leq N$. Then $A \leq T$ so that $A = J_1 \cap J_2 \cap \ldots \cap J_m$ where each $[T:J_i]$ is a prime-power, say $p_i^{e_i}$, and $[T:J_i]$, $[T:J_j]$ are coprime for $i \neq j$. Then since $N \lhd T$, NJ_i is a subgroup for all i and $[N:N \cap J_i] = [NJ_i:J_i]$. This last index divides $[T:J_i]$ and so each $[N:N \cap J_i]$ is a power of p_i and for $i \neq j$, $[N:N \cap J_i]$ and $[N:N \cap J_j]$ are coprime. Finally $(N \cap J_1) \cap (N \cap J_2) \cap \ldots \cap (N \cap J_m) = N \cap A = A$. Thus each subgroup A of N is an intersection of subgroups of pairwise coprime prime-power indices, hence N satisfies condition (4).

Continuing the proof that (4) implies (5), suppose G satisfies (4) and A, B are subnormal in G with $A \leq B$, and suppose $d \,|\, [B:A]$. Then B satisfies (4) by repeated application of the previous paragraph. If $B < G$, then B satisfies (5) by induction; and (since A and B are subnormal in B) we're done. If $B = G$, then since G satisfies condition (4) it also satisfies condition (2) since we've already shown equivalence of (4) and (2). Hence there is a subgroup C between A and B ($= G$) such that $[C:A] = d$.

(5)$\Rightarrow$(1) Let G satisfy (5). First note that G is supersolvable by virtue of Theorem 5.1 (condition (22)). Secondly we note that the class of groups satisfying condition (5) is clearly quotient-closed; consequently every proper quotient group of G belongs to $\mathcal{Y}$ by induction. Let $H \leq G$ with $q \,|\, [G:H]$, q a prime. Let p denote the largest prime divisor of $|G|$. Then by supersolvability, G has a normal Sylow p-subgroup, call it S, of order, say, p^e.

We distinguish two cases. Suppose first that $q \neq p$. We claim that $G = SN_G(H \cap S)$. Let $|G| = mp^e$ where $(m, p) = 1$. Since S is nilpotent, $H \cap S$ is subnormal in S and hence is subnormal in G. Since m divides $[G:H \cap S]$, condition (5) yields a subgroup L such that $H \cap S \leq L$ and $[L:H \cap S] = m$. Thus $|L| = mp^b$ where $|H \cap S| = p^b$. Supersolvability of L implies existence of a normal Sylow p-subgroup which must be $H \cap S$. Thus $L \leq N_G(H \cap S)$ and so m divides $|N_G(H \cap S)|$. Hence $|G| = mp^e$ divides $|SN_g(H \cap S)|$ proving $G = SN_G(H \cap S)$ as claimed. Consequently $[G:N_G(H \cap S)]$ is a power of p, say p^a.

Let C be a Hall p'-subgroup of H. Since $H \cap S \lhd H$, we have $H = (H \cap S)C$. Now $C \leq H \leq N_G(H \cap S)$ so C is contained in some Hall p'-subgroup K of $N_G(H \cap S)$. Comparing the orders of $N_G(H \cap S)$ (recall it has index p^a) and G, shows that K is also a Hall p'-subgroup of G. Since

$H \cap S \triangleleft N_G(H \cap S)$, $R = (H \cap S)K$ is a subgroup. Then $R/H \cap S$ $\cong K/K \cap (H \cap S) \cong K \cong G/S$. Thus $R/H \cap S \in \mathcal{Y}$ by induction. We next note that $H \leq R$ (because $H = (H \cap S)C$ and so $H/H \cap S$ is a subgroup of $R/H \cap S$. Now $[G:R]$, being a divisor of $[G:K]$, has order a power of p and is thus coprime to q. Then $q|[G:H] = [G:R][R:H]$ implies $q|[R:H]$ and so $q|[R/H \cap S : H/H \cap S]$. This and $R/H \cap S \in \mathcal{Y}$ produce a subgroup $W/H \cap S$ such that $[W/H \cap S : H/H \cap S] = q$. Hence $[W:H] = q$, completing the proof in the case $q \neq p$.

Now suppose $q = p$. Let $x \in Z(S)$ have order p. Then $H \cap \langle z \rangle$ is either $\langle z \rangle$ or $\{1\}$. If $H \cap \langle z \rangle = \langle z \rangle$, then $\langle z \rangle$ is a proper normal subgroup under H (normal in G because $|N_G(\langle z \rangle)| = |G|$ by an argument analogous to that which showed $G = SN_G(H \cap S)$ in the first case). Hence $G/\langle z \rangle \in \mathcal{Y}$ which is sufficient to produce a subgroup W with $[W:H] = p$. Finally, if $H \cap \langle z \rangle = \{1\}$, then $W = H\langle z \rangle$ is the required subgroup.

7. The Class $\mathbb{S}\mathcal{Y}$

DEFINITION The group G belongs to $\mathbb{S}\mathcal{Y}$ iff all subgroups of G belong to $\mathcal{Y}$.

It is immediate from the definition that $\mathbb{S}\mathcal{Y}$ is subgroup-closed. Quotient-closure follows quickly from quotient-closure of the class $\mathcal{Y}$. Finally $S_3 \times \mathbb{Z}_3$, the same counterexample that showed $\mathcal{Y}$ is not product-closed, shows $\mathbb{S}\mathcal{Y}$ is not product-closed since both S_3 and $\mathbb{Z}_3$ belong to $\mathbb{S}\mathcal{Y}$ but $S_3 \times \mathbb{Z}_3$ does not.

THEOREM 7.1 *For a finite solvable group G, the following equivalent:*
(1) $G \in \mathbb{S}\mathcal{Y}$.
(2) $G \in \mathbb{S}\mathcal{X}$.
Proof: By Theorem 5.3 of Chapter 4.

8. LM-groups

DEFINITION G is an **LM-group** iff for every pair of subgroups A, B of G with A maximal in $\langle A, B \rangle$, $A \cap B$ is maximal in B.

This class is subgroup-closed, quotient-closed, and product-closed (Proposition 4.1 and Corollary 4.5 of Chapter 4).

We now consider characterizations. One of these will be an analogue of a characterization of supersolvable groups. We need the following definition.

DEFINITION The subgroup H is **LM-embedded** in the group G iff $H \lhd G$ and for all chief factors R/S of G with $R \leq H$, $|R/S|$ is prime and $|\text{Aut}_G(R/S)|$ is either 1 or a prime.

Proofs of the following 3 facts about LM-embedded subgroups are analogous to results about supersolvably embedded subgroups—see Chapter 1, Section 7.

(i) If H, $K \lhd G$ with $K \leq H$ and if H is LM-*embedded in* G, *then* H/K *is* LM-*embedded in* G/K.

(ii) If H, K are LM-*embedded in* G, *then so is* HK.

Because of (ii), every group G has a largest LM-embedded subgroup, namely $\text{LME}(G) = \langle H : H \text{ is LM-embedded in } G \rangle$.

(iii) If $N \lhd G$ and $N \leq \text{LME}(G)$, *then* $\text{LME}(G)/N = \text{LME}(G/N)$.

THEOREM 8.1 *For a finite solvable group G the following are equivalent*:
(1) G *is an* LM-*group.*
(2) *For every pair A, B of distinct maximal subgroups of some subgroup H of G, $A \cap B$ is maximal in both A and B.*
(3) G *is supersolvable and* $\text{Aut}_G(H/K)$ *is either trivial or of prime order for all chief factors H/K of G.*
(4) $G = \text{LME}(G)$.
Proof: Equivalence of (1) and (2) was shown in Proposition 4.2 of Chapter 4. Equivalence of (1) and (3) is the content of Theorem 4.4 of Chapter 4.

(3) iff (4) Suppose condition (3) holds. Let N be a minimal normal subgroup of G. Then $|N|$ is prime (since G is supersolvable) and $|\text{Aut}_G(N)|$ is either 1 or a prime. Thus N is LM-embedded in G and so $N \leq \text{LME}(G)$. By (iii) above, $\text{LME}(G)/N = \text{LME}(G/N)$. But induction applied to the group G/N (we can do this because the class of groups satisfying condition (3) is quotient-closed by, say, equivalence of (3) and (1)) yields that $G/N = \text{LME}(G/N)$. Hence $\text{LME}(G)/N = G/N$ so $\text{LME}(G) = G$ showing that (3) implies (4). Conversely if (4) holds, we first note that $\text{LME}(G) \leq \text{SE}(G)$ so that $G = \text{SE}(G)$ and hence G is supersolvable. Thus the first part of (3) holds. Moreover, the rest of (3) holds as well, by definition of $\text{LME}(G)$ being LM-embedded in G. Thus (4) implies (3).

Appendixes

Appendix A: Extended Sylow Theory

Let π denote a set of primes and π' its complement in the set of all primes; should π consist of a single prime p we denote these sets p and p' respectively. An integer n is called a π-**number** whenever every prime divisor of n lies in π. A subgroup H of a group G is called a π-**subgroup** of G whenever $|H|$ is a π-number; H is called a **Hall π-subgroup** of G when H is a π-subgroup of G and $[G:H]$ is a π'-number. A Hall π'-subgroup of the group G is also called a **Hall π-complement** of G. When π consists of exactly one prime p a Hall π-subgroup of G is, of course, a Sylow p-subgroup of G and a Hall π-complement of G is a **Sylow p-complement** of G. The following theorem of P. Hall plays an important role in the theory of solvable groups.

THEOREM 1   *Let G be a finite solvable group and π a set of primes. Then*
(1) *G contains a Hall π-subgroup and any two Hall π-subgroups of G are conjugate in G, and*
(2) *Each π-subgroup of G lies in at least one Hall π-subgroup of G.*
Proof: We proceed by induction on $|G|$. Let N be a minimal normal subgroup of G with $|N| = p^a$ where p is a prime. By induction G/N has a Hall π-subgroup H/N, moreover, any two Hall π-subgroups of G/N are conjugate in G/N. We consider separately the cases $p \in \pi$ and $p \notin \pi$.

Case 1: $p \in \pi$. In this case H is a π-subgroup of G and, since $[G/N : H/N] = [G:H]$, it is easily seen that H is in fact a Hall π-subgroup of G. Were L any other π-subgroup of G then, by induction $LN/N \leq (H/N)^{Nx}$ for some x in G and therefore $L \leq H^x$. Should L also be a Hall π-subgroup of G, then $|L| = |H| = |H^x|$. This and $L \leq H^x$ give $L = H^x$.

Case 2: $p \notin \pi$. In this situation, should H be a proper subgroup of G, then we may apply induction to conclude that H has a Hall π-subgroup H_1 which must also be a Hall π-subgroup of G. Were L any other π-subgroup of G, then, by induction LN/N lies in H^x/N for some x in G and hence $L \leq H^x$; Since H_1^x is a Hall π-subgroup of H^x, we may apply induction to conclude that L lies in H_1^{xy} for some $y \in H^x$. Moreover, were L also a Hall subgroup of G then, as in Case 1, L would coincide with H_1^{xy}.

We assume for the remainder of the proof that $G = H$ and, without loss of generality, that $G \neq N$. We also observe that under these conditions N is a Sylow p-subgroup of G. Let T/N be a minimal normal subgroup of G/N with $|T/N| = q^r$ for some prime $q \neq p$. Let Q be a Sylow q-subgroup of T, so that $T = NQ$. By the Frattini argument, $G = TN_G(Q) = NN_G(Q)$. Should $G = N_G(Q)$, then we may apply the argument of Case 1 using Q for N. We therefore assume that $G \neq N_G(Q)$. Since $N \cap N_G(Q)$ must be a normal subgroup of G (It's normal in both $N_G(Q)$ and in N—the latter because N is Abelian), it follows that $N \cap N_G(Q) = \{1\}$. Thus $[G:N] = |N_G(Q)|$ and $N_G(Q)$ is a Hall π-subgroup of G. Set $H_1 = N_G(Q)$ and let L be any π-subgroup of G. Since $G = NH_1$, $LN = LN \cap NH_1 = N(H_1 \cap LN)$. Thus $H_1 \cap LN$ is a Hall π-subgroup of LN. If $LN \neq G$, then by induction $L \leq (H_1 \cap LN)^x \leq H_1^x$ for some x in LN. Should $LN = G$, then $T = N(L \cap T)$ and in this situation $L \cap T = Q^x$ for some x in T. Then $T \triangleleft G$ implies $L \cap T \triangleleft L$ so that $L \leq N_G(Q^x) = H_1^x$. Again, if L is also a Hall π-subgroup of G, the argument of Case 1 can be used to show that $L = H_1^x$.

It should be noted that the only finite groups which satisfy the conclusion of Theorem 1 are the finite solvable groups. The proof of this, however, uses Burnside's two-prime theorem and is considerably more difficult than Theorem 1 itself.

LEMMA 2 *Let H and K be subgroups of relatively prime index in the group G. Then $G = HK$ and $[G : H \cap K] = [G:H][G:K]$. Moreover, if H and K are Hall subgroups of G, so is $H \cap K$.*
Proof: $|G| \geq |HK| = |H| \cdot [K : H \cap K]$ and hence $[G : H] \geq [K : H \cap K]$. Thus $[G : H \cap K] = [G : K][K : H \cap K] \leq [G : K][G : H]$. Since $[G : H]$, $[G : K]$ both divide $[G : H \cap K]$, it follows that $[G:H][G:K]$ must also divide $[G : H \cap K]$. Hence $[G:H][G:K] \leq [G : H \cap K]$ and therefore $[G:H][G:K] = [G : H \cap K]$. This last equality implies that the preceding inequalities are all equalities so that $G = HK$ as well.

Should H and K be Hall subgroups of G, then any prime p which divides $|H \cap K|$ could not divide $[G:H]$ nor $[G:K]$. Hence p would not divide $[G: H \cap K]$ and therefore $H \cap K$ would also be a Hall subgroup of G.

THEOREM 3 *For each solvable group G there is a set, Σ, of Hall subgroups of G such that*

(1) *for each set of primes π, Σ contains exactly one Hall π-subgroup of G, and*

(2) *any two Hall subgroups in Σ permute.*

Proof: For each prime p dividing the order of G, select a Sylow p-complement G^p (existence by Theorem 1). For each set of primes π, set $G_\pi = \bigcap \{ G^p : p \notin \pi \}$. By repeated use of Lemma 2 it easily follows that G_π is a Hall π-subgroup of G. We now set $\Sigma = \{ G_\pi \}$, so that Σ contains exactly one Hall π-subgroup of G for each set of primes π.

For any two G_π and G_ρ in Σ, $G_\pi G_\rho \leq \bigcap \{ G^p : p \notin \pi \cup \rho \} = G_{\pi \cup \rho}$. Then G_π and G_ρ are subgroups of $G_{\pi \cup \rho}$ of relatively prime index. Lemma 2 implies $G_{\pi \cup \rho} = G_\pi G_\rho$ so that indeed G_π and G_ρ permute.

A set Σ of Hall subgroups of the group G which satisfies conditions (1) and (2) of Theorem 3 is called a **Sylow system** of the group G. It is clear from the proof of Theorem 3 that a Sylow system is completely determined by the Sylow complements in the system and that any complete set of Sylow complements determines a Sylow system. We can also deduce from that theorem that any Hall subgroup of a Sylow system is simply the product of appropriate Sylow subgroups from the system. Hence a Sylow system will be completely determined by the Sylow subgroups in the system, however, a complete set of Sylow subgroups of G (i.e., one for each prime divisor of $|G|$) will only determine a Sylow system when they are pairwise permutable. A complete set of pairwise permutable Sylow subgroups is called a **Sylow basis** of the group.

THEOREM 4 *Let G be a solvable group. Then the following statements are valid.*

(1) *Any two Sylow systems of G are conjugate in G (i.e., if Σ and $\tilde{\Sigma}$ are two Sylow systems of G, there is an $x \in G$ such that $\tilde{\Sigma} = \Sigma^x = \{ G_\pi^x : G_\pi \in \Sigma \}$).*

(2) *If H is a subgroup of G and Σ_H is a Sylow system of H, then there is a Sylow system Σ of G with $\Sigma_H = \Sigma \cap H = \{ G_\pi \cap H : G_\pi \in \Sigma \}$.*

(3) *If N is a normal subgroup of G and Σ is a Sylow system of G, then $\Sigma N / N = \{ G_\pi N / N : G_\pi \in \Sigma \}$ is a Sylow system of G/N.*

Proof:

(1) As a Sylow system is completely determined by its Sylow complements, it is only necessary to find an $x \in G$, such that $\tilde{G}^p = (G^p)^x$ for each prime p dividing the order of G and Sylow p-complements $\tilde{G}^p \in \tilde{\Sigma}$ and $G^p \in \Sigma$.

Let $\Gamma = \{S_1, S_2, \ldots, S_r\}$ and $\tilde{\Gamma} = \{\tilde{S}_1, \tilde{S}_2, \ldots, \tilde{S}_r\}$ be the Sylow comple-ments in Σ and $\tilde{\Sigma}$ respectively and select $x \in G$ so that $\tilde{\Gamma} \cap \Gamma^x$ is as large as possible (here $\Gamma^x = \{S_1^x, S_2^x, \ldots, S_r^x\}$). As any two Sylow p-complements are conjugate in G (Theorem 1), $\tilde{\Gamma} \cap \Gamma^x \neq \emptyset$. So (after rela-belling if necessary) we can write $\tilde{\Gamma} \cap \Gamma^x = \{\tilde{S}_1, \tilde{S}_2, \ldots, \tilde{S}_t\}$. If $t = r$, there is nothing to show, thus we assume $t < r$. Let $\tilde{P}_{t+1}$ be the Sylow p_{t+1}-subgroup of G in $\tilde{\Sigma}$, where $[G : S_{t+1}] = p_{t+1}^a$. From Lemma 2 we know that

$$G = (\tilde{P}_{t+1})(S_{t+1})^x$$

and, since $\tilde{S}_{t+1}$ is conjugate to $(S_{t+1})^x$, there is a $y \in \tilde{P}_{t+1}$ with $(S_{t+1})^{xy} = \tilde{S}_{t+1}$. But $y \in \tilde{P}_{t+1} \leq \tilde{S}_i$ for $i \leq t$ and hence

$$(\Gamma)^{xy} = \{S_1^{xy}, S_2^{xy}, \ldots, S_t^{xy}, S_{t+1}^{xy}, \ldots, S_r^{xy}\}$$
$$= \{\tilde{S}_1^y, \tilde{S}_2^y, \ldots, \tilde{S}_t^y, \tilde{S}_{t+1}, \ldots, S_r^{xy}\}$$
$$= \{\tilde{S}_1, \tilde{S}_2, \ldots, \tilde{S}_t, \tilde{S}_{t+1}, \ldots, S_r^{xy}\}.$$

So $\Gamma^{xy} \cap \tilde{\Gamma} \supset \Gamma^x \cap \tilde{\Gamma}$ which is contrary to the choice of x. Therefore $t = r$ and $\tilde{\Gamma} = \Gamma^x$.

(2) Let Σ_H be a Sylow system of the subgroup H of G. For each prime p let H^p be a Sylow p-complement of H in Σ_H, and G^p a Sylow p-complement of G which contains H^p. The Sylow p-complements G^p of G determine a Sylow system Σ of G and $H \cap \Sigma = \Sigma_H$.

(3) The homomorphic images of permuting Hall subgroups are again permuting Hall subgroups.

A Sylow system Σ of the group G is said to **reduce** into the subgroup H whenever $H \cap \Sigma$ is a Sylow system of H. For a fixed subgroup H of the group G Theorem 4 implies the existence of Sylow systems which reduce into H, however, a particular Sylow system need not reduce into H.

Appendix B: Chief Factors, Centralizers, and Induced Automorphisms

DEFINITION H/K is a **chief factor** of the group G if $H, K \triangleleft G$, $K \leq H$, and H/K is a minimal normal subgroup of G/K.

If H/K is a chief factor of G and $|H/K|$ is some power of the prime p, then H/K is called a **p-chief factor** of G. We note that in solvable groups every chief factor is a p-chief factor for some prime p.

DEFINITION   If H/K is a chief factor of G, then the **centralizer** of H/K in G, denoted $C_G(H/K)$ is defined as

$$C_G(H/K) = \{\, x : x^{-1}h^{-1}xh \in K \text{ for all } h \in H \,\}.$$

We next define induced automorphisms. Let H/K be a chief factor of G. Then the elements of G induce automorphisms of H/K as follows: For all $x \in G$, let $f_x : H/K \to H/K$ be given by

$$f_x(Kh) = Kxhx^{-1}.$$

It is easily seen that f_x is an automorphism of H/K; it is called the **automorphism induced by** x.

We denote $\{\, f_x : x \in G \,\}$ by $\mathrm{Aut}_G(H/K)$. For any two elements $x, y \in G$, and all $Kh \in H/K$, $f_x \circ f_y(Kh) = f_x(Kyhy^{-1}) = Kxyhy^{-1}x^{-1} = f_{xy}(Kh)$. Thus $f_x \circ f_y = f_{xy}$. This shows $\mathrm{Aut}_G(H/K)$ is a subgroup of $\mathrm{Aut}(H/K)$.

THEOREM 1   If H/K is a chief factor of G, then $C_G(H/K) \triangleleft G$ and $G/C_G(H/K) \cong \mathrm{Aut}_G(H/K)$.

Proof: Define $\alpha : G \to \mathrm{Aut}_G(H/K)$ by $\alpha(x) = f_x$. As just shown above, $f_x \circ f_y = f_{xy}$ for all $x, y \in G$; so α is a homomorphism. Hence $G/\ker \alpha \cong \mathrm{Aut}_G(H/K)$ so it suffices to show $\ker \alpha = C_G(H/K)$. Let $c \in C_G(H/K)$. Then for all $Kh \in H/K$, $f_c(Kh) = Kchc^{-1} = Kchc^{-1}h^{-1}h$. But $c \in C_G(H/K)$ implies $c^{-1}h^{-1}ch \in K$ and so $chc^{-1}h^{-1}$ also belongs to K. Thus $f_c(Kh) = Kh$ for all $Kh \in H/K$ so f_c is the identity of $\mathrm{Aut}_G(H/K)$; that is, $c \in \ker \alpha$. This shows $C_G(H/K)$ is contained in $\ker \alpha$. Proof of the reverse containment is similar, so $\ker \alpha = C_G(H/K)$ as required.

THEOREM 2   If A/B is a chief factor of G and $N \triangleleft G$ such that $N \leq B \leq A$, then $\mathrm{Aut}_{G/N}((A/N)/(B/N)) \cong \mathrm{Aut}_G(A/B)$.

Proof: Let $\eta : G \to G/N$ be the natural map. It is easily seen that $\eta[C_G(A/B)] \leq C_{G/N}((A/N)/(B/N))$. This means that the map $\eta^* : G/C_G(A/B) \to (G/N)/C_{G/N}((A/N)/(B/N))$ given by $\eta^*(C_G(A/B)x) = C_{G/N}((A/N)/(B/N))\eta(x)$ is well-defined. It is also clearly a homomorphism and is easily seen to be bijective. Thus we get $G/C_G(A/B) \cong (G/N)/C_G((A/N)/(B/N))$. Now apply Thoerem 1 to conclude the result.

THEOREM 3   If $A, B \triangleleft G$ and if AB/B, $A/A \cap B$ are both chief factors of G, then $\mathrm{Aut}_G(AB/B) \cong \mathrm{Aut}_G(A/A \cap B)$.

Proof: This will follow immediately from Theorem 1 once it is shown that $C_G(AB/B) = C_G(A/A \cap B)$.

Let $x \in C_G(AB/B)$. Then for all $a \in A$, $x^{-1}a^{-1}xa \in B$. But $a \in A \triangleleft G$ also implies $x^{-1}a^{-1}xa \in A$. Thus $x^{-1}a^{-1}xa \in A \cap B$ for all $a \in A$ and so

$x \in C_G(A/A \cap B)$, showing that $C_G(AB/B) \leq C_G(A/A \cap B)$. Now let $x \in C_G(A/A \cap B)$. Then for all $t \in AB$, $t = ab$ for some $a \in A$, $b \in B$. Then $x^{-1}t^{-1}xt = x^{-1}b^{-1}a^{-1}xab = (x^{-1}b^{-1}x)(x^{-1}a^{-1}xa)b$. But all three factors of this product belong to B (the first because $b \in B \lhd G$; the second because $x \in C_G(A/A \cap B)$). Thus $x^{-1}t^{-1}xt \in B$ for all $t \in AB$ and so $x \in C_G(AB/B)$ showing $C_G(A/A \cap B) \leq C_G(AB/B)$.

DEFINITIONS Let H/K be a chief factor of G and let A be any subgroup of G. Then
(1) A **covers** H/K iff $HA = KA$.
(2) A **avoids** H/K iff $H \cap A = K \cap A$.

It is straightforward to show that A covers H/K iff $K(H \cap A) = H$ and that A avoids H/K iff $KA \cap H = K$.

We note two things about covering and avoidance: (1) It is not possible for a subgroup A to both cover and avoid the chief factor H/K (since the equations $HA = KA$ and $H \cap A = K \cap A$ imply $H = K$). (2) A subgroup A need neither cover nor avoid a chief factor (for example, let $G = A_4 \times \mathbb{Z}_2$; $H = V \times \{1\}$, where V is the 4-element subgroup of A_4; $K = \{1\}$; and $A = B \times \mathbb{Z}_2$, where B is any 2-element subgroup of A_4; then $HA = V \times \mathbb{Z}_2$, order 8 while $KA = B \times \mathbb{Z}_2$, order 4 so $HA \neq KA$; and $H \cap A = B \times \{1\}$, order 2 while $K \cap A = \{1\}$ so $H \cap A \neq K \cap A$).

Because of (2), it is of interest to have sufficient conditions on a subgroup A which will ensure that A either covers or avoids every chief factor. We present two such conditions: namely that A be normal, or that it be maximal.

THEOREM 4 *Let H/K be a chief factor of G and let $A \lhd G$. Then either*
(1) A covers H/K; $H \cap A/K \cap A \cong H/K$; $H \cap A/K \cap A$ is a chief factor of G; and $\mathrm{Aut}_G(H \cap A/K \cap A) \cong \mathrm{Aut}_G(H/K)$.
(2) A avoids H/K; $HA/KA \cong H/K$; HA/KA is a chief factor of G; and $\mathrm{Aut}_G(HA/KA) \cong \mathrm{Aut}_G(H/K)$.
Proof: The subgroup $H \cap KA$ is normal in G and moreover $K \leq H \cap KA \leq H$. This means that either $H \cap KA = H$ or $H \cap KA = K$.

In the first case we have $H \leq KA$ so HA/KA, i.e., A covers H/K. Note that $KA \cap H = K(A \cap H)$ (either by the modular law or an easily given direct proof). Thus in this first case we also have $H = K(A \cap H)$ so that $H/K = K(A \cap H)/K \cong A \cap H/A \cap K$. To show that $A \cap H/A \cap K$ is a chief factor, suppose $B \lhd G$ with $A \cap K \leq B \leq A \cap H$. Then $KB \lhd G$ with $K \leq KB \leq H$. Hence either $KB = K$ or $KB = H$. If $KB = K$, then $B \leq K$ so $B = A \cap K$. If $KB = H$, then $B = A \cap H$ (Let $x \in A \cap H$. Then $x \in H = KB$ means that $x = kb$ for some $k \in K$ and $b \in B$. Consequently x, $b \in A$ imply $k \in A$, so $k \in A \cap K \leq B$, so $x \in B$). Thus either $B = A \cap K$ or $B = A \cap H$ so $A \cap H/A \cap K$ is indeed a chief factor. Finally apply Theorem 3 to conclude that $\mathrm{Aut}_G(H \cap A/K \cap A) \cong \mathrm{Aut}_G(H/K)$. This completes all parts of (1).

In the second case ($H \cap KA = K$) we again note that $KA \cap H = K(A \cap H)$ so that $K(A \cap H) = K$. Thus $A \cap H \leq K$ so $A \cap H = A \cap K$, i.e., A avoids H/K. Now $HA = H(KA)$ so $HA/KA = H(KA)/KA \cong H/H \cap KA = H/K$. To show that HA/KA is a chief factor, suppose $B \triangleleft G$ with $KA \leq B \leq HA$. Then $H \cap B \triangleleft G$ with $K \leq H \cap B \leq H$. Hence either $H \cap B = K$ or $H \cap B = H$. If $H \cap B = K$, then $B = KA$. (Let $b \in B$. Then $b \in HA$ so $b = ha$ for some $h \in H, a \in A$. Since $b, a \in B$, we have $h \in B$ so that $h \in H \cap B = K$. Thus $b = ha \in KA$.) If $H \cap B = H$, then $H \leq B$ so $B = HA$. Thus either $B = KA$ or $B = HA$ so HA/KA is indeed a chief factor. Finally we apply Theorem 3 to conclude that $\text{Aut}_G(HA/KA) \cong \text{Aut}_G(H/K)$. This completes all parts of (2).

THEOREM 5 *If H/K is a chief factor of G and A is a maximal subgroup of G, then A either covers or avoids H/K.*
Proof: Suppose A does not cover H/K, i.e., $HA \neq KA$. Since H is not contained in A ($H \leq A$ would imply $HA = A = KA$), we have $G = HA$ by maximality of A. Consequently, $G \neq KA$, which implies, again by maximality of A, that $K \leq A$. Thus $K \leq H \cap A \leq H$. Now $H/K \triangleleft G/K$ implies $H/K \cap A/K \triangleleft A/K$ and furthermore $H/K \cap A/K$ is also normal in H/K because H/K is Abelian. These facts and $G/K = (H/K)(A/K)$ imply $H/K \cap A/K \triangleleft G/K$, i.e., $H \cap A/K \triangleleft G/K$ so $H \cap A \triangleleft G$. But H/K is a chief factor and $H \cap A \neq H$ (lest $H \leq A$) so we conclude $H \cap A = K$. Thus A avoids H/K.

Appendix C: Various Subgroups

1   The Frattini Subgroup $\Phi(G)$

The intersection of all maximal subgroups of a group G is called the **Frattini subgroup** of G and is denoted by $\Phi(G)$. It is always characteristic in G.

LEMMA 1.1 *If $N \triangleleft G$, then $\Phi(N) \leq \Phi(G)$.*
Proof: If the lemma is false, then there is a maximal subgroup M of G such that $\Phi(N) \not\leq M$. Since $\Phi(N)$ is characteristic in N, it is normal in G. Hence the choice of M implies $G = \Phi(N)M$ and so $N = N \cap G = N \cap \Phi(N)M = \Phi(N)(N \cap M)$. We claim that $N \cap M = N$. If, to the contrary, $N \cap M < N$, then let N_1 be a maximal subgroup of N containing $N \cap M$,

so that $N = \Phi(N)N_1$. Since $\Phi(N) \le N_1$, it follows that $N = N_1$. This shows $N \cap M = N$ as claimed, so $\Phi(N) \le N \le M$. This violates the choice of M and therefore $\Phi(N) \le \Phi(G)$.

LEMMA 1.2 *If $N/\Phi(G)$ is a nilpotent normal subgroup of $G/\Phi(G)$, then N is nilpotent.*

Proof Let P be a Sylow p-subgroup of N. Then $\Phi(G)P/\Phi(G)$ is a Sylow p-subgroup of the nilpotent group $N/\Phi(G)$, from which it follows that $\Phi(G)P \lhd G$. If $N_G(P) < G$, then there is a maximal subgroup M of G with $N_G(P) \le M$. Now P is a Sylow p-subgroup of $\Phi(G)P$, so we get that $G = \Phi(G)PN_G(P)$ by the Frattini argument ([Scott 1964, 6.2.4]). Hence $G \le M$, a contradiction. It follows that P is a normal subgroup of G and thus also is a normal subgroup of N. Therefore every Sylow subgroup of N is normal in N and hence N is nilpotent.

When $N = \Phi(G)$ in the above lemma, we get

COROLLARY 1.3 $\Phi(G)$ *is nilpotent for any group* G.

2 The Fitting Subgroup Fit(G)

The product AB of two normal nilpotent subgroups A and B of a group G is again a nilpotent subgroup ([Scott 1964, 7.4.1]). This means that every group G contains a maximal nilpotent normal subgroup which is called the **Fitting subgroup** of G and is denoted by Fit(G). It is always characteristic in G.

LEMMA 2.1 *For any group* G,
(a) $\Phi(G) \le \mathrm{Fit}(G)$.
(b) $\mathrm{Fit}(G)/\Phi(G)$ *is Abelian.*
(c) $\Phi(G/\Phi(G)) = \{1\}$.

Proof: (a) follows from Corollary 1.3.

As for (b), let F denote Fit(G). Then F is nilpotent, so each maximal subgroup of F is normal and hence contains F'. Thus $F' \le \Phi(F)$. But $\Phi(F) \le \Phi(G)$ by Lemma 1.1. Conclude $F' \le \Phi(G)$ from which it follows that $F/\Phi(G)$ is Abelian.

To prove (c), let $M_1, M_2, \ldots, M_k$ be the maximal subgroups of G. Then the list of maximal subgroups of $G/\Phi(G)$ is exactly $M_1/\Phi(G)$, $M_2/\Phi(G), \ldots, M_k/\Phi(G)$. Hence $\Phi(G/\Phi(G)) = M_1/\Phi(G) \cap M_2/\Phi(G) \cap \ldots \cap M_k/\Phi(G) = (M_1 \cap M_2 \cap \ldots \cap M_k)/\Phi(G)$ which is equal to $\Phi(G)/\Phi(G) = \{1\}$.

LEMMA 2.2   *For any group G, $\mathrm{Fit}(G/\Phi(G)) = \mathrm{Fit}(G)/\Phi(G)$.*
Proof: Let $N/\Phi(G) = \mathrm{Fit}(G/\Phi(G))$ so that $N \lhd G$. By Lemma 1.2, N is nilpotent and hence $N \le \mathrm{Fit}(G)$. Since quotient groups of nilpotent groups are nilpotent; it follows that $\mathrm{Fit}(G)/\Phi(G)$ is a normal nilpotent subgroup of $G/\Phi(G)$. Hence $\mathrm{Fit}(G)/\Phi(G) \le \mathrm{Fit}(G/\Phi(G))$ so that $\mathrm{Fit}(G) \le N$ and therefore $N = \mathrm{Fit}(G)$.

LEMMA 2.3   *If G is solvable and $\Phi(G) = \{1\}$, then $\mathrm{Fit}(G)$ is the direct sum of (Abelian) minimal normal subgroups of G.*
Proof: Let L be maximal among all subgroups of $\mathrm{Fit}(G)$ which can be expressed as the direct sum of minimal normal subgroups of G. Note then that $L \lhd G$. We will first show that there is a subgroup S of G with $LS = G$ and $L \cap S = \{1\}$.

Let S be a minimal element of the set $\{T : LT = G, \ T \le G\}$. Then $S \cap L \lhd G$ ($S \cap L \lhd S$ because $L \lhd G$; $S \cap L \lhd L$ because L is Abelian; hence $S \cap L \lhd LS = G$). If $S \cap L \ne \{1\}$, then since $\Phi(G) = \{1\}$, there is a maximal subgroup M of G such that $G = M(S \cap L)$. We now see that $S = S \cap G = S \cap M(S \cap L) = (S \cap M)(S \cap L)$ and, since M does not contain $S \cap L$, that $S \cap M < S$. However, $G = SL = (S \cap M)(S \cap L)L = (S \cap M)L$. This contradicts the choice of S and we conclude that $S \cap L = \{1\}$.

We next consider the subgroup $S \cap \mathrm{Fit}(G)$. It is normal in both S and in $\mathrm{Fit}(G)$ (the latter because $\Phi(G) = \{1\}$ implies $\mathrm{Fit}(G)$ is Abelian by Lemma 2.1) hence it is normal in G. If $S \cap \mathrm{Fit}(G) \ne \{1\}$, then there is an Abelian minimal normal subgroup H of G with $H \le S \cap \mathrm{Fit}(G)$. As $S \cap L = \{1\}$, we conclude that $L < L \oplus H$. This contradicts the choice of L and therefore $S \cap \mathrm{Fit}(G) = \{1\}$. It now follows that $L = \mathrm{Fit}(G)$ and so $\mathrm{Fit}(G)$ is the direct sum of (Abelian) minimal normal subgroups of G.

Combining Lemmas 2.2 and 2.3 with 2.1(c) yields the result

THEOREM 2.4   *In the solvable group G, $\mathrm{Fit}(G)/\Phi(G)$ is the direct sum of (Abelian) minimal normal subgroups of $G/\Phi(G)$.*

The following theorem gives a very useful and important characterization of $\mathrm{Fit}(G)$.

THEOREM 2.5 $\mathrm{Fit}(G) = \bigcap \{C_G(H/K) : H/K \text{ is a chief factor of } G\}$.
Proof: Let D denote the right-hand side of the above equation. Note that $D \lhd G$. If $\{1\} = H_0 < H_1 < H_2 < \ldots < H_r = G$ is a chief series of G, then

(*) $\{1\} = D \cap H_0 \le D \cap H_1 \le D \cap H_2 \le \ldots \le D \cap H_r = D$

is a normal series for D. Since D is contained in each $C_G(H_i/H_{i-1})$, $[D \cap H_i, D] \le D \cap [H_i, D] \le D \cap H_{i-1}$ for $i = 1, 2, \ldots, r$. Thus (*) is a central series for D. Hence D is nilpotent and therefore $D \le \mathrm{Fit}(G)$.

To show the reverse inclusion, let H/K be a chief factor of G. Since H/K is a minimal normal subgroup of G/K, and since $\text{Fit}(G)K/K \lhd G/K$, we get either $H/K \cap \text{Fit}(G)K/K = \{1\}$ or $H/K \leq \text{Fit}(G)K/K$. In the first case $H \cap \text{Fit}(G) \leq K$ and so $\text{Fit}(G) \leq C_G(H/K)$. In the second case, nilpotency of $\text{Fit}(G)K/K$ implies H/K meets $Z(\text{Fit}(G)K/K)$ nontrivially. Since H/K is a minimal normal subgroup of G/K, we get $H/K \leq Z(\text{Fit}(G)K/K)$ so again $\text{Fit}(G) \leq C_G(H/K)$. Since this has been shown for an arbitrary chief factor H/K, it follows that $\text{Fit}(G) \leq D$.

THEOREM 2.6 *If G is a solvable group, then $C_G(\text{Fit}(G)) \leq \text{Fit}(G)$.*
Proof: For brevity set $F = \text{Fit}(G)$ and $C = C_G(F)$. Suppose the theorem fails so that $CF \neq F$. Then CF/F is a nontrivial normal subgroup of G/F, hence contains a minimal normal subgroup Q/F of G/F. Solvability implies Q/F is Abelian so we have $Q' \leq F < Q \leq CF$. Now $(Q \cap C)F = Q$, as is easily seen. On the other hand $Q \cap C \lhd G$ and $Q \cap C$ is nilpotent because $[Q \cap C, Q \cap C, Q \cap C] \leq [Q', C] \leq [F, C] = \{1\}$. Hence $Q \cap C \leq F$ so that $(Q \cap C)F = F$. This and $(Q \cap C)F = Q$ yields the contradiction $F = Q$ completing the proof.

3 $O_p(G)$, $O_{p'}(G)$, and $O_{p',p}(G)$

The subgroups $O_p(G)$ and $O_{p'}(G)$ can be thought of as analogues of the Fitting subgroup for a group G, as all three are defined in the same sort of way—namely as joins of normal subgroups which share a certain property. Letting p denote a prime throughout, we define

$$O_p(G) = \langle A : A \lhd G \text{ and } A \text{ is a } p\text{-group} \rangle,$$
$$O_{p'}(G) = \langle A : A \lhd G \text{ and } A \text{ is a } p'\text{-group} \rangle.$$

It is evident then that $O_p(G)$ and $O_{p'}(G)$ share the defining property of their normal constituents—i.e., $O_p(G)$ is a normal p-subgroup of G and $O_{p'}(G)$ is a normal p'-subgroup.

Since p-groups are nilpotent, we immediately have

$$O_p(G) \leq \text{Fit}(G).$$

It is also easy to see the following useful characterization for $O_p(G)$.

$$O_p(G) = \bigcap \{ S : S \text{ is a Sylow } p\text{-subgroup of } G \}.$$

LEMMA 3.1 *Let H/K be a p-chief factor of the group G. Then*
(1) $O_{p'}(G)K \cap H = K$ (i.e., $O_{p'}(G)$ avoids H/K).
(2) $O_{p'}(G)H/O_{p'}(G)K \cong H/K$.
(3) $O_{p'}(G)H/O_{p'}(G)K$ is a p-chief factor of G.

Proof: By Theorem 4 of Appendix B $O_{p'}(G)$ either covers or avoids H/K. But it cannot cover H/K for then $H/K \approx O_{p'}(G) \cap H/O_{p'}(G) \cap K$ which is impossible as H/K is a nontrivial p-group and $O_{p'}(G) \cap H/O_{p'}(G) \cap K$ is a p'-group.

Thus $O_{p'}(G)$ avoids H/K proving (1). (2) and (3) now follow from Theorem 4 of Appendix B.

We now define the subgroup $O_{p', p}(G)$ of G by means of the following equation:

$$O_{p', p}(G)/O_{p'}(G) = O_p(G/O_{p'}(G))$$

(in other words $O_{p', p}(G) = \{x : O_{p'}(G)x \in O_p(G/O_{p'}(G))\}$).

THEOREM 3.2 *If G is solvable, then $O_{p', p}(G) = \bigcap\{C_G(H/K) : H/K$ is a p-chief factor of $G\}$.*

Proof: Let D denote the intersection of the centralizers in the above equation. Note that $D \lhd G$. First we show $O_{p', p}(G) \leq D$ as follows. Let $x \in O_{p', p}(G)$ so that (letting Θ denote $O_{p'}(G)$ for notational convenience) we get

$$\Theta x \in O_p(G/\Theta) \leq \mathrm{Fit}(G/\Theta)$$
$$\leq \bigcap\{C_{G/\Theta}((A/\Theta)/(B/\Theta)) : (A/\Theta)/(B/\Theta) \text{ is a chief factor of } G/\Theta\},$$

the last inclusion by Theorem 2.5. Then for any p-chief factor H/K of G, $\Theta H/\Theta K$ is a chief factor of G by Lemma 3.1 and so $(\Theta H/\Theta)/(\Theta K/\Theta)$ is a chief factor of G/Θ. Thus

$$\Theta x \in C_{G/\Theta}((\Theta H/\Theta)/(\Theta K/\Theta)).$$

This and $\Theta K \cap H = K$ (Lemma 3.1) yield $x \in C_G(H/K)$. Since this holds for arbitrary p-chief factors H/K, $O_{p', p}(G) \leq D$.

To show the reverse inclusion we first note   we may assume $\Theta = \{1\}$. This is because in any group G, $O_{p'}(G/\Theta) = \{1\}$ which implies $O_{p', p}(G/\Theta) = O_p(G/\Theta)$. Hence if Θ is not trivial, then

$$O_{p', p}(G/\Theta) = \bigcap\{C_{G/\Theta}((H/\Theta)/(K/\Theta)) : (H/\Theta)/(K/\Theta)$$

is a p-chief factor of $G/\Theta\}$

by induction on the order of G. But then

$$O_{p', p}(G)/\Theta = \bigcap\{C_G(H/K) : H/K \text{ is a } p\text{-chief factor of } G \text{ and } \Theta \leq K\}/\Theta.$$

Thus $D/\Theta \leq O_{p', p}(G)/\Theta$ so $D \leq O_{p', p}(G)$ as desired.

Hence we will assume $\Theta = \{1\}$ in the remainder of the proof. This means $O_{p', p}(G) = O_p(G)$ so we must show $D \leq O_p(G)$. Since $D \lhd G$, it suffices to show that D is a p-group. Assume to the contrary that there is some $x \in D$ of prime order q where $q \neq p$, and seek a contradiction. Let H be a minimal normal subgroup of G such that $H \leq O_p(G)$. Then $x \in D$ implies Hx belongs to $C_{G/H}((A/H)/(B/H))$ for all p-chief factors (A/H) $/(B/H)$ of G/H. Consequently, $Hx \in O_{p', p}(G/H)$ by induction. This means that the coset $O_{p'}(G/H)Hx$ has order some power of p, say p^n. Hence $(Hx)^{p^n} \in O_{p'}(G/H)$ and since p^n and q are coprime, we further conclude $Hx \in O_{p'}(G/H)$.

Denote $O_{p'}(G/H)$ by W/H and denote Fit$(D \cap W)$ by L. Since H is an Abelian normal subgroup of $D \cap W$, $H \leq L$. Two cases arise: $H = L$ or $H < L$. In the first case we note that $x \in C_G(H)$ since H is a p-chief factor of G. Also $x \in D \cap W$ so that $x \in C_{D \cap W}(H)$. Finally, solvability of G yields $C_{D \cap W}(H) \leq H$ by Theorem 2.6 and the fact that $H = L$. Thus we get $x \in H$ so q divides $|H|$, a contradiction.

Suppose now that $H < L$ holds. Since both $O_p(G)H/H$ and L/H are nilpotent normal subgroups, so is $O_p(G)L/H$. But $O_p(G)L \leq D \leq C_G(H)$ (the last inclusion because H is a p-chief factor of G) which means $H \leq Z(O_p(G)L)$. Hence $O_p(G)L$ is an extension of its center by a nilpotent group so $O_p(G)L$ is nilpotent. Thus $L \leq O_p(G)L \leq$ Fit(G). But $L/H \leq W/H$ means that L/H is a p'-group; and moreover a nontrivial one since $H < L$. Thus there exists some $y \in L$ of prime order r where $r \neq p$. Then $y \in L \leq$ Fit(G) implies $O_r(G) \neq \{1\}$ and hence $\Theta \neq \{1\}$, the desired contradiction.

4   The p-Frattini Subgroup $\Phi_p(G)$

We next consider an analogue of the Frattini subgroup. For G a group and p a prime we define

$$\Phi_p(G) = \bigcap \{ M : M \text{ is a maximal subgroup of } G \text{ and } [G : M] \text{ is a power of } p \}$$

and call $\Phi_p(G)$ the p-**Frattini subgroup**. This subgroup is always characteristic (automorphisms take maximal subgroups into maximal subgroups and also preserve indices).

THEOREM 4.1 *If G is solvable, then $O_{p'}(G) \leq \Phi_p(G)$ and furthermore $O_{p'}(G)$ is a normal p-complement for $\Phi_p(G)$.*
Proof: If M is a maximal subgroup of G having index of power of p, then G cannot equal $O_{p'}(G)M$ lest $[G:M] = [O_{p'}(G):O_{p'}(G) \cap M]$, a p'-number. Thus $O_{p'}(G) \leq M$ for each such M and so $O_{p'}(G) \leq \Phi_p(G)$.

Now let R be a Hall p'-subgroup of $\Phi_p(G)$. Then $O_{p'}(G) \leq R$. By a straightforward extension of the Frattini argument from Sylow subgroups to Hall subgroups we got $G = \Phi_p(G)N_G(R)$. If R is not normal in G, then $N_G(R) \leq M$ for some maximal subgroup M. But $[G:N_G(R)] = [\Phi_p(G): \Phi_p(G) \cap N_G(R)]$ which divides $[\Phi_p(G):R]$ which is a power of p. Hence $[G:M]$ is a power of p and so $\Phi_p(G) \leq M$. This gives the contradiction $G \leq M$ and so we have shown $R \triangleleft G$. Hence $R = O_{p'}(G)$ completing the proof.

THEOREM 4.2 *If G is solvable, then $\Phi_p(G)/O_{p'}(G) = \Phi(G/O_{p'}(G))$.*
Proof: That $\Phi(G/O_{p'}(G)) \leq \Phi_p(G)/O_{p'}(G)$ is clear. Conversely, we must show $\Phi_p(G)$ is contained in every M such that $M/O_{p'}(G)$ is a maximal subgroup of $G/O_{p'}(G)$. For all such M, $O_{p'}(G) \leq M$ and $O_{p'}(G) \leq \Phi_p(G)$ (Theorem 4.1) give $O_{p'}(G) \leq \Phi_p(G) \cap M$ so that $[\Phi_p(G) : \Phi_p(G) \cap M]$ divides $[\Phi_p(G) : O_{p'}(G)]$ which is a power of p (Theorem 4.1 again). Hence $[\Phi_p(G)M : M]$ is a power of p, which ensures $\Phi_p(G) \leq M$ as required.

THEOREM 4.3 *If G is solvable, then*
(i) Fit $(G/\Phi_p(G)) = O_{p',\,p}(G)/\Phi_p(G)$.
(ii) Fit$(G/\Phi_p(G))$ *is the direct sum of minimal normal p-subgroups of* $G/\Phi_p(G)$.
Proof: The group Fit$(G/O_{p'}(G))$ is clearly a p-group so

$$O_{p',\,p}(G)/O_{p'}(G) = \mathrm{Fit}(G/O_{p'}(G)).$$

Then

$$\mathrm{Fit}(G/\Phi_p(G)) \cong \mathrm{Fit}((G/O_{p'}(G))/(\Phi_p(G)/O_{p'}(G)))$$
$$= \mathrm{Fit}((G/O_{p'}(G))/\Phi(G/O_{p'}(G)))$$

(by Theorem 4.2) which is equal to $\mathrm{Fit}(G/O_{p'}(G))/\Phi(G/O_{p'}(G))$ (by Lemma 2.2) which is equal to

$$(O_{p',\,p}(G)/O_{p'}(G))/(\Phi_p(G)/O_{p'}(G))$$

which establishes the isomorphism

$$\mathrm{Fit}(G/\Phi_p(G)) \cong O_{p',p}(G)/\Phi_p(G).$$

Since the group on the left-hand side of this isomorphism is nilpotent, so is the group on the right-hand side. Thus $O_{p',p}(G)/\Phi_p(G) \leq \mathrm{Fit}(G/\Phi_p(G))$. This and the isomorphism establish (i).

Since $\Phi(G/\Phi_p(G))$ is trivial, Lemma 2.3 yields that $\mathrm{Fit}(G/\Phi_p(G))$ is the direct sum of minimal normal subgroups of $G/\Phi_p(G)$. To establish (ii) then, it suffices to show that any minimal normal subgroup, say $A/\Phi_p(G)$, of $G/\Phi_p(G)$ is a p-group. Since A is not contained in $\Phi_p(G)$, there is some maximal subgroup M of G of index a power of p such that A is not contained in M. Then $G = AM$ so that $[A : A \cap M]$, and hence $[A : \Phi_p(G)]$, is a power of p.

5 $O^p(G)$ and the Nilpotent Residual

For each prime p and group G we define

$$O^p(G) = \langle S : S \text{ is a Sylow } q\text{-subgroup of } G, \text{ where } q \neq p \rangle.$$

Since any automorphism of G will carry a Sylow q-subgroup into another such, $O^p(G)$ is characteristic in G.

THEOREM 5.1

(a) $G/O^p(G)$ is a p-group.

(b) If H is a normal subgroup of G such that G/H is a p-group, then $O^p(G) \leq H$.

(c) $O^p(G) \cap \{H : H \lhd G \text{ and } G/H \text{ is a } p\text{-group}\}$.

Proof: For any prime $q \neq p$, $O^p(G)$ contains (all and hence) some Sylow q-subgroup Q, so $q \nmid |G/O^p(G)|$. This proves (a).

Let H meet the hypotheses of (b). Let S be any Sylow q-subgroup of G, $q \neq p$. Then $HQ/H \cong Q/Q \cap H$, which is a q-group. Thus HQ/H is a q-subgroup of the p-group G/H so that $HQ = H$, implying $Q \leq H$. Thus $O^p(G) \leq H$ completing the proof of (b).

By (a), $\cap\{H : H \lhd G \text{ and } G/H \text{ is a } p\text{-group}\}$ is contained in $O^p(G)$. By (b), $O^p(G)$ is contained in each $H \lhd G$ such that G/H is a p-group, hence $O^p(G) \leq \cap\{H : H \lhd G \text{ and } G/H \text{ is a } p\text{-group}\}$. This proves (c).

Next we define

$$\gamma_\infty(G) = \cap\{H : H \lhd G \text{ and } G/H \text{ is nilpotent}\}$$

and call it the **nilpotent residual** of G.

THEOREM 5.2

(a) $G/\gamma_\infty(G)$ is nilpotent.

(b) If H is a normal subgroup of G such that G/H is nilpotent, then $\gamma_\infty(G) \leq H$.

(c) $\gamma_\infty(G) = \cap\{O^p(G) : p \text{ divides } |G|\}$.

(d) $\gamma_\infty(G) \cap \{\gamma_i(G) : i = 1, 2, \ldots \}$.

Proof: Letting $H_1, H_2, \ldots, H_r$ denote the normal subgroups of G whose quotient groups are nilpotent, we have $G/\gamma_\infty(G) = G/H_1 \cap \ldots \cap H_r$, which can be embedded in $G/H_1 \times G/H_2 \times \ldots \times G/H_r$, which is nilpotent. This proves (a).

Part (b) is immediate.

To prove (c) we first note that $G/O^p(G)$ is a p-group (Theorem 5.1) and hence nilpotent. Thus $\gamma_\infty(G) \leq O^p(G)$, for each prime p and so $\gamma_\infty(G) \leq \cap\{O^p(G) : p \text{ divides } |G|\}$. Conversely, consider any normal subgroup H such that G/H is nilpotent. Then for each prime p dividing $|G/H|$, G/H factors as $G/H = S_p/H \oplus T_p/H$ where S_p/H is a (normal) Sylow p-subgroup of G/H and T_p/H is a (normal) Hall p'-subgroup of G/H. Clearly $\cap\{T_p/H : p \text{ divides } |G/H|\}$ is trivial, i.e., $\cap\{T_p : p \text{ divides } |G/H|\} \leq H$. Since each G/T_p is a p-group, $O^p(G) \leq T_p$ by Theorem 5.1(b). Thus $\cap\{O^p(G) : p \text{ divides } |G/H|\} \leq \cap\{T_p : p \text{ divides } |G/H|\} \leq H$. Hence $\cap\{O^p(G) : p \text{ divides } |G|\} \leq H$ for all H such that G/H is nilpotent. Conclude $\cap\{O^p(G) : p \text{ divides } |G|\} \leq \gamma_\infty(G)$ completing the proof of (c).

To show (d) we first note that $G/\gamma_i(G)$ is nilpotent for all i and so $\gamma_\infty(G) \le \gamma_i(G)$ by (b); thus $\gamma_\infty(G) \le \bigcap\{\gamma_i(G) : i = 0, 1, 2, \ldots\}$. On the other hand $G/\gamma_\infty(G)$ is nilpotent, of class say, n so that $\gamma_n(G) \le \gamma_\infty(G)$ (an easy induction argument shows that $\gamma_n(G) \le H$ for any normal subgroup H such that G/H is class-n nilpotent). Hence $\bigcap\{\gamma_i(G) : i = 0, 1, 2, \ldots\}$ $\le \gamma_n(G) \le \gamma_\infty(G)$.

6 The Hypercenter $Z^*(G)$

The **hypercenter** of a group G is defined as the union of the ascending chain of subgroups $Z(G) \le Z_2(G) \le \ldots$. It should always be kept in mind that for finite groups the inclusions in this chain cannot all be proper: Thus $Z^*(G) = Z_k(G)$ for some k.

THEOREM 6.1 $Z^*(G/Z(G)) = Z^*(G)/Z(G)$.
Proof: It is easy to establish by induction that

$$Z_i(G/Z(G)) = Z_{i+i}(G)/Z(G) \qquad i = 0, 1, 2, \ldots .$$

Now let $Z(G)x \in Z^*(G/Z(G))$. Thus $Z(G)x \in Z_i(G/Z(G))$ for some i and so $x \in Z_{i+1}(G) \le Z^*(G)$ by the above display. This shows $Z^*(G/Z(G)) \le Z^*(G)/Z(G)$. Conversely, if $Z(G)x \in Z^*(G)/Z(G)$, then $x \in Z^*(G)$ implies $x \in Z_j(G)$ for some j. Hence $x \in Z_{j+1}(G)$ as well, so $Z(G)x \in Z_j(G/Z(G)) \le Z^*(G/Z(G))$ by the above display.

THEOREM 6.2 *Let G by any group. Let A be a nilpotent subgroup of G. Then $Z^*(G)A$ is also nilpotent.*
Proof: If $Z(G)$ is trivial, then so is $Z^*(G)$, and hence $Z^*(G)A = A$ is nilpotent. We may then assume that $Z(G) \ne \{1\}$. By induction, $Z^*(G/Z(G))(Z(G)A/Z(G))$ is nilpotent because $Z(G)A/Z(G)$ is. Applying Theorem 6.1 yields nilpotency of $(Z^*(G)/Z(G))(Z(G)A/Z(G))$ $= Z^*(G)A/Z(G)$. But since $Z(G) \le Z(Z^*(G)A)$, we may conclude that $Z^*(G)A/Z(Z^*(G)A)$ is nilpotent. Hence so too is $Z^*(G)A$.

The next theorem follows from a result of Baer (Theorem 2'' of [Baer 1953]). We give a simple independent proof due to Humphreys and Johnson.

THEOREM 6.3 *Let P be a normal p-subgroup of G such that $|G/C_G(P)|$ is a power of p. Then $P \le Z^*(G)$.*
Proof: Let S by a Sylow p-subgroup of G containing P. Our hypothesis ensures that $G = C_G(P)S$. Hence $[P, G] = [P, S]$ and since $P \triangleleft S$, $P_1 = [P, G] < P$. Since $C_G(P_1) \ge C_G(P)$, we can apply the same argument to P_1, and so $P_2 = [P_1, G] < P_1$. We continue this process until reaching $P_k = \{1\}$.

An easy induction proof shows $P_{k-i} \leq Z_i(G)$ for all i. In particular, $P_1 \leq Z_{k-1}(G)$. Hence $P \leq Z_k(G) \leq Z^*(G)$.

As an application of Theorem 6.3 we give the following corollary which will be used in the book several times.

COROLLARY 6.4 *If H/K is a p-chief factor of G, then $G/C_G(H/K)$ has no nontrivial normal p-subgroups.*
Proof: If $K \neq \{1\}$, then by induction $(G/K)/C_{G/K}(H/K)$ has no nontrivial normal p-subgroups. But $C_{G/K}(H/K) = C_G(H/K)/K$ so $G/C_G(H/K) \cong (G/K)/(C_G(H/K)/K)$ has no nontrivial normal p-subgroups.

We may therefore assume that $K = \{1\}$. Suppose $B/C_G(H)$ is a normal p-subgroup of $G/C_G(H)$. Then $H \triangleleft B$ and, since $C_G(H) \leq C_B(H)$, $B/C_B(H)$ is a p-group. Apply the theorem to conclude that $H \leq Z^*(B)$. Thus $H \leq Z_k(B)$ for some k. Now $H, B \triangleleft G$ imply $[H, B] \triangleleft G$. Then since $[H, B] \leq H$, we have either $[H, B] = H$ or $[H, B] = \{1\}$. The first case leads to a contradiction: $H \leq Z_k(B)$ implies $H = [H, B] \leq Z_{k-1}(B)$ implies $H = [H, B] \leq Z_{k-2}(B)$ ... implies $H \leq Z_0(B) = \{1\}$. Thus we have $[H, B] = \{1\}$ instead. Hence $B \leq C_G(H)$ so $B/C_G(H)$ is trivial.

References

Agrawal, R. K. 1976: Generalized center and hypercenter of a finite group, Proc. Amer. Math. Soc. 58, 13–21.

Baer, R. 1953: Group elements of prime power index, Trans. Amer. Math. Soc. 75, 20–47.

Baer, R. 1957: Classes of finite groups and their properties, Illinois J. Math. 1, 115–187.

Baer, R. 1958: Sylowturmgruppen, Math. Z. 69, 239–246.

Baer, R. 1959: Supersoluble immersion, Canad. J. Math. 11, 353–369.

Baskaran, S. 1973: Highly non-supersolvable CLT groups, Rendi-conti-Atti della Accademia Nazionale dei Lincei, 54, 224–227.

Basmaji, B. G. 1969: Monomial representations and metabelian groups, Nagoya Math. J. 35, 99–107.

Berman, S. D. 1957: Groups of which all representations are monomial (Ukrainian), Dopovidi Akad. Nauk. Ukrain RSR, 539–542.

Berman, S. D. 1959: Representations of groups of order 2^m over an arbitrary field of characteristic 0 (Ukrainian), Dopovidi Akad. Nauk. Ukrain RSR, 243–246.

Berkovic, J. G. and Engels, F. 1979: Some estimates for the degree of a finite group, Notices Amer. Math. Soc. 26, A-605.

Blessenohl, D. and Gaschütz, W. 1970: Über normale Schunk- und Fitting-klassen, Math. Zeit. 118, 1–8.

Bray, H. G. 1968: A note on CLT groups, Pacific Journal of Mathematics, 27, 229–231.

Brewster, B. and Ottaway, M. 1976: The supersolvable residual of an $\mathcal{L}_1$-group, Math. Proc. Camb. Phil. Soc. 80, 447–450.

Bryant, R. M., Bryce, R. A., and Hartley, B. 1970: The formation generated by a finite group, Bull. Austr. Math. Soc. (2), 347–357.

Buckley, J. 1970: Finite groups whose minimal subgroups are normal, Math. Z. 116, 15–17.

Burnside, W. 1955: *Theory of Finite Groups,* Dover, New York.

Carter, R. W. 1959: On a class of finite solvable groups, Proc. London Math. Soc. (3) 9, 623–640.

Carter R. W. 1961: Nilpotent self-normalizing subgroups of solvable groups, Math. Z. 75, 136–139.

Carter, R. W. and Hawkes, T. O. 1967: The $\mathcal{F}$-normalizers of a finite solvable group, J. of Algebra 5, 175–202.

Chen, M. S. 1977: A remark of finite groups, Tamkang J. Math 8, 105–109.

Cooper, C. D. H. 1971: Subgroups of a supersolvable group, The Amer. Math. Monthly, 78, 1007.

Curtis, C. W. and Reiner, I. 1962: *Representation Theory of Finite Groups and Associative Algebras,* Interscience, New York.

Dade, E. 1973: Normal subgroups of M-groups need not be M-groups, Math. Z. 133, 313–317.

Deskins, W. E. 1963: On quasinormal subgroups of finite groups, Math. Z. 82, 125–132.

Deskins, W. E. 1968: A characterization of finite supersolvable groups, The Amer. Math. Monthly, 75, 180–182.

Dixon, J. D. 1967: The Fitting subgroup of a linear solvable group, J. Austral. Math. Soc. 7, 417–424.

Dixon, J. D. 1968: The solvable length of a solvable linear group, Math. Z. 107, 151–158.

Dixon, J. D. 1971: *The Structure of Linear Groups,* Van Nostrand Reinhold, New York.

Djoković, D. Z. and Melzan, J. 1974: Imprimitive irreducible complex characters of the symmetric group, Math. Z. 138, 219–224.

Dornhoff, L. 1967: M-groups and 2-groups, Math. Z. 100, 226–256.

Dornhoff, L. 1970: Jordan's theorem for solvable groups, Proc. Amer. Math. Soc. 24, 533–537.

Dornhoff, L. 1971: *Group Representation Theory,* Marcel Dekker, New York.

Feit, W. 1967: *Characters of Finite Groups,* Benjamin, New York.

Fischer, B. 1966: Klassen konjugierter Untergruppen in endlicher auflösbaren Gruppen, dissertation, University of Frankfurt.

Fischer, B. Gaschütz, W., and Hartley, B. 1967: Injectoren endlicher auflösbaren Gruppen, Math. Z. 102, 337–339.

Fisher, R. U. 1974: The polycyclic length of linear and finite polycyclic groups, Can. J. Math. 26, 1002–1009.

Friesen, D. R. 1971: Products of normal supersolvable subgroups, Proc. Amer. Math. Soc. 30, 46–48.

Gaschütz, W. 1963: Zur theorie der endlichen auflösbaren Gruppen, Math. Z. 80, 300–305.

Gaschütz, W. and Lubeseder, U. 1963: Kennzeichnung gesättigter Formationen, Math. Z. 82, 198–199.

Gorenstein, D. 1968: *Finite Groups,* Harper and Row, New York; reprinted by Chelsea, New York, 1980.

Hall, M. 1959: *The Theory of Groups,* Macmillan, New York.

Hall, P. 1963: On non-strictly simple groups, Proc. Camb. Phil. Soc. 59, 531–553.

Hartley, B. 1969: On Fischer's dualization of formation theory, Proc. London Math. Soc. (3) 19, 193–207.

Hoffman, K. and Kunze, R. 1971: *Linear Algebra,* 2nd ed., Prentice-Hall, Englewood Cliffs.

Holmes, C. V. 1966: A characterization of finite nilpotent groups, Amer. Math. Monthly 73, 1113–1114.

Humphreys, J. F. 1974: On groups satisfying the converse of Lagrange's theorem, Proc. Camb. Phil. Soc. 75, 25–32.

Humphreys, J. F. and Johnson, D. L. 1973: On Lagrangian groups, Trans. Amer. Math. Soc. 180, 291–300.

Huppert, B. 1953: Monomiale Darstellung endlicher Gruppen, Nagoya Math. J. 6, 93–94.

Huppert, B. 1954: Normalteiler und maximale Untergruppen endlicher Gruppen, Math. Z. 60, 409–434.

Huppert, B. 1967: *Endliche Gruppen* I, Springer-Verlag, New York.

Inagaki, N. 1965: On groups with nilpotent commutator subgroups, Nagoya Math. J. 25, 205–210.

Isaacs, I. M. 1976: *Character Theory of Finite Groups,* Academic Press, New York.

Ito, N. 1951: Note on (LM)-groups of finite order, Kodai Math. Sem. Reports, 1–6.

Ito, N. 1954: On monomial representations of finite groups, Osaka Math. J. 6, 119–127.

Iwasawa, K. 1941a: Über die endlichen Gruppen und die Verbände ihrer Unter-gruppen, J. Fac. Sci. Univ. Tokyo 4, 171–199.

Iwasawa, K. 1941b: Über die Struktur der endlichen Gruppen, deren echte Untergruppen sämtlich nilpotent sind, Proc. Phys.-Math. Soc. Japan 23, 1–4.

Jacobson, N. 1953: *Lectures in Abstract Algebra* II, Van Nostrand, New York; reprinted by Springer-Verlag, New York.

Johnson, D. L. 1971a: A note on supersolvable groups, Can. J. Math. 23, 562–564.

Johnson, D. L. 1971b: Minimal permutation representations of finite groups, Amer. J. Math. 93, 897–903.

Kappe, W. 1969: Über gruppen theoretische Eigenschaften die sich auf *t*-Prod-ukte ubertragen, Acta. Aci. Math. Szeged 30, 277–284.

Kegel, O. H. 1962: Sylow Gruppen und Subnormalteiler endlicher Gruppen, Math. Z. 78, 205–221.

Kegel, O. H. 1967: On Huppert's characterization of finite supersolvable groups, in *Proceedings of the International Conference on the Theory of Groups* (Canberra 1965), Gordon and Breach, New York.

Kerber, A. 1970: Zur theorie der M-gruppen, Math. Z. 115, 4–6.

Kramer, O.-U. 1976: Über Durchschnitte von Untergruppen endlicher auflösbar-er Gruppen, Math. Z. 148, 88–97.

Laue, H. 1978: Dualization of saturation for locally defined formations, J. Algebra 52, 347–353.

Lausch, H. 1973: On normal Fitting classes, Math. Z. 130, 67–72.

Ljubic, M. 1979: On the logarithmic property of the degree of a finite group, Dokl. Akad. Nauk SSSR 247; Soviet Math Dokl. 20, 798–801.

Lubeseder, U. 1963: Formationsbildungen in endlicher auflösbaren Gruppen, dissertation, Kiel.

Mac Lane, S. and Birkhoff, G. 1968: *Algebra,* Macmillan, New York.

Maier, R. and Schmid, P. 1973: The embedding of quasinormal subgroups in finite groups, Math. Z. 131, 269–272.

Mann, A. 1968: On $\mathfrak{F}$-normalizers and $\mathfrak{F}$-covering subgroups, Proc Amer. Math. Soc. 19, 1159–1160.

Mann, A. 1971: Injectors and normal subgroups of finite groups, Israel J. Math. 9, 554–558.

McCarthy, D. 1970: Sylow's theorem is a sharp partial converse to Lagrange's theorem, Math. Z. 113, 383–384.

McCarthy, D. 1971: A survey of partial converses to Lagrange's theorem on finite groups, Transactions of the NY Academy of Sciences, ser II, 33, 586–594.

McLain, D. 1957: The existence of subgroups of a given order in finite groups, Proc. Camb. Phil. Soc. 53, 278–285.

Mukherjee, N. P. 1972: The hyperquasicenter of a finite group, Proc. Amer. Math. Soc. 32, 24–28.

Niven, I. and Zuckerman, H. S. 1980: *An Introduction to the Theory of Numbers,* 4th ed., John Wiley and Sons, New York.

Ore, O. 1939: Contributions to the theory of finite groups, Duke Math. J. 5, 431–460.

Pazderski, G. 1959: Die Ordnung, zu denen nur Gruppen mit gegebener Eigenschaft gehören, Arch. Math. 10, 331–343.

Redei, L. 1967: *Algebra,* Pergamon Press, New York.

Sah, C. -H. 1957: On a generalization of finite nilpotent groups, Math. Z. 68, 189–205.

Sastry, N. S. and Deskins, W. E. 1978: Influence of normality conditions on almost minimal subgroups, J. Algebra 52, 364–377.

Schmidt, O. 1924: Uber gruppen, deren sämtliche Teiler spezielle sind, Dec. Math. 31, 366–372.

Schunk, H. 1967: $\mathcal{H}$-Untergruppen in endlicher auflösbaren Gruppen, Math. Z. 97, 326–330.

Scott, W. 1964: *Group Theory,* Prentice-Hall, Englewood Cliffs.

Seitz, G. 1968: M-groups and the supersolvable residual, Math. Z. 110, 101–122.

Seitz, G. and Wright, C. 1969: On finite groups whose Sylow subgroups are modular or quaternion-free, J. Algebra 13, 374–381.

Slepova, L. 1977: Radical formations, Math. Notes 21, 485–486.

Suzuki, M. 1956: Structure of a group and the structure of its lattice of subgroups, Springer-Verlag, New York.

Troccolo, J. 1975: $\mathfrak{F}$-projectors of finite solvable groups, Proc. Amer. Math. Soc. 48, 33–38.

Venzke, P. 1977: System quasinormalizers in finite solvable groups, J. Algebra 44, 160–168.

Venzke, P. 1979: A contribution to the theory of finite supersolvable groups, J. Algebra 57, 567–579.

Walls, G. L. 1971: Formations and $\mathcal{F}$-covering subgroups, Master's thesis, University of Utah.

Ward, H. N. 1969: An extension of a theorem on monomial groups, Amer. Math. Monthly 76, 534—535.

Weinstein, M. 1977: *Examples of Groups,* Polygonal Publishing House, Passaic.

Wielandt, H. 1967/68: Factors of groups, Indam, Rome, Symposia Mathematica 1, 187—194.

Wielandt, H. and Huppert, B. 1960: *Arithmetical and Normal Structure of Finite Groups,* Proc. Sym. in Pure Math., AMS, Providence.

Woodall, D. R. 1974: An exchange theorem for bases of matroids, J. Combinatorial Theory, ser B., 16, 227—228.

Wright, D. 1976: Degrees of minimal embeddings for some direct products, Amer. J. Math. 97, 897—903.

Yokoyama, A. 1975, 1976: Finite solvable groups whose F-hypercenter contains all minimal subgroups I, Arch. Math. 26, 123—130; II, Arch. Math. 27, 572—575.

Zappa, G. 1940: Remark on a recent paper of O. Ore, Duke Math. J. 6, 511—512.

Notation

Script

$\mathcal{C}, \mathcal{F}, \mathcal{F}(p), \mathcal{G}, \mathcal{M}, \mathcal{R}$	denote arbitrary classes
$\mathcal{Q}$	class of Abelian groups
$\mathcal{BNCLT}$	class of BNCLT groups
$\mathcal{CLT}$	class of CLT groups
$\hat{\mathcal{F}}$	see page 150
$\mathcal{F}_p$	see page 148
$\mathcal{G}_\pi$	class of solvable π-groups
$\mathcal{M}_c\mathcal{C}$	class of McCarthy groups
$\mathcal{N}$	class of nilpotent groups
$\mathcal{NCLT}$	class of non-CLT groups
$\mathcal{QCLT}$	class of groups all of whose quotients are CLT groups
$\mathcal{S}$	class of solvable groups
$\mathcal{SN}$	class of semi-nilpotent groups
$\mathcal{SS}$	class of supersolvable groups
$\mathcal{STP}$	class of groups satisfying the Sylow tower property
$\mathcal{SY}$	class of groups all of whose subgroups are $\mathcal{Y}$-groups
$\mathcal{X}$	class of groups all of whose primitive subgroups have prime-power index
$\mathcal{Y}$	class of groups G such that for all $H \leqslant G$ and for all primes q dividing $[G:H]$, there exists some K such that $H \leqslant K \leqslant G$ and $[K:H] = q$

Greek

$\gamma_i(\ )$	$\gamma_i(G)$ denotes the ith term of the lower central series for G; defined inductively by $\gamma_0(G) = G$, $\gamma_{i+1}(G) = [\gamma_i(G), G]$
$\gamma_\infty(\ )$	see page 219
π	denotes an arbitrary set of primes
$\rho_\mathcal{F}(\ )$	see page 173
$\rho_\mathcal{F}{}^*(\ )$	see page 151
$\Phi(\ )$	see page 212
$\Phi_p(\ )$	see page 217
ψ	see page 94

Latin

$\mathrm{Aut}_G(\ )$	see page 210
$C_G(\ )$	see page 210
$C_G{}^*(\ )$	see page 33
$\mathrm{Core}_G(\ )$	$\mathrm{Core}_G(H)$ is the intersection of the subgroups $x^{-1}Hx$ for all $x \in G$
$C_p(\ )$	see page 162
$\exp(\ ,\ )$	see page 69
$\mathrm{Fit}(\ )$	see page 213

Other

Index

QUEEN MARY
COLLEGE
LIBRARY

WITHDRAWN
FROM STOCK
QMUL LIBRARY